CAMBRIDGE MONOGRAPHS ON MATHEMATICAL PHYSICS

THE INTERACTING BOSON MODEL

THE INTERACTING BOSON MODEL

F. IACHELLO
Yale University

A. ARIMA
University of Tokyo

The right of the University of Cambridge to print and sell all manner of books was granted by Henry VIII in 1534. The University has printed and published continuously since 1584.

CAMBRIDGE UNIVERSITY PRESS
Cambridge
New York New Rochelle
Melbourne Sydney

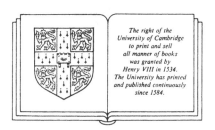

CAMBRIDGE UNIVERSITY PRESS
Cambridge, New York, Melbourne, Madrid, Cape Town, Singapore, São Paulo

Cambridge University Press
The Edinburgh Building, Cambridge CB2 2RU, UK

Published in the United States of America by Cambridge University Press, New York

www.cambridge.org
Information on this title: www.cambridge.org/9780521302821

First published 1987
This digitally printed first paperback version 2006

A catalogue record for this publication is available from the British Library

Library of Congress Cataloguing in Publication data
Iachello, F.
The interacting boson model.
(Cambridge monographs on mathematical physics;)
1. Bosons—Mathematical models. 2. Nuclear models—
Mathematical models. 3. Nuclear structure—Mathematical
models. I. Arima, Akito, 1930– . II. Title.
III. Series
Q793.5.B622I25 1987 539.7′21 86-18790

ISBN-13 978-0-521-30282-1 hardback
ISBN-10 0-521-30282-X hardback

ISBN-13 978-0-521-02879-0 paperback
ISBN-10 0-521-02879-5 paperback

TO IRENA, GIOVANNI AND HIROKO
for their patience and understanding

Contents

Part II: The interacting boson model-2

Part III: The interacting boson model-k

Preface

In 1974 a new model, the interacting boson model, was introduced in an attempt to describe in a unified way collective properties of nuclei. This model is rooted in the spherical shell model developed by Jensen and Mayer (Haxel, Jensen and Suess, 1949; Mayer, 1949), which is the fundamental model for describing properties of nuclei, but in addition has properties similar, and in many cases identical, to the collective model developed by Bohr and Mottelson (Rainwater, 1950; Bohr, 1951, 1952; Bohr and Mottelson, 1953) and based on the concept of shape variables. Since 1974, the interacting boson model has been the subject of many investigations and it has been extended to cover most aspects of nuclear structure. In this book, which is intended to be the first in a series of three, we give an account of some properties of the interacting boson model.

We have particularly in mind here two purposes. First, we want to present the mathematical technique used to analyze the structure of the model. This, which could go under the general name of boson calculus, is potentially of interest to a large number of researchers, since the same technique can be used (and has been used) to describe other physical systems, such as molecular structures. The mathematical framework of the model is discussed in detail in Part I, which is the part of most interest for non-nuclear physicists. The second purpose of the book is to collect in a single, easily accessible reference all the formulas that have been developed throughout the years to account for collective properties of nuclei. These formulas can be used by experimentalists to compare their data with the predictions of the model.

In this book, we cover only properties of nuclei with an even number of protons and neutrons. The second volume is intended to cover properties of nuclei with an even number of protons and an odd number of neutrons, or vice-versa. The model describing these cases is usually called the interacting boson–fermion model. Finally, the third volume is intended to cover the microscopic structure and justification of the model.

It is our pleasure to express our gratitude to *all* the researchers who have contributed to the development of the model. Although we mention here only those with whom we have been in closer contact throughout the years, we intend this to be truly an occasion for thanking all persons who have worked in this area of physics.

First of all we thank Igal Talmi, without whom many aspects of the model, in particular the microscopic aspect, would not have been developed. We also thank J. Phillip Elliott, whose pioneering work on the application of group theory to nuclear structure provided us with much inspiration, Herman Feshbach, whose input in the early stages led to the present version of the model and D. Allan Bromley for his constant interest and encouragement. Thirdly, we want to express special thanks to Takaharu Otsuka and Olaf Scholten for their endless help. Together with them we thank Alex Dieperink, Joe Ginocchio, Kris Heyde, Bruce Barrett, Da Hsuan Feng, Stuart Pittel, Pieter van Isacker, Roelof Bijker, Alejandro Frank and Onno van Roosmalen.

Among the many experimenters who have analyzed the predictions of the model and confronted them with the data, we want again to express special thanks to those with whom we have mostly interacted throughout the years. First of all, we thank Richard Casten, whose enthuasiasm and interest has been crucial in developing several aspects of the model. We also want to thank Jolie Cizewski, Adrian Gelberg, Peter von Brentano, Jean Vervier, Michel Vergnes, David Warner and John Wood.

Finally, we wish to thank Marguerite Scalesse and Janet DeFillippo for their patience in coping with this very difficult typescript.

New Haven, Connecticut
December 1985

Part I

The interacting boson model-1

1
Operators

1.1 Introduction

The interacting boson model-1 originated from early ideas of Feshbach and Iachello (Iachello, 1969; Feshbach and Iachello, 1973, 1974), who in 1969 described some properties of light nuclei in terms of interacting bosons, and from the work of Janssen, Jolos and Dönau (1974), who in 1974 suggested a description of collective quadrupole states in nuclei in terms of a SU(6) group. The latter description was subsequently cast into a different mathematical form by us (Arima and Iachello, 1975) with the introduction of an s-boson, which made the SU(6), or rather U(6), structure more apparent. The success of this phenomenological approach to the structure of nuclei has led to major developments in the understanding of nuclear structure.

The major new development was the realization that the bosons could be interpreted as nucleon pairs (Arima *et al.*, 1977) in much the same way as Cooper pairs in the electron gas (Cooper, 1956). This provided a framework for a microscopic description of collective quadrupole states in nuclei and stimulated a large number of theoretical investigations. An immediate consequence of this interpretation was that, since one expected both neutron and proton pairs, one was led to consider a model with two types of bosons, proton bosons and neutron bosons. In order to make the distinction between proton and neutron bosons more apparent, the resulting model was called the interacting boson model-2, while the original version retained the name of interacting boson model-1.

Subsequently, the model was further expanded by introducing explicitly unpaired fermions, thus allowing one to treat odd–even nuclei (Iachello and Scholten, 1979). Of this extension there exist now two versions, called the interacting boson–fermion model-1 and -2. In recent years, yet more extensions have been developed, including mixing of configurations, giant resonances, etc. As a result, there is hardly any aspect of nuclear structure that has not been touched by it.

The purpose of this book is to present the basic mathematical framework of the interacting boson models-1 and -2. We thus leave out a description of odd–even nuclei and of the microscopic approaches to the model. These will form the subject of subsequent books. Our aim here is to draw together work on the subject which is scattered in several journals. Because of the lack of space, we shall not be able to cover all publications that have appeared on the subject but select some which fit the homogeneity of the presentation.

1.2 Boson operators

In the interacting boson model, collective excitations of nuclei are described by bosons. An appropriate formalism to describe the situation is provided by second quantization. One thus introduces boson creation (and annihilation) operators of multipolarity l and z-component m, $b^\dagger_{l,m}(b_{l,m})$. A boson model is specified by the number of boson operators that are introduced. In the interacting boson model-1, it is assumed (Arima and Iachello, 1975) that low-lying collective states of nuclei can be described in terms of a monopole boson with angular momentum and parity, $J^P=0^+$, called s, and a quadrupole boson with $J^P=2^+$, called d. Thus, the building blocks of this model are:

$$\begin{cases} s^\dagger, d^\dagger_\mu & (\mu=0,\pm1,\pm2), \\ s, d_\mu & (\mu=0,\pm1,\pm2). \end{cases} \tag{1.1}$$

The reason why only these bosons are kept has to do with the microscopic interpretation of the bosons and will be discussed at a later stage.

The operators (1.1) satisfy Bose commutation relations:

$$\begin{gathered} [s,s^\dagger]=1; \quad [s,s]=[s^\dagger,s^\dagger]=0; \\ [d_\mu,d^\dagger_{\mu'}]=\delta_{\mu\mu'}; \quad [d_\mu,d_{\mu'}]=[d^\dagger_\mu,d^\dagger_{\mu'}]=0; \\ [s,d^\dagger_\mu]=[s,d_\mu]=[s^\dagger,d^\dagger_\mu]=[s^\dagger,d_\mu]=0. \end{gathered} \tag{1.2}$$

We shall, at times, use a more compact notation for the boson operators (1.1), i.e. either

$$b^\dagger_{l,m}; \quad b_{l,m}; \quad (l=0,2; \ -l\leqslant m\leqslant l), \tag{1.3}$$

or

$$b^\dagger_\alpha; \quad b_\alpha; \quad (\alpha=1,\ldots,6), \tag{1.4}$$

with

$$b_1=s, \quad b_2=d_{+2}, \quad b_3=d_{+1}, \quad b_4=d_0, \quad b_5=d_{-1}, \quad b_6=d_{-2}. \tag{1.5}$$

The commutation relations (1.2) are written then as

$$[b_{l,m}, b^\dagger_{l',m'}] = \delta_{ll'}\delta_{mm'};$$
$$[b_{l,m}, b_{l',m'}] = [b^\dagger_{l,m}, b^\dagger_{l',m'}] = 0, \tag{1.6}$$

or as

$$[b_\alpha, b^\dagger_{\alpha'}] = \delta_{\alpha\alpha'}; \quad [b_\alpha, b_{\alpha'}] = [b^\dagger_\alpha, b^\dagger_{\alpha'}] = 0. \tag{1.7}$$

For applications, one needs to construct with the boson operators spherical tensors $T^{(k)}_\kappa$ of degree k, in the sense of Racah, i.e. operators that transform as basis vectors of a $(2k+1)$ dimensional representation $R^{(k)}_{\kappa\kappa'}$ of the rotation group

$$\mathscr{R} T^{(k)}_\kappa \mathscr{R}^{-1} = \sum_{\kappa'} T^{(k)}_{\kappa'} R^{(k)}_{\kappa'\kappa}. \tag{1.8}$$

The creation operators can be defined as such. However, the annihilation operators are not. In order to construct spherical tensors, one introduces the operators

$$\tilde{b}_{l,m} = (-)^{l+m} b_{l,-m}. \tag{1.9}$$

Equation (1.9) gives, when applied to the present case,

$$\tilde{s} = s, \quad \tilde{d}_\mu = (-)^\mu d_{-\mu}. \tag{1.10}$$

Although there is no need to do so, we will keep for consistency the tilde over the operator s when forming tensors.

With spherical tensors one can form tensor products. We use the notation (De Shalit and Talmi, 1963)

$$T^{(k)}_\kappa = [T^{(k_1)} \times T^{(k_2)}]^{(k)}_\kappa \tag{1.11}$$

for the tensor product of $T^{(k_1)}$ and $T^{(k_2)}$, i.e. the square bracket indicates

$$[T^{(k_1)} \times T^{(k_2)}]^{(k)}_\kappa = \sum_{\kappa_1\kappa_2} (k_1\kappa_1 k_2\kappa_2 | k\kappa) T^{(k_1)}_{\kappa_1} T^{(k_2)}_{\kappa_2} \tag{1.12}$$

where the symbol $(k_1\kappa_1 k_2\kappa_2|k\kappa)$ denotes a Clebsch–Gordon coefficient. A particular case of tensor product is the scalar product. We shall denote this by a round bracket, defined as (De Shalit and Talmi, 1963)

$$(U^{(k)} \cdot V^{(k)}) = (-)^k (2k+1)^{\frac{1}{2}} [U^{(k)} \times V^{(k)}]^{(0)}_0. \tag{1.13}$$

This product can also be rewritten as

$$(U^{(k)} \cdot V^{(k)}) = \sum_\kappa (-)^\kappa U^{(k)}_\kappa V^{(k)}_{-\kappa}. \tag{1.14}$$

1.3 Basis states

Basis states can be constructed by repeated application of boson operators on a boson vacuum $|0\rangle$,

$$\mathscr{B}: b_\alpha^\dagger b_{\alpha'}^\dagger \dots |0\rangle. \tag{1.15}$$

Here too it is convenient to construct states of good angular momentum by coupling the boson operators appropriately,

$$\mathscr{B}: [b_l^\dagger \times b_{l'}^\dagger \times \cdots]_M^{(L)} |0\rangle. \tag{1.16}$$

However, the total angular momentum alone is not, in general, sufficient to characterize the states uniquely, since there may be several states with the same L for a given number of bosons, N. A way to find appropriate labels is discussed in Chapter 2. Most numerical calculations are performed by using the so-called spherical basis in which the s and d bosons are explicitly separated

$$\mathscr{B}^{(I)}: s^{\dagger n_s} [d^{\dagger n_d}]_{v n_\Delta L M_L} |0\rangle. \tag{1.17}$$

These are states with total boson number

$$N = n_s + n_d \tag{1.18}$$

where n_s and n_d are the numbers of s and d bosons. The numbers v, n_Δ are additional quantum numbers to be discussed in Chapter 2.

1.4 Physical operators

1.4.1 The Hamiltonian operator

In order to calculate properties of the physical system described by the bosons, one writes all operators in terms of boson operators. We begin by considering the Hamiltonian operator. If one assumes that the operator H conserves the total boson number N, one can write

$$H = E_0 + \sum_{\alpha\beta} \varepsilon_{\alpha\beta} b_\alpha^\dagger b_\beta + \sum_{\alpha\beta\gamma\delta} \tfrac{1}{2} u_{\alpha\beta\gamma\delta} b_\alpha^\dagger b_\beta^\dagger b_\gamma b_\delta + \cdots. \tag{1.19}$$

Here E_0 is a c-number, the term with $b^\dagger b$ represents one-body contributions, the next one two-body contributions, etc. The presence of the interaction terms gives the name 'interacting boson models' to this type of models. We stress the fact that it is a basic assumption of the interacting boson model that the interactions in (1.19) are number conserving, i.e. in each term the number of creation operators equals the number of annihilation operators. Other boson models, for example those arising from boson expansion techniques, do not have this assumption.

When written as in Eq. (1.19), the fact that the Hamiltonian is a scalar operator under rotations is not evident. To make it apparent, it is convenient to rewrite H as

$$H = E_0 + \sum_l \varepsilon_l (b_l^\dagger \cdot \tilde{b}_l) + \sum_{L,ll'l''l'''} \tfrac{1}{2} u_{ll'l''l'''}^{(L)} [[b_l^\dagger \times b_{l'}^\dagger]^{(L)} \times [\tilde{b}_{l''} \times \tilde{b}_{l'''}]^{(L)}]_0^{(0)} + \cdots. \tag{1.20}$$

In most calculations, only up to two-body terms have been retained. When written explicitly in terms of s and d bosons, Eq. (1.20) reads

$$\begin{aligned} H = {} & E_0 + \varepsilon_s (s^\dagger \cdot \tilde{s}) + \varepsilon_d (d^\dagger \cdot \tilde{d}) \\ & + \sum_{L=0,2,4} \tfrac{1}{2}(2L+1)^{\frac{1}{2}} c_L [[d^\dagger \times d^\dagger]^{(L)} \times [\tilde{d} \times \tilde{d}]^{(L)}]_0^{(0)} \\ & + \frac{1}{\sqrt{2}} v_2 [[d^\dagger \times d^\dagger]^{(2)} \times [\tilde{d} \times \tilde{s}]^{(2)} + [d^\dagger \times s^\dagger]^{(2)} \times [\tilde{d} \times \tilde{d}]^{(2)}]_0^{(0)} \\ & + \tfrac{1}{2} v_0 [[d^\dagger \times d^\dagger]^{(0)} \times [\tilde{s} \times \tilde{s}]^{(0)} + [s^\dagger \times s^\dagger]^{(0)} \times [\tilde{d} \times \tilde{d}]^{(0)}]_0^{(0)} \\ & + u_2 [[d^\dagger \times s^\dagger]^{(2)} \times [\tilde{d} \times \tilde{s}]^{(2)}]_0^{(0)} + \tfrac{1}{2} u_0 [[s^\dagger \times s^\dagger]^{(0)} \times [\tilde{s} \times \tilde{s}]^{(0)}]_0^{(0)}. \end{aligned} \tag{1.21}$$

Here one has also used the fact that the Hamiltonian is an Hermitian operator, i.e. $H^\dagger = H$. There are thus two one-body terms, specified by the parameters ε_s, ε_d, and seven two-body terms, specified by the parameters c_L ($L=0,2,4$), v_L ($L=0,2$), u_L ($L=0,2$), to this order.

1.4.2 *Transition operators*

Operators inducing electromagnetic transitions of multipolarity L can also be written in terms of boson operators as

$$T^{(L)} = t_0^{(0)} \delta_{L0} + \sum_{\alpha\beta} t_{\alpha\beta}^{(L)} b_\alpha^\dagger b_\beta + \cdots. \tag{1.22}$$

Again, since the operators must transform as tensors of rank L under rotations, it is more convenient to introduce coupled tensors as

$$T_\mu^{(L)} = t_0^{(0)} \delta_{L0} + \sum_{ll'} t_{ll'}^{(L)} [b_l^\dagger \times \tilde{b}_{l'}]_\mu^{(L)} + \cdots. \tag{1.23}$$

Usually, one stops at one-body terms. One can again write down explicitly the operators (1.23) in terms of s and d bosons as

$$\begin{aligned} T_0^{(E0)} &= \gamma_0 + \alpha_0[s^\dagger \times \tilde{s}]_0^{(0)} + \beta_0[d^\dagger \times \tilde{d}]_0^{(0)}, \\ T_\mu^{(M1)} &= \beta_1[d^\dagger \times \tilde{d}]_\mu^{(1)}, \\ T_\mu^{(E2)} &= \alpha_2[d^\dagger \times \tilde{s} + s^\dagger \times \tilde{d}]_\mu^{(2)} + \beta_2[d^\dagger \times \tilde{d}]_\mu^{(2)}, \\ T_\mu^{(M3)} &= \beta_3[d^\dagger \times \tilde{d}]_\mu^{(3)}, \\ T_\mu^{(E4)} &= \beta_4[d^\dagger \times \tilde{d}]_\mu^{(4)}. \end{aligned} \tag{1.24}$$

In writing Eq. (1.24) we have taken into account the fact that the s and d bosons have positive parity. Thus, for example, the operator with multipolarity one has positive parity and therefore corresponds to an M1 operator (the E1 operator would have negative parity). Equation (1.24) shows also that, if one stops at one-body operators with s and d bosons, one can construct transition operators with multipolarity up to four. The constants γ_0, α_L $(L=0,2)$, β_L $(L=0,\ldots,4)$ are parameters specifying the magnitude and scale of the corresponding operators.

1.4.3 Choice of phases

The operators (1.21) and (1.24) have been chosen to be Hermitian and real. In principle, there is a phase ambiguity due to the fact that the operators s and d could undergo a gauge transformation (Van Isacker, Frank and Dukelsky, 1985)

$$\begin{aligned} s &\to s\,\mathrm{e}^{-\mathrm{i}\phi}, \quad s^\dagger \to s^\dagger\,\mathrm{e}^{+\mathrm{i}\phi}, \\ d &\to d\,\mathrm{e}^{-\mathrm{i}\psi}, \quad d^\dagger \to d^\dagger\,\mathrm{e}^{+\mathrm{i}\psi}. \end{aligned} \tag{1.25}$$

This gauge transformation would have no effect on terms having an equal number of creation and annihilation operators of the same type. However, under this transformation, the first term in the E2 operator of Eq. (1.24) would change to

$$[d^\dagger \times \tilde{s} + s^\dagger \times \tilde{d}]^{(2)} \to \mathrm{e}^{\mathrm{i}(\psi-\phi)}[d^\dagger \times \tilde{s} + \mathrm{e}^{-2i(\psi-\phi)} s^\dagger \times \tilde{d}]^{(2)}. \tag{1.26}$$

One thus has to decide which gauge (i.e. what choice of phases) one selects. We shall denote the choice of the previous subsections, $(\psi=\phi=0)$, the standard choice and adopt it throughout. A choice $\psi-\phi=\pi/2$ would lead to a transition operator of the form

$$T_\mu^{(E2)} = \alpha_2 \mathrm{i}[d^\dagger \times \tilde{s} - s^\dagger \times \tilde{d}]_\mu^{(2)} + \beta_2[d^\dagger \times \tilde{d}]_\mu^{(2)}. \tag{1.27}$$

This operator is still Hermitian, but it is now complex. The effects of non-

Table 1.1. *List of three-body terms in the Hamiltonian operator, Eq. (1.29)*

i	i'	i''	j	j'	j''	k
2	2	2	2	2	2	6, 4, 3, 2, 0
2	2	2	2	2	0	4, 2, 0
2	2	2	2	0	0	2
2	2	2	0	0	0	0
2	2	0	2	2	0	4, 2, 0
2	2	0	2	0	0	2
2	2	0	0	0	0	0
2	0	0	2	0	0	2
0	0	0	0	0	0	0

standard choices in the description of properties of nuclei have been investigated by Van Isacker, Frank and Dukelsky (1985).

1.4.4 Higher-order terms

Although most calculations up to now have been done using the operators (1.21) and (1.24), it has appeared to be necessary, in some cases, to introduce higher-order terms. In this subsection, we enumerate all cubic terms in the Hamiltonian and all quadratic terms in the transition operators. This subsection can be skipped if the reader is not interested in detailed fits to nuclear spectra.

We begin by writing the three-body terms in the Hamiltonian

$$H_3 = \sum_{\alpha\beta\gamma\delta\varepsilon\eta} w_{\alpha\beta\gamma\delta\varepsilon\eta} b_\alpha^\dagger b_\beta^\dagger b_\gamma^\dagger b_\delta b_\varepsilon b_\eta, \tag{1.28}$$

in coupled tensor form

$$H_3 = \sum_{\substack{k \\ i,i',i'' \\ j,j',j''}} w_{ii'i'',jj'j''}^{(k)} [[b_i^\dagger \times b_{i'}^\dagger \times b_{i''}^\dagger]^{(k)} \times [\tilde{b}_j \times \tilde{b}_{j'} \times \tilde{b}_{j''}]^{(k)}$$
$$+ [b_{j''}^\dagger \times b_{j'}^\dagger \times b_j^\dagger]^{(k)} \times [\tilde{b}_{i''} \times \tilde{b}_{i'} \times \tilde{b}_i]^{(k)}]_0^{(0)}, \tag{1.29}$$

and list in Table 1.1 all independent three-body terms. There are 17 possible three-body terms in the Hamiltonian operators. The effects of some of these on the spectra of nuclei have been investigated by Casten *et al.* (1985).

Similarly, one can consider all possible two-body terms in the transition operators

$$T_2^{(L)} = \sum_{\alpha\beta\gamma\delta} t^{(L)}_{\alpha\beta\gamma\delta} b^\dagger_\alpha b^\dagger_\beta b_\gamma b_\delta. \tag{1.30}$$

These, when written in coupled tensor form, read

$$T^{(L)}_{2,\mu} = \sum_{\substack{k,i,i'\\k',j,j'}} t^{(L)}{}_{k,i,i',k',j,j'} [[b^\dagger_i \times b^\dagger_{i'}]^{(k)} \times [\tilde{b}_j \times \tilde{b}_{j'}]^{(k')}$$

$$+ [b^\dagger_{j'} \times b^\dagger_j]^{(k')} \times [\tilde{b}_{i'} \times \tilde{b}_i]^{(k)}]^{(L)}_\mu. \tag{1.31}$$

All possible terms are listed in Table 1.2. To this order, one can now have operators up to multipolarity eight. As an example, we write down explicitly in terms of s and d bosons, the M1 operator including two-body terms

$$T^{(\mathrm{M1})}_\mu = \beta_1 [d^\dagger \times \tilde{d}]^{(1)}_\mu + \tilde{\alpha}_1 [[d^\dagger \times d^\dagger]^{(4)} \times [\tilde{d} \times \tilde{d}]^{(4)}]^{(1)}_\mu$$

$$+ \tilde{\gamma}_1 [[d^\dagger \times d^\dagger]^{(2)} \times [\tilde{d} \times \tilde{d}]^{(2)}]^{(1)}_\mu$$

$$+ \tilde{\delta}_1 [[d^\dagger \times d^\dagger]^{(2)} \times [\tilde{d} \times \tilde{s}]^{(2)} + [d^\dagger \times s^\dagger]^{(2)} \times [\tilde{d} \times \tilde{d}]^{(2)}]^{(1)}_\mu$$

$$+ \tilde{\eta}_1 [[d^\dagger \times s^\dagger]^{(2)} \times [\tilde{s} \times \tilde{d}]^{(2)}]^{(1)}_\mu. \tag{1.32}$$

In this case, there are four two-body terms, $\tilde{\alpha}_1, \tilde{\gamma}_1, \tilde{\delta}_1, \tilde{\eta}_1$.

1.4.5 Independent parameters

If the Hamiltonian and transition operators are considered within a space of fixed number of bosons, N, the number of independent terms is reduced. The operators affected by this reduction are the Hamiltonian and the operator $T(\mathrm{E0})$. Consider first the Hamiltonian H. Since

$$\hat{N} = \hat{n}_\mathrm{s} + \hat{n}_\mathrm{d}, \tag{1.33}$$

(where we have denoted by a caret ˆ the operators corresponding to the total number, N, and the number of s and d bosons, n_s and n_d) one can eliminate $\hat{n}_\mathrm{s}$ from the expression (1.21). For example, the term $\varepsilon_\mathrm{s}(s^\dagger \cdot \tilde{s})$ can be written as

$$\varepsilon_\mathrm{s}(s^\dagger \cdot \tilde{s}) = \varepsilon_\mathrm{s}\hat{n}_\mathrm{s} = \varepsilon_\mathrm{s}(\hat{N} - \hat{n}_\mathrm{d}) = \varepsilon_\mathrm{s}\hat{N} - \varepsilon_\mathrm{s}(d^\dagger \cdot \tilde{d}). \tag{1.34}$$

Within the space of a given N, the operator $\hat{N}$ is diagonal with eigenvalue

Table 1.2. *List of two-body terms in the transition operators, Eq. (1.31)*

Operator	i	i'	j	j'	k	k'
E0	2	2	2	2	4	4
					2	2
					0	0
	2	2	2	0	2	2
	2	2	0	0	0	0
	2	0	2	0	2	2
	0	0	0	0	0	0
M1	2	2	2	2	4	4
					2	2
	2	2	2	0	2	2
	2	0	2	0	2	2
E2	2	2	2	2	0	2
					2	2
					2	4
					4	4
	2	2	2	0	0	2
					2	2
					4	2
	2	2	0	0	2	0
	2	0	2	0	2	2
	2	0	0	0	2	0
M3	2	2	2	2	2	2
					2	4
					4	4
	2	2	2	0	2	2
					4	2
	2	0	2	0	2	2
E4	2	2	2	2	0	4
					2	2
					2	4
					4	4
	2	2	2	0	2	2
					4	2
	2	0	2	0	2	2
M5	2	2	2	2	2	4
					4	4
	2	2	2	0	4	2
E6	2	2	2	2	2	4
					4	4
	2	2	2	0	4	2
M7	2	2	2	2	4	4
E8	2	2	2	2	4	4

N. Elimination of $\hat{n}_s$ then yields

$$H = E'_0 + \varepsilon'(d^\dagger \cdot \tilde{d})$$

$$+ \sum_{L=0,2,4} \tfrac{1}{2}(2L+1)c'_L[[d^\dagger \times d^\dagger]^{(L)} \times [\tilde{d} \times \tilde{d}]^{(L)}]_0^{(0)}$$

$$+ \frac{1}{\sqrt{2}} v_2[[d^\dagger \times d^\dagger]^{(2)} \times [\tilde{d} \times \tilde{s}]^{(2)} + [d^\dagger \times s^\dagger]^{(2)} \times [\tilde{d} \times \tilde{d}]^{(2)}]_0^{(0)}$$

$$+ \tfrac{1}{2}v_0[[d^\dagger \times d^\dagger]^{(0)} \times [\tilde{s} \times \tilde{s}]^{(0)} + [s^\dagger \times s^\dagger]^{(0)} \times [\tilde{d} \times \tilde{d}]^{(0)}]_0^{(0)}, \tag{1.35}$$

$$E'_0 = E_0 + \varepsilon_s N + \tfrac{1}{2}u_0 N(N-1),$$

with

$$\varepsilon' = (\varepsilon_d - \varepsilon_s) + \frac{1}{\sqrt{5}} u_2(N-1) - u_0(N-1), \tag{1.36}$$

$$c'_L = c_L + u_0 - \frac{2}{\sqrt{5}} u_2.$$

Since E_0 itself is, if one stops at two-body terms in the Hamiltonian, a quadratic function of N,

$$E_0 = E_{00} + E_{01}N + \tfrac{1}{2}E_{02}N(N-1), \tag{1.37}$$

one can write

$$E'_0 = E_{00} + (E_{01} + \varepsilon_s)N + \tfrac{1}{2}(E_{02} + u_0)N(N-1). \tag{1.38}$$

The quantity E'_0 is the same for all states with fixed N (a given nucleus). It does not contribute to the calculation of excitation energies. These depend only on the six independent parameters ε', c'_L $(L=0,2,4)$, v_L $(L=0,2)$. Conversely, the quantity E'_0 plays a role in the calculation of absolute energies (for example, binding energies). For these, three extra independent parameters are needed, E_{00}, E'_{01}, E'_{02}, where

$$\begin{aligned} E'_{01} &= E_{01} + \varepsilon_s, \\ E'_{02} &= E_{02} + u_0. \end{aligned} \tag{1.39}$$

The other operator in which some terms can be eliminated is the E0 operator of Eq. (1.24). This operator can be rewritten as

$$T_0^{(\mathrm{E0})} = \gamma'_0 + \beta'_0[d^\dagger \times \tilde{d}]_0^{(0)} \tag{1.40}$$

where

$$\begin{aligned} \gamma'_0 &= \gamma_0 + \alpha_0 N, \\ \beta'_0 &= \beta_0 - \sqrt{5}\,\alpha_0. \end{aligned} \tag{1.41}$$

Again, the constant term γ_0' does not contribute to transitions. Since γ_0 is itself a linear function of N (if one stops at one-body terms in the transition operators),

$$\gamma_0 = \gamma_{00} + \gamma_{01} N, \tag{1.42}$$

one can write

$$\gamma_0' = \gamma_{00} + (\gamma_{01} + \alpha_0) N. \tag{1.43}$$

Thus, diagonal matrix elements of the E0 operator depend on two more parameters γ_{00} and γ_{01}' in addition to β_0'. Here

$$\gamma_{01}' = \gamma_{01} + \alpha_0. \tag{1.44}$$

1.4.6 Other realizations

The conservation of total boson number allows one to introduce other realizations of the Hamiltonian and transition operators. Particularly important is the realization introduced by Janssen, Jolos and Dönau (1974). This realization is based on the so-called Holstein–Primakoff (Holstein and Primakoff, 1940) realization of the algebra of bilinear products of boson operators, to be discussed in Chapter 2. In this realization, the s boson operators can be eliminated by appropriate replacements. For example, s and $s^\dagger$ operators can be eliminated using

$$s \text{ or } s^\dagger \rightarrow \sqrt{[N - (d^\dagger \cdot \tilde{d})]}. \tag{1.45}$$

However, since these operators contain explicitly d bosons, their ordering becomes important. When applied to the Hamiltonian, H, of Eq. (1.35), the replacement procedure yields

$$\begin{aligned} H^{\mathrm{HP}} = E_0' + \varepsilon'(d^\dagger \cdot \tilde{d}) \\ &+ \sum_{L=0,2,4} \tfrac{1}{2}(2L+1)^{\frac{1}{2}} c_L' [[d^\dagger \times d^\dagger]^{(L)} \times [\tilde{d} \times \tilde{d}]^{(L)}]_0^{(0)} \\ &+ \frac{1}{\sqrt{2}} v_2 [[d^\dagger \times d^\dagger]^{(2)} \times \tilde{d}]_0^{(0)} \sqrt{[N - (d^\dagger \cdot \tilde{d})]} \\ &+ \sqrt{[N - (d^\dagger \cdot \tilde{d})]} [d^\dagger \times [\tilde{d} \times \tilde{d}]^{(2)}]_0^{(0)} \\ &+ \frac{1}{\sqrt{2}} v_0 [[d^\dagger \times d^\dagger]_0^{(0)} \sqrt{[N - (d^\dagger \cdot \tilde{d})]} \sqrt{[N - (d^\dagger \cdot \tilde{d}) - 1]} \\ &+ \sqrt{[N - (d^\dagger \cdot \tilde{d})]} \sqrt{[N - (d^\dagger \cdot \tilde{d}) - 1]} [\tilde{d} \times \tilde{d}]_0^{(0)}]. \end{aligned} \tag{1.46}$$

This Hamiltonian contains only d boson operators, whose number is not conserved, but bounded by $n_d \leqslant N$. Since the number is bounded, the corresponding model is called the truncated quadrupole model (TQM).

The only other operator affected by the replacement (1.45) is the E2 transition operator which becomes

$$T_\mu^{(E2)\,HP} = \alpha_2[d_\mu^\dagger \sqrt{[N-(d^\dagger\cdot\tilde{d})]} + \sqrt{[N-(d^\dagger\cdot\tilde{d})]}\tilde{d}_\mu] + \beta_2[d^\dagger \times \tilde{d}]_\mu^{(2)}. \tag{1.47}$$

From the mathematical point of view, the two realizations are equivalent and produce identical results. The realization (1.24), (1.35) and (1.40), is sometimes called the Schwinger realization (Schwinger, 1965), since it is based on a realization of the algebra of bilinear products of boson operators introduced originally by Schwinger. A third realization, called the Dyson realization (Dyson, 1956) is also sometimes used. The interrelations among the three realizations have been discussed recently in detail by Kyrchev and Paar (1986).

1.4.7 Multipole expansion

In addition to those described above, there are several other equivalent ways of writing the Hamiltonian and transition operators. Two forms often used are

$$H = E_0' + \varepsilon_d \hat{n}_d + a_0(\hat{\mathscr{P}}^\dagger\cdot\hat{\mathscr{P}}) + a_1(\hat{L}\cdot\hat{L}) + a_2(\hat{Q}\cdot\hat{Q}) + a_3(\hat{U}\cdot\hat{U}) + a_4(\hat{V}\cdot\hat{V}), \tag{1.48}$$

and

$$H = E_0' + \varepsilon \hat{n}_d + a_0(\hat{\mathscr{P}}^\dagger\cdot\hat{\mathscr{P}}) + a_1(\hat{L}\cdot\hat{L}) + a_2(\hat{Q}\cdot\hat{Q}) + a_3(\hat{U}\cdot\hat{U}) + a_5\hat{n}_d^2. \tag{1.49}$$

Since the operators appearing in these equations,

$$\begin{aligned}
\hat{n}_d &= (d^\dagger\cdot\tilde{d}),\\
\hat{\mathscr{P}} &= \tfrac{1}{2}(\tilde{d}\cdot\tilde{d}) - \tfrac{1}{2}(\tilde{s}\cdot\tilde{s}),\\
\hat{L} &= (10)^{\frac{1}{2}}[d^\dagger\times\tilde{d}]^{(1)},\\
\hat{Q} &= [d^\dagger\times\tilde{s} + s^\dagger\times\tilde{d}]^{(2)} - \tfrac{1}{2}(7^{\frac{1}{2}})[d^\dagger\times\tilde{d}]^{(2)},\\
\hat{U} &= [d^\dagger\times\tilde{d}]^{(3)},\\
\hat{V} &= [d^\dagger\times\tilde{d}]^{(4)},
\end{aligned} \tag{1.50}$$

are called multipole operators, the corresponding expressions (1.48) and (1.49) are often referred to as multipole expansions. The transition operators, $T^{(L)}$, can also be written in terms of the operators (1.50) plus the operator

$$\hat{Q}' = [d^\dagger \times \tilde{d}]^{(2)}, \tag{1.51}$$

as

$$\begin{aligned} T^{(\mathrm{E0})} &= \gamma'_0 + \beta''_0 \hat{n}_\mathrm{d}, \\ T^{(\mathrm{M1})} &= g'\hat{L}, \\ T^{(\mathrm{E2})} &= \alpha_2 \hat{Q} + \alpha'_2 \hat{Q}', \\ T^{(\mathrm{M3})} &= \beta_3 \hat{U}, \\ T^{(\mathrm{E4})} &= \beta_4 \hat{V}. \end{aligned} \tag{1.52}$$

1.4.8 Consistent-Q formalism

Finally, another parametrization used in some calculations is (Warner and Casten, 1983)

$$H = E'_0 + \varepsilon \hat{n}_\mathrm{d} + c_1(\hat{L}\cdot\hat{L}) + c_2(\hat{Q}^\chi\cdot\hat{Q}^\chi) + c_3(\hat{U}\cdot\hat{U}) + c_4(\hat{V}\cdot\hat{V}), \tag{1.53}$$

where the operators $\hat{n}_\mathrm{d}$, $\hat{L}$, $\hat{U}$ and $\hat{V}$ are the same as in Eq. (1.50), but

$$\hat{Q}^\chi = [d^\dagger \times \tilde{s} + s^\dagger \times \tilde{d}]^{(2)} + \chi[d^\dagger \times \tilde{d}]^{(2)}. \tag{1.54}$$

In this parametrization, the transition operators are given still by Eq. (1.52), except for the E2 operator which is given by

$$T^{(\mathrm{E2})} = \alpha_2 \hat{Q}^\chi. \tag{1.55}$$

This parametrization reduces the number of parameters by one, since the same quadrupole operator is used in the Hamiltonian and in the transition operator, and it is suggested by microscopic considerations.

1.4.9 Transfer operators

Another set of operators of practical interest is formed by transfer operators. In the interacting boson model, the bosons are interpreted as nucleon pairs. One may thus attempt to describe in terms of boson operators transfer reactions, in which a pair of nucleons is added or removed. These operators will be denoted by $P_+^{(L)}$ for addition and $P_-^{(L)}$ for removal. If one retains only one-body operators, one can write schematically

$$\begin{aligned} P_+^{(L)} &= \sum_\alpha p_\alpha^{(L)} b_\alpha^\dagger, \\ P_-^{(L)} &= \sum_\alpha p_\alpha^{(L)} b_\alpha. \end{aligned} \tag{1.56}$$

A more explicit form is

$$P^{(L)}_{+,m} = p_l b^{\dagger}_{l,m},$$
$$P^{(L)}_{-,m} = p_l \tilde{b}_{l,m}. \tag{1.57}$$

Introducing s and d bosons,

$$P^{(0)}_{+,0} = p_0 s^{\dagger}; \quad P^{(0)}_{-,0} = p_0 \tilde{s}$$
$$P^{(2)}_{+,\mu} = p_2 d^{\dagger}_{\mu}; \quad P^{(2)}_{-,\mu} = p_2 \tilde{d}_{\mu}. \tag{1.58}$$

To this order, only transfer with $L=0$ and $L=2$ can be calculated.

1.4.10 Higher-order terms in the transfer operators

Higher-order terms in transfer operators can be written as

$$P^{(L)}_{3,+} = \sum_{\alpha\beta\gamma} q^{(L)}_{\alpha\beta\gamma} b^{\dagger}_{\alpha} b^{\dagger}_{\beta} b_{\gamma},$$
$$P^{(L)}_{3,-} = \sum_{\alpha\beta\gamma} q^{(L)}_{\alpha\beta\gamma} b^{\dagger}_{\gamma} b_{\beta} b_{\alpha}. \tag{1.59}$$

In coupled tensor form, Eq. (1.59) can be rewritten as

$$P^{(L)}_{3,+,\mu} = \sum_{k,l,l',l''} q^{(L)}_{k,l,l',l''} [[b^{\dagger}_{l} \times b^{\dagger}_{l'}]^{(k)} \times \tilde{b}_{l''}]^{(L)}_{\mu},$$
$$P^{(L)}_{3,-,\mu} = \sum_{k,l,l',l''} q^{(L)}_{k,l,l',l''} [b^{\dagger}_{l''} \times [\tilde{b}_{l'} \times \tilde{b}_{l}]^{(k)}]^{(L)}_{\mu}. \tag{1.60}$$

A list of all possible cubic terms is given in Table 1.3. To this order, one can have transfer operators up to multipolarity six. As an example, we write down explicitly the transfer addition operator with $L=0$, in terms of s and d bosons, including cubic terms

$$P^{(0)}_{+,0} = p_0 s^{\dagger} + q_0 [[d^{\dagger} \times d^{\dagger}]^{(0)} \times \tilde{s}]^{(0)}_0 + q_2 [[d^{\dagger} \times d^{\dagger}]^{(2)} \times \tilde{d}]^{(0)}_0$$
$$+ q'_2 [[d^{\dagger} \times s^{\dagger}]^{(2)} \times \tilde{d}]^{(0)}_0 + q'_0 [[s^{\dagger} \times s^{\dagger}]^{(0)} \times \tilde{s}]^{(0)}_0. \tag{1.61}$$

There are, in this case, four additional terms, q_0, q_2, q'_0, q'_2, in addition to p_0. Some of these can be eliminated or rewritten by using the conservation of boson number (1.18).

1.4.11 Consistent-Q formalism for transfer operators

Transfer operators can also be written using a consistent-Q formalism. For example, in this formalism, the $L=0$ transfer addition operator, (1.61), is written as

$$P^{(0)}_{+,0}=s^{\dagger}(p_0+\tilde{q}_0\hat{N}+\tilde{q}'_0\hat{n}_d)+\tilde{q}'_2(d^{\dagger}\times\hat{Q}^{\chi})^{(0)}_0. \tag{1.62}$$

1.4.12 Ω formalism for transfer operators

Microscopic considerations have suggested another form of the transfer operators, usually used in calculations. This form is of the type

$$\begin{aligned} P^{(0)}_{+,0}&=p_0 s^{\dagger}\sqrt{(\Omega-N+\delta_0\hat{n}_d)}\\ P^{(0)}_{-,0}&=p_0\sqrt{(\Omega-N+\delta_0\hat{n}_d)}\tilde{s}\\ P^{(0)}_{+,\mu}&=p_2 d^{\dagger}_{\mu}\sqrt{(\Omega-N+\delta_0\hat{n}_d)}\\ P^{(0)}_{-,\mu}&=p_2\sqrt{(\Omega-N+\delta_0\hat{n}_d)}\tilde{d}_{\mu} \end{aligned} \tag{1.63}$$

and it is often referred to as the Ω form since the quantity Ω is related, in the microscopic theory, to the pair degeneracy of the shell.

1.4.13 Other realizations

As in the case of the Hamiltonian and transition operators, other realizations of the transfer operators are possible. Kyrchev and Paar (1983) have developed an algorithm by means of which two-nucleon transfer reaction strengths can be evaluated within the framework of the Holstein–Primakoff realization.

1.5 Transition densities

In electron-scattering experiments, one measures the spatial dependence of the matrix elements of the electromagnetic transition operators. In order to analyze these experiments one needs a generalization of the transition operators (1.24) to include this spatial dependence. This can be achieved by making the numbers, γ_0, α_0, α_2, β_L ($L=0, \ldots, 4$), functions of a spatial coordinate, r. Thus, one writes (Dieperink *et al.*, 1978; Dieperink, 1979; Iachello, 1981),

$$\begin{aligned} T^{(E0)}_0(r)&=\gamma_0(r)+\alpha_0(r)[s^{\dagger}\times\tilde{s}]^{(0)}_0+\beta_0(r)[d^{\dagger}\times\tilde{d}]^{(0)}_0,\\ T^{(M1)}_{\mu}(r)&=\beta_1(r)[d^{\dagger}\times\tilde{d}]^{(1)}_{\mu},\\ T^{(E2)}_{\mu}(r)&=\alpha_2(r)[d^{\dagger}\times\tilde{s}+s^{\dagger}\times\tilde{d}]^{(2)}_{\mu}+\beta_2(r)[d^{\dagger}\times\tilde{d}]^{(2)}_{\mu},\\ T^{(M3)}_{\mu}(r)&=\beta_3(r)[d^{\dagger}\times\tilde{d}]^{(3)}_{\mu},\\ T^{(E4)}_{\mu}(r)&=\beta_4(r)[d^{\dagger}\times\tilde{d}]^{(4)}_{\mu}. \end{aligned} \tag{1.64}$$

The matrix elements of these operators, usually called transition densities, are then obtained in terms of the functions,

$$\gamma_0(r), \quad \alpha_0(r), \quad \alpha_2(r), \quad \beta_L(r) \quad (L=0,\ldots,4).$$

Table 1.3. *List of possible cubic terms in the transfer operators*

Operator (L)	l	l'	k	l''
0	2	2	0	0
			2	2
	2	0	2	2
	0	0	0	0
1	2	2	2	2
	2	0	2	2
2	2	2	0	2
			2	0
			2	2
			4	2
	2	0	2	0
			2	2
	0	0	0	2
3	2	2	2	2
			4	2
	2	0	2	2
4	2	2	2	2
			4	2
	2	0	2	2
5	2	2	4	2
6	2	2	4	2

2
Algebras

2.1 Introduction

Having defined operators $b_\alpha^\dagger$ and space $\mathscr{B}$, one could now proceed and find properties of the corresponding physical system by direct numerical methods. However, it turns out that a great deal of information about the system can be obtained by exploiting its algebraic properties. This is one of the main advantages of models of this type and this chapter will be devoted to the study of the algebraic properties of the interacting boson model-1. The reader of this chapter, which is of fundamental importance for an understanding of the model, is expected to have a rudimentary knowledge of the theory of Lie algebras.

Besides shedding light on the structure of the solutions, the study of the algebraic properties of the model allows one to construct bases in which the numerical diagonalization of the Hamiltonian and the evaluation of the matrix elements of relevant operators can be done. This is obviously the ultimate goal if one wants to compare the results of the calculations with experiments.

2.2 Boson algebra

Consider the set of bilinear products of the boson creation ($b_\alpha^\dagger$) and annihilation (b_α) operators

$$\mathscr{G}: G_{\alpha\beta} = b_\alpha^\dagger b_\beta, \quad (\alpha, \beta = 1, \ldots, 6). \tag{2.1}$$

These operators satisfy the commutation relations

$$[G_{\alpha\beta}, G_{\gamma\delta}] = G_{\alpha\delta}\delta_{\beta\gamma} - G_{\gamma\beta}\delta_{\delta\alpha}. \tag{2.2}$$

Operators X satisfying the commutation relations

$$[X_a, X_b] = \sum_c C_{ab}^c X_c, \tag{2.3}$$

Table 2.1. *Admissible Lie algebras*

Name	Label	Cartan label
[Special] Unitary	[S]U(n)	$A_{(n-1)}$
[Special] Orthogonal	[S]O(n) n=odd	$B_{(n-1)/2}$
[Special] Orthogonal	[S]O(n) n=even	$D_{(n/2)}$
Symplectic	Sp(n) n=even	$C_{(n/2)}$
Exceptional	G_2, F_4, E_6, E_7, E_8	G_2, F_4, E_6, E_7, E_8

with $C^c_{ab} = -C^c_{ba}$, together with the Jacobi identity

$$[[X_a, X_b], X_c] + [[X_b, X_c], X_a] + [[X_c, X_a], X_b] = 0, \tag{2.4}$$

are said to form a Lie algebra (Wybourne, 1974).

The $6 \times 6 = 36$ operators (2.1) satisfy the commutation relations of the unitary algebra in six dimensions, $u(6)$. Associated with each algebra there are groups of transformations. These are usually denoted by capital letters, U(6). In this chapter, we shall not make use of groups, but only of algebras. Following standard notation we will, however, not distinguish between a group, U(6), and its algebra, $u(6)$, but denote both by a capital letter, U(6).

The quantities C^c_{ab} appearing in (2.3) are called Lie structure constants. For the algebra g of (2.1), the Lie structure constants are either zero or one. For applications to nuclear physics the form (2.1) is not very convenient. It is more appropriate to use a coupled form, in the sense of Racah,

$$G^{(k)}_{\kappa}(l, l') = [b^\dagger_l \times \tilde{b}_{l'}]^{(k)}_{\kappa}, \quad (l, l' = 0, 2). \tag{2.5}$$

The commutation relations of the operators (2.5) are

$$[G^{(k)}_{\kappa}(l, l'), G^{(k')}_{\kappa'}(l'', l''')]$$
$$= \sum_{k'', \kappa''} (2k+1)^{\frac{1}{2}}(2k'+1)^{\frac{1}{2}}(k\kappa k'\kappa' | k''\kappa'')(-)^{k-k'}$$
$$\times \Bigg[(-)^{k+k'+k''} \begin{Bmatrix} k & k' & k'' \\ l''' & l & l' \end{Bmatrix} \delta_{l'l''} G^{(k'')}_{\kappa''}(l, l''')$$
$$- \begin{Bmatrix} k & k' & k'' \\ l'' & l' & l \end{Bmatrix} \delta_{ll'''} G^{(k'')}_{\kappa''}(l'', l') \Bigg]. \tag{2.6}$$

In this expression, the quantity in curly brackets denotes a Wigner $6j$-symbol (De Shalit and Talmi, 1963). The operators G are also called generators of the corresponding group. The 36 operators (2.5) are thus the generators of U(6) in Racah's form. Their explicit expression in terms of s and d bosons is

$$
\begin{array}{lr}
G_0^{(0)}(s,s)=[s^\dagger \times \tilde{s}]_0^{(0)} & 1 \\
G_0^{(0)}(d,d)=[d^\dagger \times \tilde{d}]_0^{(0)} & 1 \\
G_\mu^{(1)}(d,d)=[d^\dagger \times \tilde{d}]_\mu^{(1)} & 3 \\
G_\mu^{(2)}(d,d)=[d^\dagger \times \tilde{d}]_\mu^{(2)} & 5 \\
G_\mu^{(3)}(d,d)=[d^\dagger \times \tilde{d}]_\mu^{(3)} & 7 \\
G_\mu^{(4)}(d,d)=[d^\dagger \times \tilde{d}]_\mu^{(4)} & 9 \\
G_\mu^{(2)}(d,s)=[d^\dagger \times \tilde{s}]_\mu^{(2)} & 5 \\
G_\mu^{(2)}(s,d)=[s^\dagger \times \tilde{d}]_\mu^{(2)} & 5 \\
 & \overline{} \\
 & 36.
\end{array}
\tag{2.7}
$$

Each k-tensor has $(2k+1)$ independent components whose number is indicated to the right.

2.3 Boson subalgebras

For purposes of classification of states and of construction of a basis, one needs to construct all possible subalgebras of the algebra of U(6). In general, given an algebra $\mathscr{g}$, any subset $\mathscr{h}$ of $\mathscr{g}$ closed with respect to commutation is called a subalgebra. This means that if

$$X \in \mathscr{g}, \quad Y \in \mathscr{h}, \quad X \supset Y, \tag{2.8}$$

then

$$[Y_i, Y_j] = \sum_k C_{ij}^k Y_k. \tag{2.9}$$

For applications to nuclear physics, we want states characterized by a good value of the angular momentum. This has the consequence that we must always include, as a subalgebra of U(6), the algebra of three-dimensional rotations, which we denote by O(3), since this algebra is orthogonal. The search for appropriate subalgebras is simplified by the fact that all admissible Lie algebras have been classified by Cartan (Wybourne, 1974). This classification is reproduced in Table 2.1.

The letter S is introduced to denote algebras that correspond to special transformations, i.e. transformations with determinant = +1. For applications in nuclear physics, the letter S is not essential when dealing

Table 2.2. *Number of generators of Lie groups*

Group	Number
U(n)	n^2
SU(n)	n^2-1
O(n)	$\frac{1}{2}n(n-1)$
Sp(n)	$\frac{1}{2}n(n+1)$
G_2	14
F_4	52
E_6	78
E_7	133
E_8	248

with orthogonal groups, since all the orthogonal algebras we use correspond to special transformations. We shall therefore omit it, in order to avoid confusion. For unitary algebras (and groups) the letter S is instead essential since a unitary group and its corresponding special unitary group differ by one generator (i.e. the unitary group U(n) has one more generator than its corresponding group SU(n)). The number of generators of the admissible Lie groups is shown in Table 2.2, and we shall pay particular attention to the letter S for unitary groups, since this problem has in the past been mistreated in the literature. In the particular case we are treating in this book, it is convenient to search for appropriate subalgebras in the coupled form (2.7). This will insure automatically the inclusion of O(3) as a subalgebra, since the three operators $G_\mu^{(1)}(d,d)$ are the generators of the O(3) algebra.

2.3.1 *Boson subalgebra I*

A subalgebra of U(6) can be obtained by considering the 25 operators

$$
\begin{array}{ll}
G_0^{(0)}(d,d)=[d^\dagger\times\tilde{d}]_0^{(0)} & 1\\
G_\mu^{(1)}(d,d)=[d^\dagger\times\tilde{d}]_\mu^{(1)} & 3\\
G_\mu^{(2)}(d,d)=[d^\dagger\times\tilde{d}]_\mu^{(2)} & 5\\
G_\mu^{(3)}(d,d)=[d^\dagger\times\tilde{d}]_\mu^{(3)} & 7\\
G_\mu^{(4)}(d,d)=[d^\dagger\times\tilde{d}]_\mu^{(4)} & 9\\
 & \overline{}\\
 & 25.
\end{array}
\tag{2.10}
$$

These operators close under the algebra U(5). The ten operators

$$\begin{array}{ll} G_\mu^{(1)}(d,d)=[d^\dagger \times \tilde{d}]_\mu^{(1)} & 3 \\ G_\mu^{(3)}(d,d)=[d^\dagger \times \tilde{d}]_\mu^{(3)} & 7 \\ & \overline{} \\ & 10 \end{array} \tag{2.11}$$

form a subalgebra of U(5), the orthogonal algebra in five dimensions, O(5). The three operators

$$G_\mu^{(1)}(d,d)=[d^\dagger \times \tilde{d}]_\mu^{(1)} \qquad 3 \tag{2.12}$$

close under the algebra of O(3), the rotation algebra. Finally, the single component

$$G_0^{(1)}(d,d)=[d^\dagger \times \tilde{d}]_0^{(1)} \qquad 1 \tag{2.13}$$

generates the algebra O(2) of rotations around the z-axis. This yields a possible chain of algebras,

$$U(6) \supset U(5) \supset O(5) \supset O(3) \supset O(2), \quad (I) \tag{2.14}$$

which we shall denote by I.

2.3.2 *Boson subalgebra II*

While the first chain of algebras, I, can be derived almost by inspection, the second chain is more difficult to obtain, since it involves linear combinations of the operators in (2.7). Consider the operators

$$\begin{array}{ll} G_0^{(0)}(s,s)+\sqrt{5}\, G_0^{(0)}(d,d)=[s^\dagger \times \tilde{s}]_0^{(0)}+\sqrt{5}\,[d^\dagger \times \tilde{d}]_0^{(0)} & 1 \\ G_\mu^{(1)}(d,d)=[d^\dagger \times \tilde{d}]_\mu^{(1)} & 3 \\ G_\mu^{(2)}(d,s)+G_\mu^{(2)}(s,d)\pm\tfrac{1}{2}\sqrt{7}\, G_\mu^{(2)}(d,d) & \\ \quad =[d^\dagger \times \tilde{s}+s^\dagger \times \tilde{d}]_\mu^{(2)}\pm\tfrac{1}{2}\sqrt{7}\,[d^\dagger \times \tilde{d}]_\mu^{(2)} & 5 \\ & \overline{} \\ & 9. \end{array} \tag{2.15}$$

By using the commutation relations (2.6) one can show that the nine operators (2.15) close under commutation and form the algebra of U(3). This occurs for both signs ($\pm$) in (2.15). The operator $G_\mu^{(2)}(d,s)+G_\mu^{(2)}(s,d)\pm\frac{1}{2}\sqrt{7}\, G_\mu^{(2)}(d,d)$ is proportional to the electric quadrupole operator. In a geometric picture, this is in turn related to quadrupole deformations of the nucleus, as it will be shown later. The two signs correspond to positive (+) and negative (−) quadrupole moments. Since most nuclei have negative quadrupole moments, the negative sign is usually taken in (2.15). Furthermore, the operator $G_0^{(0)}(s,s)+\sqrt{5}\,G_0^{(0)}(d,d)$ is the total number operator, $\hat{N}$. For applications to nuclei, where the total N is conserved, the eigenvalues of $\hat{N}$ are fixed. Thus, rather than the algebra U(3) it is more convenient to consider the algebra in which the operator $\hat{N}$ has been deleted. This is formed by the eight operators

$$\begin{aligned} &G_\mu^{(1)}(d,d)=[d^\dagger\times\tilde{d}]_\mu^{(1)} && 3\\ &G_\mu^{(2)}(d,s)+G_\mu^{(2)}(s,d)-\tfrac{1}{2}\sqrt{7}\,G_\mu^{(2)}(d,d)\\ &\qquad=[d^\dagger\times\tilde{s}+s^\dagger\times\tilde{d}]_\mu^{(2)}-\tfrac{1}{2}\sqrt{7}\,[d^\dagger\times\tilde{d}]_\mu^{(2)} && 5\\ & && \overline{8}, \end{aligned} \tag{2.16}$$

and turns out to be the algebra SU(3) of special unitary transformations in three dimensions. Subalgebras of (2.16) are now that of O(3), formed by,

$$G_\mu^{(1)}(d,d)=[d^\dagger\times\tilde{d}]_\mu^{(1)} \qquad 3, \tag{2.17}$$

and that of O(2), formed by

$$G_0^{(1)}(d,d)=[d^\dagger\times\tilde{d}]_0^{(1)} \qquad 1. \tag{2.18}$$

Thus, a second possible chain of subalgebras is

$$\mathrm{U}(6)\supset\mathrm{SU}(3)\supset\mathrm{O}(3)\supset\mathrm{O}(2), \quad \text{(II)}. \tag{2.19}$$

2.3.3 Boson subalgebra III

A third possibility is to consider the 15 operators

$$\begin{aligned} G_\mu^{(1)}(d,d)&=[d^\dagger\times\tilde{d}]_\mu^{(1)} && 3\\ G_\mu^{(3)}(d,d)&=[d^\dagger\times\tilde{d}]_\mu^{(3)} && 7\\ G_\mu^{(2)}(d,s)+G_\mu^{(2)}(s,d)&=[d^\dagger\times\tilde{s}+s^\dagger\times\tilde{d}]_\mu^{(2)} && 5\\ & && \overline{15}. \end{aligned} \tag{2.20}$$

These operators close under the algebra O(6) of orthogonal transformations in six dimensions. Deleting the five operators $G_\mu^{(2)}(d,s)+G_\mu^{(2)}(s,d)$, one obtains the ten operators

$$\begin{aligned} G_\mu^{(1)}(d,d)&=[d^\dagger\times\tilde{d}]_\mu^{(1)} && 3\\ G_\mu^{(3)}(d,d)&=[d^\dagger\times\tilde{d}]_\mu^{(3)} && 7\\ & && \overline{10} \end{aligned} \tag{2.21}$$

which form, as above, the algebra of O(5). Subalgebras of O(5) are that of O(3),

$$G_{\mu}^{(1)}(d,d)=[d^{\dagger}\times \tilde{d}]_{\mu}^{(1)} \qquad 3 \tag{2.22}$$

and O(2),

$$G_{0}^{(1)}(d,d)=[d^{\dagger}\times \tilde{d}]_{0}^{(1)} \qquad 1. \tag{2.23}$$

Thus, a third chain is

$$\mathrm{U}(6)\supset \mathrm{O}(6)\supset \mathrm{O}(5)\supset \mathrm{O}(3)\supset \mathrm{O}(2), \quad \text{(III)}. \tag{2.24}$$

One can show that the three subalgebras I, II and III are the only possible ones, if one wants to include the rotation algebra, O(3), as a subalgebra. In fact, starting from U(6) one has considered U(5), U(3) and O(6). The algebra of O(6) is isomorphic to SU(4) and thus all possibilities leading from U(6) to the rotation algebra O(3) $\approx$ SU(2) have been considered. In conclusion, there are three and only three possibilities

$$\begin{array}{lll} & \mathrm{U}(5)\supset \mathrm{O}(5)\supset \mathrm{O}(3)\supset \mathrm{O}(2), & \text{(I)} \\ & \nearrow & \\ \mathrm{U}(6) & \rightarrow \mathrm{SU}(3)\supset \mathrm{O}(3)\supset \mathrm{O}(2), & \text{(II)} \\ & \searrow & \\ & \mathrm{O}(6)\supset \mathrm{O}(5)\supset \mathrm{O}(3)\supset \mathrm{O}(2), & \text{(III)}. \end{array} \tag{2.25}$$

Before closing this section, it is worthwhile mentioning that, for operators involving linear combinations of the generators of U(6), there are phase ambiguities similar to those discussed in Sect. 1.4.3. We shall call the choice made here the 'standard' choice, and keep it throughout. Other choices have also been used (Castaños *et al.*, 1979).

2.4 Basis states

One of the main uses of the group (or algebra) chains discussed above is that they allow one to construct bases in which the Hamiltonian operator can be diagonalized. This is because one can characterize the states of the basis $\mathscr{B}$ of (1.15) by their transformation properties under the groups in (2.25). In other words, one can construct bases that transform as representations of the appropriate groups and label those states with the corresponding quantum numbers.

Tensor representations of Lie groups are labelled by a set of integer numbers $[\lambda_1, \lambda_2, \ldots]$ with $\lambda_1 \geqslant \lambda_2 \geqslant \ldots$, which are sometimes displayed

Table 2.3. *Number of integers that characterize the tensor representation of Lie groups*

Group	Number
U(n)	n
SU(n)	$n-1$
O(n), n = even	$n/2$
O(n), n = odd	$(n-1)/2$
Sp(n)	$n/2$
G_2	2
F_4	4
E_6	6
E_7	7
E_8	8

pictorially in a Young tableau (Wybourne, 1974)

$$[\lambda_1, \lambda_2, \ldots] = \begin{array}{l} \overbrace{\square\,\square\,\square}^{\lambda_1} \\ \overbrace{\square\,\square}^{\lambda_2} \\ \cdots \end{array} \qquad (2.26)$$

The Young tableau is important because it displays clearly the symmetry properties of the states. We recall in Table 2.3 the number of integers λ_1, λ_2, ... needed to characterize the representations of Lie groups. For the case of identical bosons we are treating here, the wave function must be totally symmetric. This means that we will be dealing here only with the totally symmetric representations of U(6), characterized by a single number N, the total boson number

$$[N] \equiv \overbrace{\square\,\square \ldots \square}^{N \text{ times}} . \qquad (2.27)$$

The central problem for the construction of a basis is the decomposition of the representation $[N]$ of U(6) into representations of subgroups. This is a standard group theoretic problem and we present here only its solution.

2.4.1 Chain I

Representations of the groups appearing in this chain are characterized by the quantum numbers (Arima and Iachello, 1976).

$$\begin{array}{ll} \mathrm{U}(6) & [N,0,0,0,0,0] \equiv [N] \\ \mathrm{U}(5) & (n_\mathrm{d},0,0,0,0) \equiv (n_\mathrm{d}) \\ \mathrm{O}(5) & (v,0) \equiv v \\ \mathrm{O}(3) & L \\ \mathrm{O}(2) & M_L . \end{array} \tag{2.28}$$

The values of n_d contained in a given representation $[N]$ of U(6) are

$$n_\mathrm{d} = N, \quad N-1, \quad \ldots, \quad 1, \quad 0. \tag{2.29}$$

The values of v contained in a representation (n_d) of U(5) are

$$v = n_\mathrm{d}, \quad n_\mathrm{d}-2, \quad \ldots, \quad 1 \text{ or } 0 \quad (n_\mathrm{d} = \text{odd or even}). \tag{2.30}$$

The representations of U(6), U(5) and 0(3) are all totally symmetric. When going from O(5) to O(3) a subtle problem arises: several states with the same value of the quantum number L may appear in a given representation, v, of O(5). When this occurs in general for the group reduction $G \supset G'$, one says that G is not fully reducible with respect to G'. Thus, here, O(5) is not fully reducible with respect to O(3). One needs then a further quantum number (sometimes called the missing label) to characterize the states uniquely. This quantum number can be chosen in an infinite number of ways. For classification purposes, it is convenient to choose this number as the number of boson triplets coupled to zero angular momentum, denoted by n_Δ. The algorithm to find the values of L contained in each representation v is then as follows. First, partition n_d as

$$n_\mathrm{d} = 2n_\beta + 3n_\Delta + \lambda, \tag{2.31}$$

where

$$n_\beta = (n_\mathrm{d} - v)/2. \tag{2.32}$$

Since v can take the values (2.30), n_β can take the values

$$n_\beta = 0, \quad 1, \quad \ldots, \quad n_\mathrm{d}/2 \quad \text{or} \quad (n_\mathrm{d}-1)/2; \quad (n_\mathrm{d} = \text{even or odd}). \tag{2.33}$$

Then

$$L = \lambda, \quad \lambda+1, \quad \ldots, \quad 2\lambda-2, \quad 2\lambda. \tag{2.34}$$

(Note that $2\lambda-1$ is missing!) Finally, M_L takes all integer values between $-L$ and $+L$, as usual,

$$-L \leqslant M_L \leqslant +L. \tag{2.35}$$

The basis defined in this way has the drawback of not being orthogonal. An orthogonal basis, which we shall call the Szpikowski basis (Szpikowski and Góźdź, 1980) can be obtained in the following way. For each n_d and v, let $n_{\Delta_1}, n_{\Delta_2}, \ldots$ be the values of the quantum number n_Δ with $n_{\Delta_1} < n_{\Delta_2} < n_{\Delta_3} < \cdots$. The new basis is labeled by the quantum numbers $\tilde{n}_{\Delta_1}$, $\tilde{n}_{\Delta_2}$, ..., with $\tilde{n}_{\Delta_1} < \tilde{n}_{\Delta_2} < \cdots$ and defined by

$$\begin{aligned}
&|n_d, v, \tilde{n}_{\Delta_1}, L, M_L\rangle = |n_d, v, n_{\Delta_1}, L, M_L\rangle, \\
&|n_d, v, \tilde{n}_{\Delta_2}, L, M_L\rangle = x_{21}|n_d, v, n_{\Delta_1}, L, M_L\rangle \\
&\qquad\qquad + x_{22}|n_d, v, n_{\Delta_2}, L, M_L\rangle, \\
&\ldots, \\
&|n_d, v, \tilde{n}_{\Delta_i}, L, M_L\rangle = \sum_{j=1}^{i} x_{ij}|n_d, v, n_{\Delta_j}, L, M_L\rangle.
\end{aligned} \tag{2.36}$$

The coefficients x_{ij} are obtained by the requirement

$$\langle n_d, v, \tilde{n}_{\Delta_i}, L, M_L | n_d, v, \tilde{n}_{\Delta_j}, L, M_L\rangle = \delta_{ij}. \tag{2.37}$$

This algorithm gives the values shown in Table 2.4. For $N > 4$, the appropriate values of n_d, v, $\tilde{n}_\Delta$, L can be constructed using Eqs. (2.29)–(2.34). Summarizing, the complete classification scheme for this chain is

$$\left|\begin{array}{ccccc} U(6) \supset & U(5) \supset & O(5) \supset & O(3) \supset & O(2) \\ \downarrow & \downarrow & \downarrow & \downarrow & \downarrow \\ [N], & n_d, & v, \tilde{n}_\Delta, & L, & M_L \end{array}\right\rangle. \tag{2.38}$$

2.4.2 *Chain II*

The labels needed to classify the states in this chain are (Arima and Iachello, 1978)

$$\begin{array}{ll} U(6) & [N,0,0,0,0,0] \equiv [N] \\ SU(3) & (f_1, f_2) \\ O(3) & L \\ O(2) & M_L. \end{array} \tag{2.39}$$

Instead of the Young tableau (f_1, f_2), it is customary to introduce the quantum numbers (Elliott, 1958)

$$\begin{aligned} \lambda &= f_1 - f_2, \\ \mu &= f_2, \end{aligned} \tag{2.40}$$

Table 2.4. *Classification scheme for the group chain I*

U(6)	U(5)	O(5)		O(3)
N	n_d	v	$\tilde{n}_\Delta$	L
0	0	0	0	0
1	0	0	0	0
	1	1	0	2
2	0	0	0	0
	1	1	0	2
	2	2	0	4, 2
		0	0	0
3	0	0	0	0
	1	1	0	2
	2	2	0	4, 2
		0	0	0
	3	3	0	6, 4, 3
			1	0
		1	0	2
4	0	0	0	0
	1	1	0	2
	2	2	0	4, 2
		0	0	0
	3	3	0	6, 4, 3
			1	0
		1	0	2
	4	4	0	8, 6, 5, 4
			1	2
		2	0	4, 2
		0	0	0

(Elliott numbers). The values of (λ, μ) contained in a symmetric representation $[N]$ of U(6) are given by

$$(\lambda,\mu)=(2N,0)\oplus(2N-4,2)\oplus(2N-8,4)\oplus\cdots\oplus\begin{Bmatrix}(0,N)\\(2,N-1)\end{Bmatrix}\begin{Bmatrix}N=\text{even}\\N=\text{odd}\end{Bmatrix}$$

$$\oplus(2N-6,0)\oplus(2N-10,2)\oplus\cdots\oplus\begin{Bmatrix}(0,N-3)\\(2,N-4)\end{Bmatrix}\qquad\begin{Bmatrix}N-3=\text{even}\\N-3=\text{odd}\end{Bmatrix}$$

$$\oplus(2N-12,0)\oplus(2N-16,2)\oplus\cdots\oplus\begin{Bmatrix}(0,N-6\\(2,N-7)\end{Bmatrix}\qquad\begin{Bmatrix}N-6=\text{even}\\N-6=\text{odd}\end{Bmatrix}$$

$$\oplus\cdots. \tag{2.41}$$

Again here the step from SU(3) to O(3) is not fully decomposable and one needs an additional quantum number to classify the states uniquely. The simplest choice of this additional quantum number is due to Elliott (1958). The corresponding number is called K. The values of L contained in each representation (λ, μ) are then given by the following algorithm:

$$L=K, \quad K+1, \quad K+2, \quad \ldots, \quad K+\max\{\lambda,\mu\}, \tag{2.42}$$

where

$$K=\text{integer}=\min\{\lambda,\mu\}, \quad \min\{\lambda,\mu\}-2, \quad \ldots, \quad 1 \text{ or } 0;$$
$$(\min\{\lambda,\mu\}=\text{odd or even}), \tag{2.43}$$

with the exception of $K=0$ for which

$$L=\max\{\lambda,\mu\}, \quad \max\{\lambda,\mu\}-2, \quad \ldots, \quad 1 \text{ or } 0;$$
$$(\max\{\lambda,\mu\}=\text{odd or even}). \tag{2.44}$$

Elliott's basis has the drawback of not being orthogonal. For this reason, it is convenient to introduce another basis, which we shall call the Vergados basis (Vergados, 1968). This basis can be constructed from Elliott's basis in the following way. Let $K_1, K_2, \ldots, K_n$ be the Elliott quantum numbers which occur in a given representation (λ, μ) with $K_1<K_2<\cdots<K_n$. The new basis is labeled by the quantum numbers $\tilde{\chi}_1, \tilde{\chi}_2, \ldots, \tilde{\chi}_n$, with $\tilde{\chi}_1<\tilde{\chi}_2<\cdots<\tilde{\chi}_n$ and defined by

$$|(\lambda,\mu),\tilde{\chi}_1,L,M_L\rangle=|(\lambda,\mu),K_1,L,M_L\rangle_0$$

$$|(\lambda,\mu),\tilde{\chi}_2,L,M_L\rangle=x_{21}|(\lambda,\mu),K_1,L,M_L\rangle_0+x_{22}|(\lambda,\mu),K_2,L,M_L\rangle_0$$

$$\ldots$$

$$|(\lambda,\mu),\tilde{\chi}_i,L,M_L\rangle=\sum_{j=1}^{i} x_{ij}|(\lambda,\mu),K_j,L,M_L\rangle, \tag{2.45}$$

where the states $|(\lambda,\mu),K,L,M_L\rangle_0$ are related to Elliott states $|(\lambda,\mu),\mathrm{K},\mathrm{L}, \mathrm{M_L}\rangle$ by the phase convention

$$|(\lambda,\mu),K,L,M_L\rangle_0=i^{\lambda+2\mu}|(\lambda,\mu),K,L,M_L\rangle, \tag{2.46}$$

and the coefficients x_{ij} are obtained by the requirement

$$\langle(\lambda,\mu),\tilde{\chi}_i,L,M_L|(\lambda,\mu),\tilde{\chi}_j,L,M_L\rangle=\delta_{ij}. \tag{2.47}$$

Since $\lambda+2\mu=\text{even}$ in the present case, the phase in Eq. (2.46) is either $+1$ or -1. The sequence of quantum numbers $\tilde{\chi}_1, \tilde{\chi}_2, \ldots, \tilde{\chi}_n$ is the same as $K_1, K_2, \ldots, K_n$ but the values of L contained in each $\tilde{\chi}_i$ are different from those contained in K_i. In fact, from its definition, it is clear that if a given L occurs

Table 2.5. *Classification scheme for the group chain II*

U(6)	SU(3)		O(3)
N	(λ,μ)	$\tilde{\chi}$	L
0	(0,0)	0	0
1	(2,0)	0	2, 0
2	(4,0)	0	4, 2, 0
	(0,2)	0	2, 0
3	(6,0)	0	6, 4, 2, 0
	(2,2)	0	4, 2, 0
		2	3, 2
	(0,0)	0	0
4	(8,0)	0	8, 6, 4, 2, 0
	(4,2)	0	6, 4, 2, 0
		2	5, 4, 3, 2
	(0,4)	0	4, 2, 0
	(2,0)	0	2, 0

in a given representation once, it belongs to the lowest possible $\tilde{\chi}$. If it occurs twice, it belongs to the two lowest possible $\tilde{\chi}$s, etc. The only exception is when $\tilde{\chi}=0$, for which the allowed L values are restricted to be even or odd for λ even or odd, respectively. The algorithm described above gives the values shown in Table 2.5. The complete classification for chain II is

$$\left|\begin{array}{ccccccc} \mathrm{U}(6) & \supset & \mathrm{SU}(3) & \supset \mathrm{O}(3) & \supset \mathrm{O}(2) \\ \downarrow & & \downarrow & \downarrow & \downarrow \\ {[N]}, & & (\lambda,\mu),\tilde{\chi}, & L, & M_L \end{array}\right\rangle. \tag{2.48}$$

We note here in passing that, since the missing label K can be chosen in an infinite number of ways, there are several other bases, in addition to (2.48), that have been considered in detail. Van Isacker and Lipas (1985) have studied recently some consequences of using a basis different from (2.48).

2.4.3 Chain III

The labels needed to classify the representations of the group in this chain are (Arima and Iachello, 1979)

$$\begin{array}{ll} \mathrm{U}(6) & [N,0,0,0,0,0] \equiv N \\ \mathrm{O}(6) & (\sigma,0,0,0] \equiv \sigma \\ \mathrm{O}(5) & (\tau,0) \equiv \tau \\ \mathrm{O}(3) & L \\ \mathrm{O}(2) & M_L. \end{array} \tag{2.49}$$

The values of the quantum number σ contained in a given representation N of U(6) are

$$\sigma = N, \quad N-2, \quad \ldots, \quad 1 \text{ or } 0, \quad (N = \text{odd or even}). \tag{2.50}$$

The values of the quantum number τ contained in a given representation σ of O(6) are

$$\tau = \sigma, \quad \sigma - 1, \quad \ldots, \quad 1, \quad 0. \tag{2.51}$$

The representations of U(6), O(6) and O(5) are all totally symmetric. Once more, the step from O(5) to O(3) is not fully reducible and an additional quantum number is needed. This quantum number will be called ν_Δ. The values of L are then obtained by partitioning τ as

$$\tau = 3\nu_\Delta + \lambda, \quad \nu_\Delta = 0, 1, \ldots, \tag{2.52}$$

and taking

$$L = \lambda, \quad \lambda + 1, \quad \ldots, \quad 2\lambda - 2, \quad 2\lambda. \tag{2.53}$$

(Note again that $2\lambda - 1$ is missing!) A problem similar to that discussed in Sect. 2.4.1 arises with the orthogonality of the basis. It can be dealt with in a similar way by introducing a new quantum number $\tilde{\nu}_{\Delta_i}$, and requiring that

$$\langle \sigma, \tau, \tilde{\nu}_{\Delta_i}, L, M_L | \sigma, \tau, \tilde{\nu}_{\Delta_j}, L, M_L \rangle = \delta_{ij}. \tag{2.54}$$

The algorithm discussed above gives the values shown in Table 2.6. The complete classification scheme for chain III is thus

$$\left| \begin{array}{ccccc} \mathrm{U}(6) \supset & \mathrm{O}(6) \supset & \mathrm{O}(5) \supset & \mathrm{O}(3) \supset & \mathrm{O}(2) \\ \downarrow & \downarrow & \downarrow & \downarrow & \downarrow \\ [N], & \sigma, & \tau, \tilde{\nu}_\Delta, & L, & M_L \end{array} \right\rangle. \tag{2.55}$$

2.4.4 Transformation brackets

The three bases, (2.38), (2.48) and (2.55) are complete and orthonormal. The Hamiltonian H can be diagonalized in any of them. It is also possible

Table 2.6. *Classification scheme for the group chain III*

U(6)	O(6)	O(5)		O(3)
N	σ	τ,	$\tilde{\nu}_\Delta$	L
0	0	0	0	0
1	1	1	0	2
		0	0	0
2	2	2	0	4, 2
		1	0	2
		0	0	0
	0	0	0	0
3	3	3	0	6, 4, 3
			1	0
		2	0	4, 2
		1	0	2
		0	0	0
	1	1	0	2
		0	0	0
4	4	4	0	8, 6, 5, 4
			1	2
		3	0	6, 4, 3
			1	0
		2	0	4, 2
		1	0	2
		0	0	0
	2	2	0	4, 2
		1	0	2
		0	0	0
	0	0	0	0

to go from one basis to the other by means of transformation brackets. Using these brackets one can expand the wave functions of one basis into those of another. The only transformation brackets that have explicitly been constructed in the present context are those relating the bases I and III (Arima and Iachello, 1979; Castaños *et al.*, 1979),

$$|[N], \sigma, \tau, \tilde{\nu}_\Delta, L, M_L\rangle = \sum_{n_d, v, \tilde{n}_\Delta} \zeta^{\sigma,\tau,\tilde{\nu}_\Delta}_{n_d,v,\tilde{n}_\Delta} |[N], n_d, v, \tilde{n}_\Delta, L, M_L\rangle, \qquad (2.56)$$

where $\zeta^{\sigma,\tau,\tilde{\nu}_\Delta}_{n_d,v,\tilde{n}_\Delta}$ are the (I–III) transformation brackets. In this particular case, the two chains have in common the O(5) group, in addition to O(3) and O(2). This implies that $\tilde{\nu}_\Delta = \tilde{n}_\Delta$ and $\tau = v$. Furthermore, the brackets do not depend on $\tilde{\nu}_\Delta$. Thus, they can be written as

$$|[N], \sigma, \tau, \tilde{\nu}_\Delta, L, M_L\rangle = \sum_{n_d} \zeta^{\sigma}_{n_d,\tau} |[N], n_d, \tau, \tilde{n}_\Delta, L, M_L\rangle. \qquad (2.57)$$

Explicit expressions for the brackets $\zeta^{\sigma}_{n_d,\tau}$ can be (and in some cases have been) derived. For example, the for O(6) representation $\sigma = N$, one has (Arima and Iachello, 1979)

$$\zeta^{\sigma=N}_{n_d,\tau} = \left(\frac{(N-\tau)!\,(N+3+\tau)!\,(n_d+1-\tau)!!}{(N-n_d)!\,2^{N+1}(N+1)!\,(n_d+1-\tau)!\,(n_d+3+\tau)!!} \right)^{\frac{1}{2}}, \tag{2.58}$$

with

$$n_d = \tau, \quad \tau+2, \quad \tau+4, \quad \ldots,$$
$$n_d \leqslant N. \tag{2.59}$$

Similarly, one can introduce (I–II) and (II–III) transformation brackets

$$|[N], (\lambda,\mu), \tilde{\chi}, L, M_L\rangle = \sum_{n_d,v,\tilde{n}_\Delta} \eta^{\lambda,\mu,\tilde{\chi}}_{n_d,v,\tilde{n}_\Delta} |[N], n_d, v, \tilde{n}_\Delta, L, M_L\rangle,$$

$$|[N], \sigma, \tau, \tilde{\nu}_\Delta, L, M_L\rangle = \sum_{\lambda,\mu,\tilde{\chi}} \theta^{\sigma,\tau,\tilde{\nu}_\Delta}_{\lambda,\mu,\tilde{\chi}} |[N], (\lambda,\mu), \tilde{\chi}, L, M_L\rangle, \tag{2.60}$$

and derive explicit expressions if needed.

2.5 Dynamic symmetries

In general, the Hamiltonian H must be diagonalized numerically in one of the three bases of the previous Sect. 2.4. However, insight into the nature of the solutions can be obtained by considering special circumstances in which the eigenvalue problem can be solved in closed form. These situations, called dynamic symmetries, have played a major role in the development of the interacting boson model. They arise when the Hamiltonian H can be written in terms only of certain operators $\mathscr{C}$, called Casimir (or invariant) operators, of a chain of groups

$$G \supset G' \supset \cdots, \tag{2.61}$$

i.e.

$$H = \alpha\mathscr{C}(G) + \alpha'\mathscr{C}(G') + \cdots, \tag{2.62}$$

where $\mathscr{C}(G)$ denotes a Casimir operator of the group G. Since these operators have the property of being diagonal in that chain, their eigenvalues, denoted by $\langle\mathscr{C}(G)\rangle$, provide a solution to H,

$$E = \alpha\langle\mathscr{C}(G)\rangle + \alpha'\langle\mathscr{C}(G')\rangle + \cdots. \tag{2.63}$$

2.5.1 Casimir operators

For a given algebra, $X \in g$, there exists a set of operators $\mathscr{C}$ (Casimir operators) such that

$$[\mathscr{C}, X_a]=0, \quad \text{for any } a. \tag{2.64}$$

The number of invariant operators of an algebra, called the rank of the algebra, is equal to the number of integers characterizing the tensor representations of Lie groups, given in Table 2.3.

In the notation of this book, in which the operators X are written in the coupled form (2.5), the Casimir operators are such that

$$[\mathscr{C}, \tilde{G}_\kappa^{(k)}(l, l')]=0, \quad \text{for any } k, \kappa, l, l', \tag{2.65}$$

where $\tilde{G}_\kappa^{(k)}(l, l')$ are the operators appropriate to each algebra. A trivial Casimir operator is the total number operator

$$G_0^{(0)}(s, s)+\sqrt{5}\, G_0^{(0)}(d, d)=\hat{n}_s+\hat{n}_d=\hat{N}. \tag{2.66}$$

This operator commutes with all 36 operators of U(6),

$$[\hat{N}, G_\kappa^{(k)}(l, l')]=0, \quad \text{for any } k, \kappa, l, l', \tag{2.67}$$

and it is thus a Casimir operator of U(6). Since it is linear in the generators, Eq. (2.66), it will be denoted by $\mathscr{C}_1(\text{U6})$. In general, an operator $\mathscr{C}$ can be linear, quadratic, cubic, ... in the generators. A Casimir operator containing p generators will be called of order p

$$\mathscr{C}_p=\Sigma_{\alpha_1\alpha_2\cdots\alpha_p} f_{\alpha_1\alpha_2\cdots\alpha_p} X_{\alpha_1} X_{\alpha_2} \cdots X_{\alpha_p}, \tag{2.68}$$

and denoted by $\mathscr{C}_p(G)$. For example, for the rotation group, with generators L_x, L_y, L_z, the Casimir invariant is

$$\vec{L}^2=L_x^2+L_y^2+L_z^2, \tag{2.69}$$

since

$$[\vec{L}^2, L_i]=0, \quad \text{any } i. \tag{2.70}$$

In the notation of this book, the quadratic invariant of O(3) is written as

$$G^{(1)}(d, d) \cdot G^{(1)}(d, d)=[d^\dagger \times \tilde{d}]^{(1)} \cdot [d^\dagger \times \tilde{d}]^{(1)}. \tag{2.71}$$

Since $G^{(1)}(d, d)$ is proportional to the angular momentum operator, (1.50), the operator in (2.71) is proportional to $\vec{L}^2=\vec{L} \cdot \vec{L}$. This operator is a quadratic invariant. It is worth noting that only unitary groups have linear invariants.

Casimir operators are defined up to an overall c-number factor, since, if $[\mathscr{C}, X_a]=0$, then $[\alpha\mathscr{C}, X_a]=0$. Thus, some care must be taken in their definition. We define the Casimir operators of the groups appearing in Sect. 2.3 as in Table 2.7.

Table 2.7. *Casimir operators of the groups appearing in the interacting boson model-1*

Group	Order	Casimir operator
U(6)	1	$G_0^{(0)}(s,s)+\sqrt{5}\,G_0^{(0)}(d,d)$
	2	$G^{(0)}(s,s)\cdot G^{(0)}(s,s)+G^{(2)}(s,d)\cdot G^{(2)}(d,s)$ $+G^{(2)}(d,s)\cdot G^{(2)}(s,d)+\sum_{k=0}^{4} G^{(k)}(d,d)\cdot G^{(k)}(d,d)$
U(5)	1	$\sqrt{5}\,G_0^{(0)}(d,d)$
	2	$\sum_{k=0}^{4} G^{(k)}(d,d)\cdot G^{(k)}(d,d)$
SU(3)	2	$\frac{2}{3}[2(G^{(2)}(s,d)+G^{(2)}(d,s)-\frac{1}{2}\sqrt{7}G^{(2)}(d,d))$ $\cdot(G^{(2)}(s,d)+G^{(2)}(d,s)-\frac{1}{2}\sqrt{7}\,G^{(2)}(d,d))$ $+\frac{15}{2}G^{(1)}(d,d)\cdot G^{(1)}(d,d)]$
O(6)	2	$2[G^{(2)}(s,d)+G^{(2)}(d,s)]\cdot[G^{(2)}(s,d)+G^{(2)}(d,s)]$ $+4\sum_{k=1,3} G^{(k)}(d,d)\cdot G^{(k)}(d,d)$
O(5)	2	$4\sum_{k=1,3} G^{(k)}(d,d)\cdot G^{(k)}(d,d)$
O(3)	2	$20[G^{(1)}(d,d)\cdot G^{(1)}(d,d)]$

Some of the overall factors in Table 2.7 are not essential and have been introduced to conform with standard notation. One can also introduce the operators (1.50), and rewrite the Casimir operators as

$$\begin{aligned}
\mathscr{C}_1(\mathrm{U6}) &= \hat{N},\\
\mathscr{C}_1(\mathrm{U5}) &= \hat{n}_\mathrm{d},\\
\mathscr{C}_2(\mathrm{SU3}) &= \tfrac{2}{3}[2\hat{Q}\cdot\hat{Q}+\tfrac{3}{4}\hat{L}\cdot\hat{L}],\\
\mathscr{C}_2(\mathrm{O6}) &= 2[\hat{N}(\hat{N}+4)-4\hat{\mathscr{P}}^\dagger\cdot\hat{\mathscr{P}}],\\
\mathscr{C}_2(\mathrm{O5}) &= 4[\tfrac{1}{10}\hat{L}\cdot\hat{L}+\hat{U}\cdot\hat{U}],\\
\mathscr{C}_2(\mathrm{O3}) &= 2(\hat{L}\cdot\hat{L}).
\end{aligned}\tag{2.72}$$

Other operators, sometimes found in the literature, are:

$$\begin{aligned}
\mathscr{C}(\lambda,\mu) &= \tfrac{3}{2}\mathscr{C}_2(\mathrm{SU3}),\\
\hat{C}_5 &= \tfrac{1}{12}\mathscr{C}_2(\mathrm{O5}),\\
\hat{C}_3 &= \tfrac{1}{2}\mathscr{C}_2(\mathrm{O3}),\\
\hat{C}_6 &= \mathscr{C}_2(\mathrm{O6}).
\end{aligned}\tag{2.73}$$

Table 2.8. *Eigenvalues of some Casimir operators of Lie groups*

Group	Labels	Order	$\langle\mathscr{C}\rangle$
U(n)	$[f_1,f_2,\ldots,f_n]$	1	$f=\sum_{i=1}^{n} f_i$
		2	$\sum_{i=1}^{n} f_i(f_i+n+1-2i)$
SU(n)	$[f_1,f_2,\ldots,f_{n-1},f_n=0]$	2	$\sum_{i=1}^{n}\left(f_i-\frac{f}{n}\right)\left(f_i-\frac{f}{n}+2n-2i\right)$ with $f=\sum_{i=1}^{n} f_i$
SU(3)	$(\lambda,\mu)=(f_1-f_2,f_2)$	2	$\frac{2}{3}(\lambda^2+\mu^2+\lambda\mu+3\lambda+3\mu)$
O($2n+1$)	$(f_1,f_2,\ldots,f_n)$	2	$\sum_{i=1}^{n} 2f_i(f_i+2n+1-2i)$
O($2n$)	$(f_1,f_2,\ldots,f_n)$	2	$\sum_{i=1}^{n} 2f_i(f_i+2n-2i)$
Sp($2n$)	$(f_1,f_2,\ldots,f_n)$	2	$\sum_{i=1}^{n} 2f_i(f_i+2n+2-2i)$

2.5.2 *Energy eigenvalues*

The most general Hamiltonian of the interacting boson model-1, containing up to quadratic terms, Eq. (1.21), can be rewritten as a linear combination of linear and quadratic Casimir invariants of all the groups in (2.25).

$$H=e_0+e_1\mathscr{C}_1(\mathrm{U6})+e_2\mathscr{C}_2(\mathrm{U6})+\varepsilon\mathscr{C}_1(\mathrm{U5})+\alpha\mathscr{C}_2(\mathrm{U5})+\beta\mathscr{C}_2(\mathrm{O5})$$
$$+\gamma\mathscr{C}_2(\mathrm{O3})+\delta\mathscr{C}_2(\mathrm{SU3})+\eta\mathscr{C}_2(\mathrm{O6}). \tag{2.74}$$

The form (2.74) contains as many independent parameters, nine, as the original form (1.21). It is merely a rewriting of (1.21). This Hamiltonian is not diagonalizable in closed form, since some Casimir invariants are diagonal in one chain and some in another. Analytic solutions can only be obtained whenever the Hamiltonian can be written in terms only of Casimir operators of one chain. Its eigenvalues can then be found by analyzing the eigenvalues of the corresponding Casimir invariants. These are given in Table 2.8 (Popov and Perelomov, 1967; Nwachuku and Rashid, 1977).

(i) Chain I

Consider the Hamiltonian

$$H^{(\mathrm{I})}=e_0+e_1\mathscr{C}_1(\mathrm{U6})+e_2\mathscr{C}_2(\mathrm{U6})+\varepsilon\mathscr{C}_1(\mathrm{U5})$$
$$+\alpha\mathscr{C}_2(\mathrm{U5})+\beta\mathscr{C}_2(\mathrm{O5})+\gamma\mathscr{C}_2(\mathrm{O3}). \tag{2.75}$$

This Hamiltonian is diagonal in the basis (2.38). Its eigenvalues can be found by taking the expectation value of $H^{(\mathrm{I})}$ in (2.38) and using Table 2.8. Denoting by

$$E^{(\mathrm{I})}(N, n_{\mathrm{d}}, v, \tilde{n}_\Delta, L, M_L) = \langle [N], n_{\mathrm{d}}, v, \tilde{n}_\Delta, L, M_L | H^{(\mathrm{I})} | [N], n_{\mathrm{d}}, v, \tilde{n}_\Delta, L, M_L \rangle \quad (2.76)$$

the eigenvalues, one has

$$E^{(\mathrm{I})}(N, n_{\mathrm{d}}, v, \tilde{n}_\Delta, L, M_L) = E_0 + \varepsilon n_{\mathrm{d}} + \alpha n_{\mathrm{d}}(n_{\mathrm{d}}+4) + \beta 2v(v+3) + \gamma 2L(L+1). \quad (2.77)$$

The first three terms in (2.75) have all been included in E_0, given by

$$E_0 = e_0 + e_1 N + e_2 N(N+5), \quad (2.78)$$

since they contribute only to binding energies and not to excitation energies. It should be noted that both in (2.74) and (2.75) the Casimir operator of O(2) has not been included. This term will appear only if the nucleus is placed in an external field that splits the M_L degeneracy.

In finding the eigenvalues (2.77), different combinations of Casimir invariations can be used. In the original derivation of (2.77) a different form was used (Arima and Iachello, 1976) leading to

$$\begin{aligned} E^{(\mathrm{I})}(N, n_{\mathrm{d}}, v, \tilde{n}_\Delta, L, M_L) = E_0 &+ \varepsilon' n_{\mathrm{d}} + \alpha' \tfrac{1}{2} n_{\mathrm{d}}(n_{\mathrm{d}}-1) \\ &+ \beta'[n_{\mathrm{d}}(n_{\mathrm{d}}+3) - v(v+3)] \\ &+ \gamma'[L(L+1) - 6n_{\mathrm{d}}]. \end{aligned} \quad (2.79)$$

The coefficients in (2.79) are related to those in (2.77) by

$$\begin{aligned} \gamma' &= 2\gamma, \\ \beta' &= -2\beta, \\ \alpha' &= 2\alpha + 4\beta, \\ \varepsilon' &= \varepsilon + 5\alpha + 8\beta + 12\gamma. \end{aligned} \quad (2.80)$$

Still another form that can be used is that in terms of s and d bosons,

$$H^{(\mathrm{I})} = E_0 + \varepsilon'(d^\dagger \cdot \tilde{d}) + \sum_{L=0,2,4} \tfrac{1}{2}(2L+1)^{\frac{1}{2}} c_L [[d^\dagger \times d^\dagger]^{(L)} \times [\tilde{d} \times \tilde{d}]^{(L)}]_0^{(0)}. \quad (2.81)$$

This form can be transformed to (2.75) by recoupling the terms in (2.81) and

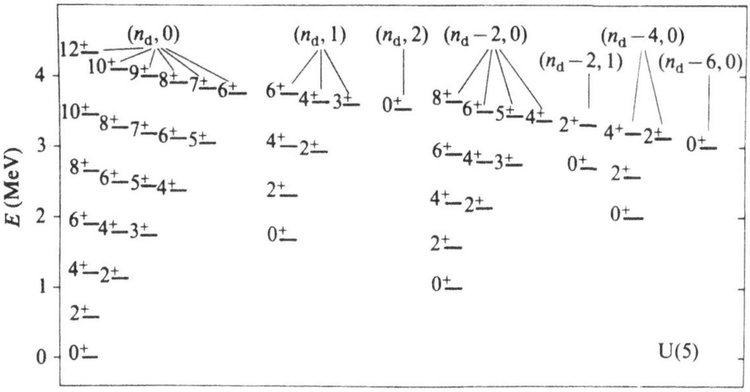

Fig. 2.1. A typical spectrum with U(5) symmetry and $N=6$. In parentheses are the values of v and $\tilde{n}_\Delta$. The angular momentum L of each state is shown to the left.

using Table 2.7. One has

$$c_4=\alpha'+8\gamma',$$

$$c_2=\alpha'-6\gamma',$$

$$c_0=\alpha'+10\beta'-12\gamma'. \tag{2.82}$$

The spectrum of states corresponding to (2.77) is shown in Fig. 2.1.

(ii) *Chain II*
Consider the Hamiltonian

$$H^{(\mathrm{II})}=e_0+e_1\mathscr{C}_1(\mathrm{U6})+e_2\mathscr{C}_2(\mathrm{U6})+\gamma\mathscr{C}_2(\mathrm{O3})+\delta\mathscr{C}_2(\mathrm{SU3}). \tag{2.83}$$

This Hamiltonian is diagonal in the basis (2.48) with eigenvalues

$$E^{(\mathrm{II})}(N,\lambda,\mu,\tilde{\chi},L,M_L)=E_0+\gamma 2L(L+1)+\delta\tfrac{2}{3}(\lambda^2+\mu^2+\lambda\mu+3\lambda+3\mu), \tag{2.84}$$

where E_0 is still given by (2.78).

Again, several linear combinations of $\mathscr{C}_2(\mathrm{SU3})$ and $\mathscr{C}_2(\mathrm{O3})$ can be used. In the article by Arima and Iachello (1978), the following form was used

$$H^{(\mathrm{II})}=E_0-\kappa'\hat{L}\cdot\hat{L}-\kappa 2\hat{Q}\cdot\hat{Q}, \tag{2.85}$$

with eigenvalues

$$E^{(\mathrm{II})}(N,\lambda,\mu,\tilde{\chi},L,M_L)=E_0+(\tfrac{3}{4}\kappa-\kappa')L(L+1)$$
$$-\kappa(\lambda^2+\mu^2+\lambda\mu+3\lambda+3\mu). \tag{2.86}$$

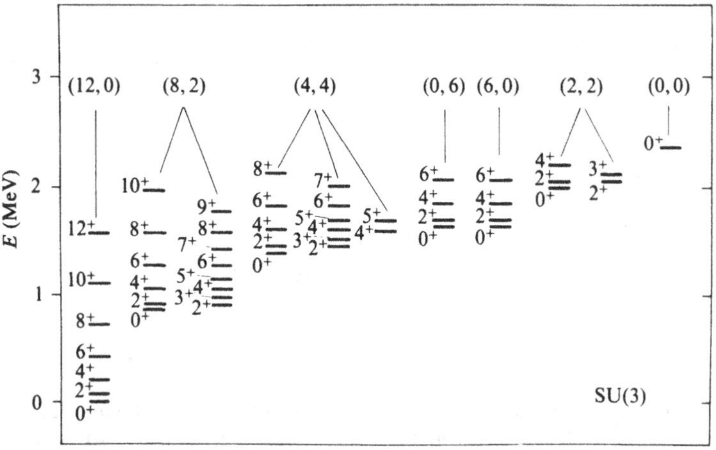

Fig. 2.2. A typical spectrum with SU(3) symmetry and $N=6$. In parentheses are the values of λ and μ. The angular momentum L of each state is shown to the left.

Equation (2.85) is obviously related to (2.83) by (2.72),

$$\kappa = -\tfrac{2}{3}\delta,$$

$$\kappa' = -\tfrac{1}{2}\delta - 2\gamma. \tag{2.87}$$

The spectrum of states corresponding to (2.84) is shown in Fig. 2.2.

(iii) Chain III

For this chain, one considers the Hamiltonian

$$H^{(\mathrm{III})} = e_0 + e_1\mathscr{C}_1(\mathrm{U6}) + e_2\mathscr{C}_2(\mathrm{U6}) + \beta\mathscr{C}_2(\mathrm{O5}) + \gamma\mathscr{C}_2(\mathrm{O3}) + \eta\mathscr{C}_2(\mathrm{O6}). \tag{2.88}$$

This Hamiltonian is diagonal in the basis (2.55) with eigenvalues

$$E^{(\mathrm{III})}(N, \sigma, \tau, \tilde{\nu}_\Delta, L, M_L) = E_0 + \beta 2\tau(\tau+3) + \gamma 2L(L+1) + \eta 2\sigma(\sigma+4), \tag{2.89}$$

where E_0 is given by (2.78). In Arima and Iachello (1979) a different combination of invariants was used, i.e.

$$H^{(\mathrm{III})} = E_0' + A\,\hat{\mathrm{P}}_6 + B\,\hat{\mathrm{C}}_5 + C\,\hat{\mathrm{C}}_3. \tag{2.90}$$

The operators $\hat{\mathrm{C}}_5$ and $\hat{\mathrm{C}}_3$ are given in (2.73), while $\hat{\mathrm{P}}_6$ is given by

$$\hat{\mathrm{P}}_6 = \hat{\mathscr{P}}^\dagger \cdot \hat{\mathscr{P}}. \tag{2.91}$$

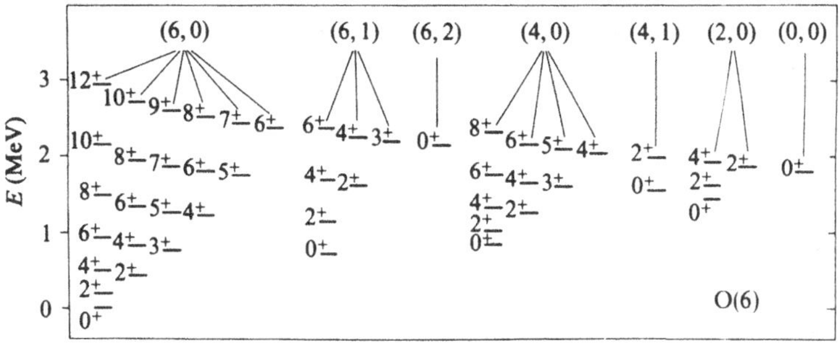

Fig. 2.3. A typical spectrum with O(6) symmetry and $N=6$. In parentheses are the values of σ and $\tilde{v}_\Delta$. The angular momentum L of each state is shown to the left.

Diagonalization of (2.90) leads to

$$E^{(\mathrm{III})}(N, \sigma, \tau, \tilde{v}_\Delta, L, M_L) = E_0' + A\tfrac{1}{4}[N(N+4) - \sigma(\sigma+4)] + B\tfrac{1}{6}\tau(\tau+3) + CL(L+1). \quad (2.92)$$

Comparing (2.92) with (2.89) yields

$$\begin{aligned} A &= -8\eta, \\ B &= 12\beta, \\ C &= 2\gamma, \\ E_0' &= E_0 - A\tfrac{1}{4}N(N+4). \end{aligned} \quad (2.93)$$

The spectrum of states corresponding to (2.89) is given in Fig. 2.3.

2.5.3 Examples of nuclear spectra with dynamic symmetries

The formulas (2.77), (2.84) and (2.89) or, conversely, (2.79), (2.86) and (2.92), can be used to analyze nuclear spectra. Extensive searches for spectra with dynamic symmetries have been performed, and several examples have been found. The regions of the periodic table where these examples are localized are shown in Fig. 2.4.

In Figs. 2.5, 2.6 and 2.7 we show three examples, one for each dynamic symmetry.

2.6 Electromagnetic transitions and moments

Whenever a dynamic symmetry exists, the wave functions of the states are given by representations of the appropriate groups, as discussed in Sect. 2.4. Matrix elements of all operators can then be calculated explicitly. In

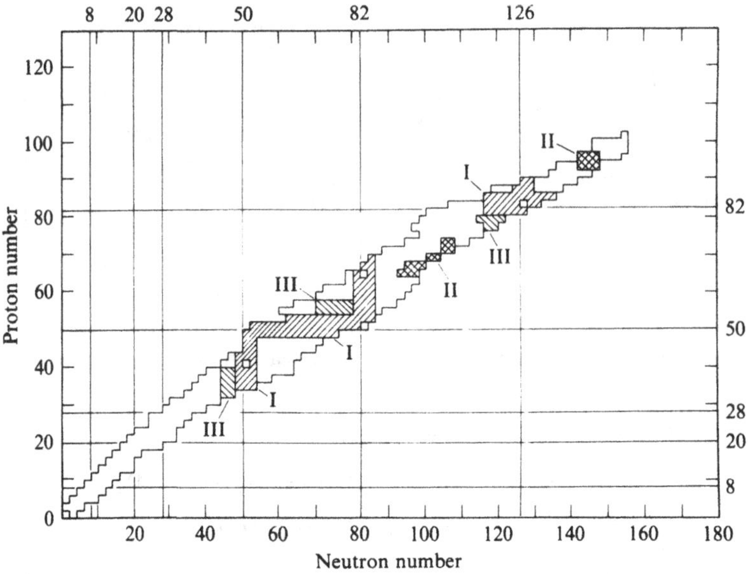

Fig. 2.4. Regions of the periodic table where examples of dynamic symmetries have been found: (I) U(5); (II) SU(3); (III) O(6).

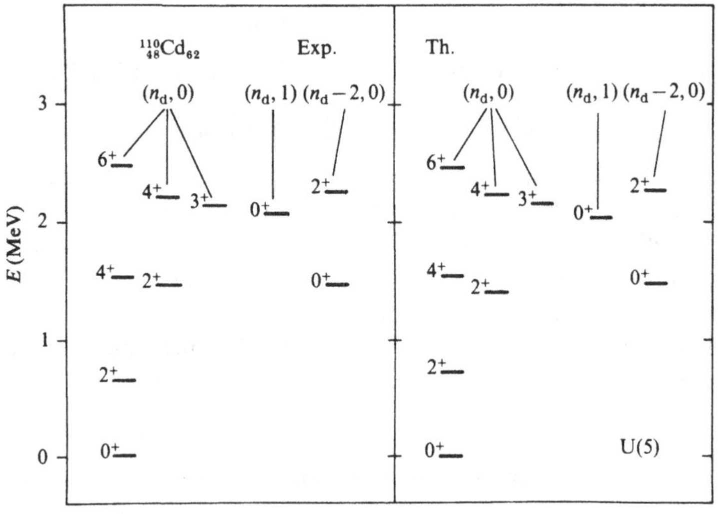

Fig. 2.5. An example of a spectrum with U(5) symmetry: $^{110}_{48}Cd_{62}$, $N=7$. The theoretical spectrum is calculated using (2.79) and (2.82) with $\varepsilon' = 722$ KeV, $c_0 = 29$ KeV, $c_2 = -42$ KeV, $c_4 = 98$ KeV.

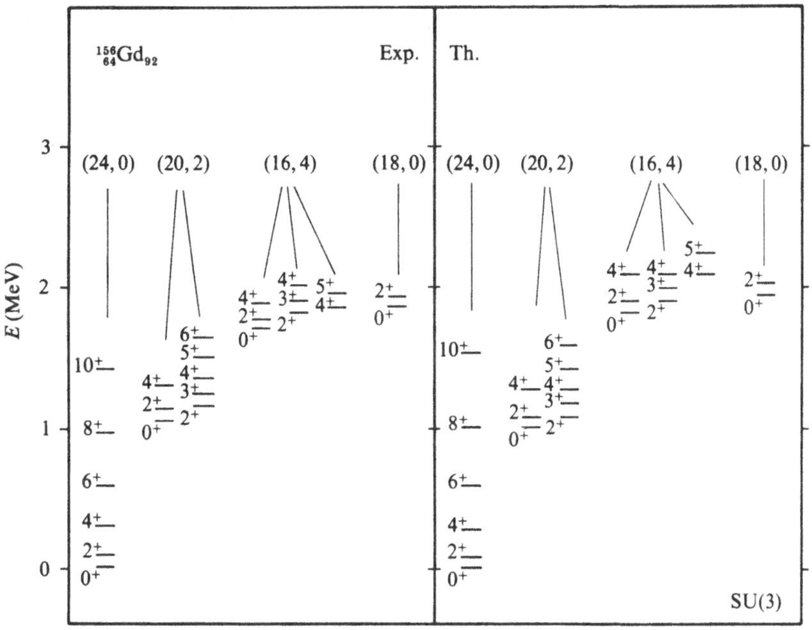

Fig. 2.6. An example of a spectrum with SU(3) symmetry: $^{156}_{64}Gd_{92}$, $N = 12$. The theoretical spectrum is calculated using (2.84) with $\gamma = 3.8$ KeV, $\delta = -20.1$ KeV.

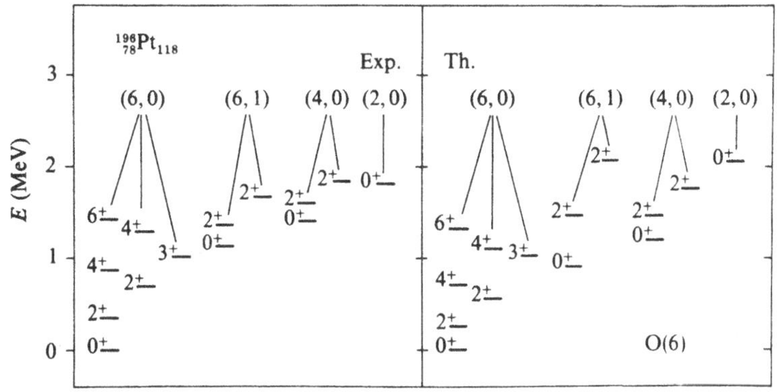

Fig. 2.7. An example of a spectrum with O(6) symmetry: $^{196}_{78}Pt_{118}$, $N = 6$. The theoretical spectrum is calculated using (2.92) with $A = 171$ KeV, $B/6 = 50$ KeV, $C = 10$ KeV.

this calculation one may, or may not, take advantage of the transformation properties of the operator under the group of interest. These transformation properties can be used in two ways: (i) to provide selection rules; and (ii) to simplify, in some cases, the explicit evaluation of the matrix elements. Calculations of matrix elements of the electromagnetic transition operators within the framework of the interacting boson model have taken advantage of the selection rules of the corresponding operators but have not made much use of the tensorial properties of the operators under the groups $G \supset G' \supset \cdots$. These properties have instead been much used in the interacting boson–fermion models and will be therefore described in detail when treating this case. We begin our discussion here by considering matrix elements of the electric quadrupole operator, since these are usually large and thus experimentally accessible.

2.6.1 *Electric quadrupole operators (E2)*

The electric quadrupole (E2) operator was written in (1.24) as

$$T_\mu^{(\mathrm{E2})} = \alpha_2 [d^\dagger \times \tilde{s} + s^\dagger \times \tilde{d}]_\mu^{(2)} + \beta_2 [d^\dagger \times \tilde{d}]_\mu^{(2)}. \tag{2.94}$$

Matrix elements of this operator can be evaluated explicitly in the three cases I, II and III.

(i) Chain I

For this chain, the evaluation of the matrix elements of the operator (2.94) splits into two parts, that of the operator $[d^\dagger \times \tilde{s} + s^\dagger \times \tilde{d}]_\mu^{(2)}$ and that of the operator $[d^\dagger \times \tilde{d}]_\mu^{(2)}$. When written explicitly in terms of s and d bosons, the wave functions (2.38) take the form (1.17),

$$|[N], n_\mathrm{d}, v, \tilde{n}_\Delta, L, M_L\rangle = \frac{1}{\mathcal{N}_\mathrm{s} \mathcal{N}_\mathrm{d}} s^{\dagger n_\mathrm{s}} [d^{\dagger n_\mathrm{d}}]_{v \tilde{n}_\Delta L M_L} |0\rangle, \tag{2.95}$$

where $\mathcal{N}_\mathrm{s}$ and $\mathcal{N}_\mathrm{d}$ are normalization constants. The wave functions thus separate into an s boson and a d boson part,

$$|[N], n_\mathrm{d}, v, \tilde{n}_\Delta, L, M_L\rangle = |n_\mathrm{s}\rangle |n_\mathrm{d}, v, \tilde{n}_\Delta, L, M_L\rangle, \tag{2.96}$$

with $n_\mathrm{s} + n_\mathrm{d} = N$. In order to evaluate matrix elements of the first term in the transition operator (2.94), we then need the matrix elements

$$\langle n_\mathrm{s}' | \tilde{s} | n_\mathrm{s} \rangle \tag{2.97}$$

and

$$\langle n_\mathrm{d}', v', \tilde{n}_\Delta', L', M_L' | d_\mu^\dagger | n_\mathrm{d}, v, \tilde{n}_\Delta, L, M_L \rangle. \tag{2.98}$$

The former are given by

$$\langle n_\mathrm{s}' | \tilde{s} | n_\mathrm{s} \rangle = \sqrt{n_\mathrm{s}}\, \delta_{n_\mathrm{s}', n_\mathrm{s} - 1}, \tag{2.99}$$

as they are in the case of a one-dimensional oscillator. The latter are rather complex and depend explicitly on the quantum numbers n_d, v, $\tilde{n}_\Delta$, L, M_L.

For electromagnetic transition rates and moments, it is sufficient to evaluate matrix elements reduced with respect to the rotation group. The relation between reduced and non-reduced matrix elements is given by the Wigner–Eckart theorem. For a tensor operator $T^{(k)}_{\kappa}$ one has (De Shalit and Talmi, 1963)

$$\langle LM_L|T^{(k)}_{\kappa}|L'M'_L\rangle = (-)^{L-M_L}\begin{pmatrix} L & k & L' \\ -M_L & \kappa & M'_L \end{pmatrix}\langle L\| T^{(k)} \| L'\rangle, \tag{2.100}$$

where the symbol in parentheses is a Wigner $3j$-symbol and the double-bar matrix element is the reduced one. Thus, we need here

$$\langle n'_{\rm d}, v', \tilde{n}'_{\Delta}, L' \| d^{\dagger} \| n_{\rm d}, v, \tilde{n}_{\Delta}, L\rangle. \tag{2.101}$$

The matrix elements (2.101) have been evaluated in closed form for a number of states. Since the operator $d^{\dagger}$ transforms under the groups $U(5) \supset O(5) \supset O(3)$ as the representation with $n_{\rm d}=1$, $v=1$, $\tilde{n}_{\Delta}=0$, $L=2$, i.e.

$$d^{\dagger} = \mathscr{R}^{[1,1,0,2]}, \tag{2.102}$$

it has the selection rules $\Delta n_{\rm d} = +1$, $\Delta v = \pm 1$, $\Delta \tilde{n}_{\Delta}=0$, $|\Delta L| \leqslant 2$. A similar argument applies to the matrix elements of the operator $\tilde{d}$ with selection rules $\Delta n_{\rm d} = -1$, $\Delta v = \pm 1$, $\Delta \tilde{n}_{\Delta}=0$, $|\Delta L| \leqslant 2$.

Using the wave functions (2.95) and the commutation relations (1.2), reduced matrix elements of the $\tilde{d}$ operator have been evaluated for a number of states. These are summarized in Table 2.9. The matrix elements of the operator $d^{\dagger}$ can be obtained from those of $\tilde{d}$ by using the relation

$$\langle n_{\rm d}, v, \tilde{n}_{\Delta}, L \| \tilde{d} \| n'_{\rm d}, v', \tilde{n}'_{\Delta}, L'\rangle = (-)^{(L'-L)}\langle n'_{\rm d}, v', \tilde{n}'_{\Delta}, L' \| d^{\dagger} \| n_{\rm d}, v, \tilde{n}_{\Delta}, L\rangle. \tag{2.103}$$

Combining the matrix elements of the operator $s^{\dagger}$ with those of $\tilde{d}$, or those of $\tilde{s}$ with $d^{\dagger}$, one can find the matrix element of the first term in (2.94). Matrix elements of the second term are more difficult to evaluate and have been derived analytically only in one case. This is the case

$$\langle n_{\rm d}, v=n_{\rm d}, \tilde{n}_{\Delta}=0, L=2n_{\rm d} \| [d^{\dagger} \times \tilde{d}]^{(2)} \| n_{\rm d}, v=n_{\rm d}, \tilde{n}_{\Delta}=0, L=2n_{\rm d}\rangle. \tag{2.104}$$

Using the reduction formula for tensor products (De Shalit and Talmi,

Table 2.9. *Reduced matrix elements of the operator $\tilde{d}$ for some low-lying states* ($\tilde{n}_\Delta = \tilde{n}'_\Delta = 0$)

n'_d, v', L'	n_d, v, L	Matrix element
$n_d+1, n_d+1, 2n_d+2$	$n_d, n_d, 2n_d$	$\sqrt{[(n_d+1)(4n_d+5)]}$
$n_d+1, n_d+1, 2n_d$	$n_d, n_d, 2n_d$	$\sqrt{\left[\frac{(4n_d+2)(4n_d+1)}{(4n_d-1)}\right]}$
$n_d+1, n_d+1, 2n_d$	$n_d, n_d, 2n_d-2$	$\sqrt{\left[\frac{(n_d-1)(4n_d+3)(4n_d+1)}{(4n_d-1)}\right]}$
$n_d+1, n_d+1, 2n_d-1$	$n_d, n_d, 2n_d$	$\sqrt{\left[\frac{(2n_d-2)(4n_d+1)}{(2n_d-1)}\right]}$
$n_d+1, n_d+1, 2n_d-1$	$n_d, n_d, 2n_d-2$	$-\sqrt{\left[\frac{3(2n_d+1)}{(n_d-1)}\right]}$
$n_d+1, n_d+1, 2n_d-1$	$n_d, n_d, 2n_d-3$	$\sqrt{\left[\frac{n_d(n_d-2)(2n_d+1)(4n_d-1)}{(n_d-1)(2n_d-1)}\right]}$
$n_d+1, n_d-1, 2n_d-2$	$n_d, n_d, 2n_d$	$\sqrt{\left[\frac{2n_d(4n_d+1)}{(2n_d+3)}\right]}$
$n_d+1, n_d-1, 2n_d-2$	$n_d, n_d, 2n_d-2$	$\sqrt{\left[\frac{2(4n_d-2)(4n_d-3)}{(2n_d+3)(4n_d-5)}\right]}$
$n_d+1, n_d-1, 2n_d-2$	$n_d, n_d, 2n_d-3$	$-\sqrt{\left[\frac{2(2n_d-4)(4n_d-3)}{(2n_d+3)(2n_d-3)}\right]}$
$n_d+1, n_d-1, 2n_d-2$	$n_d, n_d, 2n_d-4$	$\sqrt{\left[\frac{64(n_d-3)(n_d-2)(n_d-1)}{(2n_d+3)(2n_d-3)(2n_d+1)(4n_d-5)}\right]}$
$n_d+1, n_d-1, 2n_d-2$	$n_d, n_d-2, 2n_d-4$	$\sqrt{\left[\frac{(n_d-1)(2n_d+3)(4n_d-3)}{(2n_d+1)}\right]}$

1963), one has

$$\langle n_d, v=n_d, \tilde{n}_\Delta=0, L=2n_d \| [d^\dagger \times \tilde{d}]^{(2)} \| n_d, v=n_d, \tilde{n}_\Delta=0, L=2n_d\rangle$$
$$=(-)^{L+2+L}\sqrt{5}$$
$$\times \langle n_d, v=n_d, \tilde{n}_\Delta=0, L=2n_d \| d^\dagger \| n_d-1, v'=n_d-1, \tilde{n}'_\Delta=0, L'=2n_d-2\rangle$$
$$\times \langle n_d-1, v'=n_d-1, \tilde{n}'_\Delta=0, L'=2n_d-2 \| \tilde{d} \| n_d, v=n_d, \tilde{n}_\Delta=0, L=2n_d\rangle$$
$$\times \begin{Bmatrix} 2 & 2 & 2 \\ L & L & L-2 \end{Bmatrix}, \tag{2.105}$$

where the symbol in curly brackets is a Wigner $6j$-symbol. Using Table 2.9 and the value of the $6j$-symbol, one obtains

$$\langle n_{\mathrm{d}}, v=n_{\mathrm{d}}, \tilde{n}_{\Delta}=0, L=2n_{\mathrm{d}} \| [d^{\dagger}\times\tilde{d}]^{(2)} \| n_{\mathrm{d}}, v=n_{\mathrm{d}}, \tilde{n}_{\Delta}=0, L=2n_{\mathrm{d}}\rangle$$
$$=\sqrt{\left[\frac{L(2L+1)(2L+3)(L+1)}{2\times 7(2L-1)}\right]}. \tag{2.106}$$

Knowing the matrix elements, one can calculate electromagnetic transition rates and moments. Electromagnetic transition rates are governed by $B(\mathrm{E2})$ values. These are defined as usual

$$B(\mathrm{E2}; L\rightarrow L')=\frac{1}{(2L+1)}|\langle L' \| T^{(\mathrm{E2})} \| L\rangle|^{2}. \tag{2.107}$$

Quadrupole moments instead are defined by

$$Q_{L}=\langle L, M_{L}=L|\sqrt{\left[\frac{16\pi}{5}\right]}T_{0}^{(\mathrm{E2})}|L, M_{L}=L\rangle$$
$$=\sqrt{\left[\frac{16\pi}{5}\right]}\begin{pmatrix} L & 2 & L \\ -L & 0 & L\end{pmatrix}\langle L \| T^{(\mathrm{E2})} \| L\rangle. \tag{2.108}$$

From Table 2.9 and Eq. (2.106) one can calculate $B(\mathrm{E2})$ values and quadrupole moments of the set of states with $v=n_{\mathrm{d}}$, $\tilde{n}_{\Delta}=0$, $L=2n_{\mathrm{d}}$. These are given by

$$B(\mathrm{E2}; n_{\mathrm{d}}+1, v=n_{\mathrm{d}}+1, \tilde{n}_{\Delta}=0, L'=2n_{\mathrm{d}}+2 \rightarrow n_{\mathrm{d}}, v=n_{\mathrm{d}}, \tilde{n}_{\Delta}=0, L=2n_{\mathrm{d}})$$
$$=\alpha_{2}^{2}\left(\frac{L+2}{2}\right)\left(\frac{2N-L}{2}\right) \tag{2.109}$$

and

$$Q_{L}=\beta_{2}\sqrt{\left[\frac{16\pi}{5}\right]}\sqrt{\frac{1}{14}}L. \tag{2.110}$$

In particular,

$$B(\mathrm{E2}; 2_{1}^{+}\rightarrow 0_{1}^{+})=\alpha_{2}^{2}N,$$
$$Q_{2_{1}^{+}}=\beta_{2}\sqrt{\left[\frac{16\pi}{5}\right]}\sqrt{\frac{2}{7}}. \tag{2.111}$$

The structure of the $B(\mathrm{E2})$ values and quadrupole moments in the limit of large N is shown in Figs. 2.8 and 2.9.

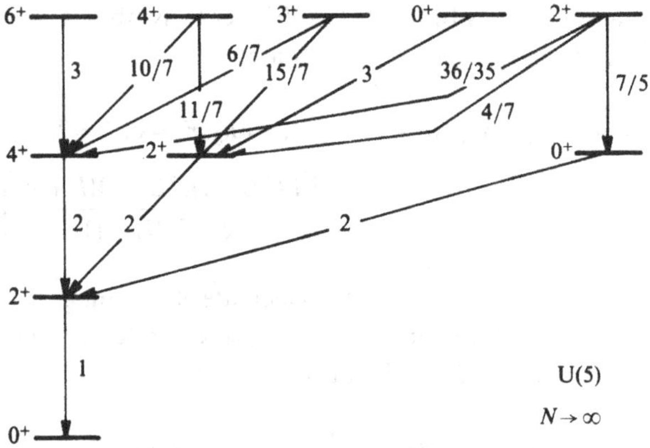

Fig. 2.8. B(E2) values, in units of $B(\mathrm{E2}; 2_1^+ \to 0_1^+)$, for the dynamic symmetry I and $N \to \infty$, for some low-lying states.

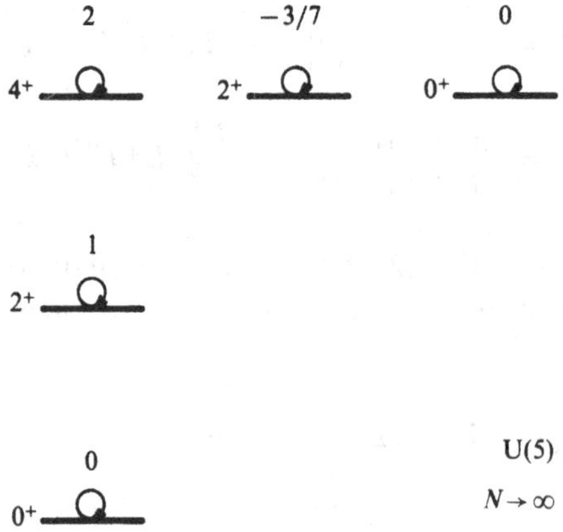

Fig. 2.9. Quadrupole moments, in units of $Q_{2_1^+}$, for the dynamic symmetry I and $N \to \infty$, for some low-lying states. The moments are actually N-independent in this case.

(ii) Chain II

For calculations in this case, it is more convenient to rewrite the transition operator (2.94) as

$$T_\mu^{(\mathrm{E2})} = \alpha_2 Q_\mu^{(2)} + \alpha_2' Q'^{(2)}_\mu, \tag{2.112}$$

where the operators $Q_\mu^{(2)}$ and $Q'^{(2)}_\mu$ are given by (1.50) and (1.51). The coefficient α_2' is related to α_2 and β_2 by

$$\alpha_2' = \beta_2 + \tfrac{1}{2}\sqrt{7}\,\alpha_2. \tag{2.113}$$

The reason why it is convenient to write the operator $T^{(E2)}$ as in (2.112) is twofold. First, it turns out empirically that the coefficient α'_2 is much smaller than α_2 in regions where the symmetry II applies. Second, since the operator $Q^{(2)}_\mu$ is a generator of SU(3), as shown in Eq. (2.18), it is relatively easy to evaluate its matrix elements in closed form. This has been done in several cases.

We begin by noting that a generator of a group has the property that it cannot connect different representations of the group. Thus, the operator $\hat{Q}^{(2)}$, when considered between states of the type $|[N],(\lambda,\mu),\tilde{\chi},L,M_L\rangle$ has the selection rules

$$\Delta\lambda=0, \quad \Delta\mu=0. \tag{2.114}$$

There are several ways in which the matrix elements of the operator $\hat{Q}^{(2)}$ can be evaluated: (i) by using the results already derived by Elliott (1958) and Vergados (1961); (ii) by evaluating separately the matrix elements of the operators $d^\dagger$ and $s^\dagger$ in the SU(3) basis and then combining them together; (iii) by expanding the SU(3) wave functions into a U(5) basis and then using the results discussed above for Chain I. Here we shall use the results of Elliott and Vergados. We note first that the Vergados basis $|[N]$, (λ,μ), $\tilde{\chi}$, L, $M_L\rangle$ can be expanded into the Elliott basis $|[N]$, (λ,μ), K, L, $M_L\rangle$ using (2.45),

$$|[N],(\lambda,\mu),\tilde{\chi},L,M_L\rangle=\sum_K x_{\tilde{\chi}K}\,|[N],(\lambda,\mu),K,L,M_L\rangle. \tag{2.115}$$

As mentioned in Sect. 2.4, the Elliott basis is not orthogonal and one thus needs the overlap integrals

$$\mathscr{I}_{KK',L}=\langle[N],(\lambda,\mu),K,L,M_L|[N],(\lambda,\mu),K',L,M_L\rangle. \tag{2.116}$$

Finally, Elliott has given the reduced matrix elements of the $\hat{Q}$ operator in his basis,

$$\begin{aligned}\langle[N],(\lambda,\mu),K',L'\|\hat{Q}^{(2)}\|[N],(\lambda,\mu),K,L\rangle=&\frac{(2L+1)}{\sqrt{8}\,c(K,L)}\\ \times\Bigg[\begin{pmatrix}L & 2 & L'\\ K & 0 & -K\end{pmatrix}&c(K,L')\{2\lambda+\mu+3+\tfrac{1}{2}L'(L'+1)-\tfrac{1}{2}L(L+1)\}\\ \times\langle[N],(\lambda,\mu),K',L'|[N],(\lambda,\mu),K,L\rangle&\\ +\sum_{\pm}\begin{pmatrix}L & 2 & L'\\ K & \pm2 & -(K\pm2)\end{pmatrix}&\left(\frac{3(2\Lambda\pm K)(2\Lambda\pm K+2)}{2}\right)^{\frac{1}{2}}c(K\pm2,L')\\ \times\langle[N],(\lambda,\mu),K',L'|[N],(\lambda,\mu),K\pm2,L'\rangle\Bigg],&\end{aligned} \tag{2.117}$$

where $\Lambda=\tfrac{1}{2}\min\{\lambda,\mu\}$.

These matrix elements require a knowledge of the coefficients $c(K,L)$ (Elliott coefficients). These, together with the expansion coefficients $x_{\tilde{\chi}K}$ and the overlap integral $\mathscr{I}_{KK',L}$, are all given by Vergados (1968).

Matrix elements of the operator $\hat{Q}$ have been evaluated explicitly for states belonging to the SU(3) representations ($\lambda=2N$, $\mu=0$) and ($\lambda=2N-4,2$). The former has $\tilde{\chi}=0$ and it is usually denoted by g-band (ground-state band). The latter has $\tilde{\chi}=0$ or $\tilde{\chi}=2$. The $\tilde{\chi}=0$ portion is usually denoted by β-band, while the $\tilde{\chi}=2$ portion is called γ-band. The reason for these names will become apparent in Chapter 3. For these bands, one has

$$
\begin{aligned}
&|[N],(2N,0),\tilde{\chi}=0,L,M_L\rangle=|[N],(2N,0),K=0,L,M_L\rangle,\\
&|[N],(2N-4,2),\tilde{\chi}=0,L,M_L\rangle=|[N],(2N-4,2),K=0,L,M_L\rangle,\\
&|[N],(2N-4,2),\tilde{\chi}=2,L,M_L\rangle\\
&\quad=-\sqrt{\left[\frac{(L-1)L(L+1)(L+2)}{4(2N-2)(2N-1)(2N-2-L)(2N-1+L)}\right]}\\
&\qquad\qquad|[N],(2N-4,2),K=0,L,M_L\rangle\\
&\quad+\sqrt{\left\{\frac{[2(2N-2)^2-L(L+1)][2(2N-1)^2-L(L+1)]}{4(2N-2)(2N-1)(2N-2-L)(2N-1+L)}\right\}}\\
&\qquad\qquad|[N],(2N-4,2),K=2,L,M_L\rangle;\\
&\qquad\qquad L=\text{even};\\
&|[N],(2N-4,2),\tilde{\chi}=2,L,M_L\rangle=|[N],(2N-4,2),K=2,L,M_L\rangle;\\
&\qquad\qquad L=\text{odd},
\end{aligned}
\tag{2.118}
$$

where we have inserted the appropriate values of the Vergados coefficients.

The overlap integrals are given by

$$
\begin{aligned}
&\langle[N],(2N,0),K=0,L,M_L|[N],(2N,0),K=0,L,M_L\rangle=1\\
&\langle[N],(2N-4,2),K=0,L,M_L|[N],(2N-4,2),K=0,L,M_L\rangle=1\\
&\langle[N],(2N-4,2),K=0,L,M_L|[N],(2N-4,2),K=2,L,M_L\rangle\\
&\quad=\sqrt{\left[\frac{(L-1)L(L+1)(L+2)}{[2(2N-2)^2-L(L+1)][2(2N-1)^2-L(L+1)]}\right]};\quad L=\text{even};
\end{aligned}
$$

Table 2.10. *Elliott coefficients c(K, L) for the representation (λ, 2)*

	$c(K,L)$
$\lambda-L=$even $K=0$	$\left[\frac{(2L+1)\phi(\lambda+1,L)}{4(\lambda+1)(\lambda+2)}b(\lambda+2,L)\right]^{\frac{1}{2}}$
$K=2$	$\left[\frac{(2L+1)\phi(\lambda+2,L)}{16(\lambda+1)(\lambda+2)}b(\lambda+2,L)\right]^{\frac{1}{2}}$
$b(\lambda,L)=\frac{\lambda(\lambda-1)(\lambda-2)\cdots(\lambda-L+1)}{(\lambda-L+1)(\lambda-L+3)\cdots(\lambda+L-1)(\lambda+L+1)}$,	$\lambda\geqslant 1$
$b(\lambda,0)=1/(\lambda+1)$	
$\phi(\lambda,L)=2(\lambda+1)^2-L(L+1)$	
$\lambda-L=$odd	
$K=2$	$\left[\frac{(2L+1)(\lambda+3)}{8(\lambda+1)}b(\lambda+1,L)\right]^{\frac{1}{2}}$

$$\langle[N],(2N-4,2),K=2,L,M_L|[N],(2N-4,2),K=2,L,M_L\rangle=1. \tag{2.119}$$

Finally, Elliott coefficients are given by

$$c(0,L)=[(2L+1)b(\lambda,L)]^{\frac{1}{2}},$$

$$b(\lambda,L)=\frac{\lambda(\lambda-1)(\lambda-2)\cdots(\lambda-L+1)}{(\lambda-L+1)(\lambda-L+3)\cdots(\lambda+L-1)(\lambda+L+1)},\quad L\geqslant 1,$$

$$b(\lambda,0)=\frac{1}{\lambda+1}. \tag{2.120}$$

for the representation $(\lambda=2N,0)$ and by Table 2.10 for the representation $(2N-4,2)$. Using these expressions, one can obtain the following $B(\mathrm{E2})$ values:

(a) $g\rightarrow g$

$$B(\mathrm{E2};(\lambda=2N,\mu=0),\tilde{\chi}=0,L'=L+2\rightarrow(\lambda=2N,\mu=0),\tilde{\chi}=0,L)$$

$$=\alpha_2^2\frac{3}{4}\frac{(L+2)(L+1)}{(2L+3)(2L+5)}(2N-L)(2N+L+3). \tag{2.121}$$

(b) $\beta \to \beta$

$$B(\mathrm{E2}; (\lambda=2N-4, \mu=2), \tilde{\chi}=0, L'=L+2 \to (\lambda=2N-4, \mu=2), \tilde{\chi}=0, L)$$

$$=\alpha_2^2 \frac{3}{4} \frac{(L+2)(L+1)}{(2L+3)(2L+5)}$$

$$\times \left\{ \left[2N+L+\frac{1}{2} \frac{(L+3)(L+4)}{2(2N-2)^2-(L+2)(L+3)} \right]^2 \right.$$

$$\left. \times \left[\frac{2(2N-2)^2-(L+2)(L+3)}{2(2N-2)^2-L(L+1)} \right] \left(\frac{2N-2-L}{2N+1+L} \right) \right\}. \quad (2.122)$$

Similarly, one can calculate quadrupole moments. For the g-band $(2N, 0)$, one obtains

$$Q_L = -\alpha_2 \sqrt{\left[\frac{16\pi}{40} \right]} \frac{L}{2L+3} (4N+3). \quad (2.123)$$

Particularly important are the values

$$B(\mathrm{E2}; 2_1^+ \to 0_1^+) = \alpha_2^2 \frac{1}{5} N(2N+3) \quad (2.124)$$

and

$$Q_{2_1^+} = -\alpha_2 \sqrt{\left[\frac{16\pi}{40} \right]} \frac{2}{7} (4N+3). \quad (2.125)$$

The structure of the $B(\mathrm{E2})$ values and quadrupole moments for this case in the limit of large N is shown in Figs. 2.10 and 2.11.

We now come to the evaluation of the matrix elements of the operator $Q_\mu'^{(2)}$. These are more difficult to evaluate and have been derived in closed form only in few cases. The technique that has been used to do this evaluation is that of calculating separately the matrix elements of the operators $d^\dagger$ and $\tilde{d}$ and of combining them together. For example, one can write

$$\langle [N], (2N, 0), \tilde{\chi}=0, L \| [d^\dagger \times \tilde{d}]^{(2)} \| [N], (2N, 0), \tilde{\chi}=0. L'=L+2 \rangle$$

$$= \sqrt{5} \sum_{L''=L, L+2} \langle [N], (2N, 0), \tilde{\chi}=0, L \| d^\dagger \| [N], (2N-2, 0), \tilde{\chi}=0, L'' \rangle$$

$$\times \langle [N-1], (2N-2, 0), \tilde{\chi}=0, L'' \| \tilde{d} \| [N], (2N, 0), \tilde{\chi}=0, L'=L+2 \rangle$$

$$\times \begin{Bmatrix} 2 & 2 & 2 \\ L'' & L & L+2 \end{Bmatrix}. \quad (2.126)$$

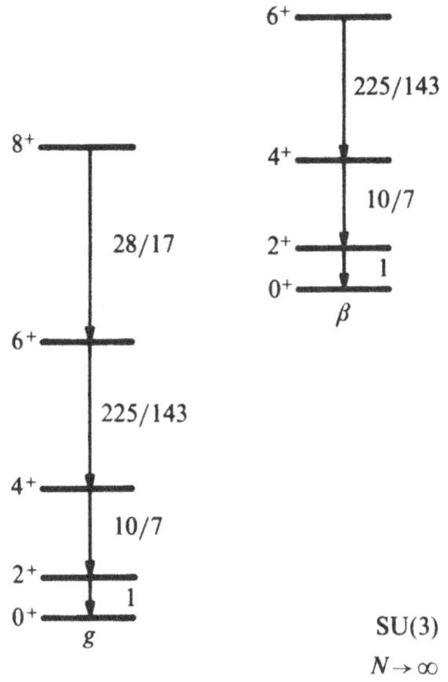

Fig. 2.10. B(E2) values in units of $B(E2; 2_1^+ \rightarrow 0_1^+)$ for the dynamic symmetry II and $N \rightarrow \infty$ for some low-lying states, assuming $\alpha'_2=0$ in (2.112).

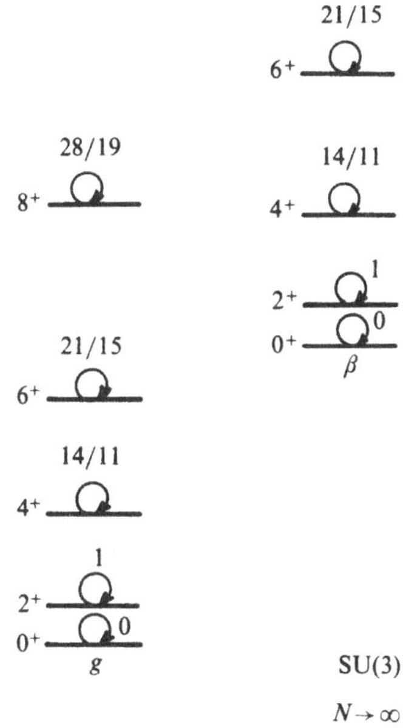

Fig. 2.11. Quadrupole moments in units of $Q_{2_1^+}$ for the dynamic symmetry II and $N \rightarrow \infty$, for some low-lying states, assuming $\alpha'_2=0$ in (2.112). These values are actually N-independent.

Using

$$\langle [N], (2N,0), \tilde{\chi}=0, L \| d^\dagger \| [N-1], (2N-2,0), \tilde{\chi}=0, L\rangle$$
$$= -\sqrt{\left[\frac{2(2N-L)(2N+L+1)L(L+1)}{3(2N-1)(2N)(2L-1)(2L+3)}\right]}\sqrt{(2L+1)}\sqrt{N},$$
$$\langle [N-1], (2N-2,0), \tilde{\chi}=0, L \| \tilde{d} \| [N], (2N,0), \tilde{\chi}=0, L+2\rangle$$
$$= \sqrt{\left[\frac{(2N+L+1)(2N+L+3)(L+1)(L+2)}{(2N-1)(2N)(2L+3)(2L+5)}\right]}\sqrt{(2L+5)}\sqrt{N}, \tag{2.127}$$
$$\langle [N], (2N,0), \tilde{\chi}=0, L \| d^\dagger \| [N-1], (2N-2,0), \tilde{\chi}=0, L+2\rangle$$
$$= \sqrt{\left[\frac{(2N-L-2)(2N-L)(L+2)(L+1)}{(2N-1)(2N)(2L+1)(2L+3)}\right]}\sqrt{(2L+1)}\sqrt{N},$$
$$\langle [N-1], (2N-2,0), \tilde{\chi}=0, L+2 \| \tilde{d} \| [N], (2N,0), \tilde{\chi}=0, L+2\rangle$$
$$= -\sqrt{\left[\frac{2}{3}\frac{(2N-L-2)(2N+L+3)(L+2)(L+3)}{(2N-1)(2N)(2L+3)(2L+7)}\right]}\sqrt{(2L+5)}\sqrt{N},$$

one then obtains

$$\langle [N], (2N,0), \tilde{\chi}=0, L \| [d^\dagger \times \tilde{d}]^{(2)} \| [N], (2N,0), \tilde{\chi}=0, L'=L+2\rangle$$
$$= -\sqrt{\left[\frac{(L+1)(L+2)}{21(2L+3)}\right]}\sqrt{[(2N-L)(2N+3)]}\,\frac{2N-2}{2N-1}. \tag{2.128}$$

These matrix elements, multiplied by α_2', can then be added to those of the operator $\hat{Q}$ multiplied by α_2, to evaluate the matrix elements of the complete operator (2.112). This calculation can be repeated for other matrix elements, if one wishes to do so. Van Isacker (1983) has used the same technique to evaluate matrix elements of the operator

$$T_\mu^{(E2)} = \alpha_2 Q_\mu^{(2)} + \theta[d^\dagger \times \tilde{s} + s^\dagger \times \tilde{d}]_\mu^{(2)}, \tag{2.129}$$

between the β, γ-bands and the g-band. As mentioned above, the first term does not contribute to these transitions since $\hat{Q}$ is a generator of SU(3). The contribution of the second term gives rise to the following $B(E2)$ values:

(a) $\beta \rightarrow g$

$$B(E2; \beta, L \rightarrow g, L-2) = \theta^2 \frac{N}{9}\frac{3(L-1)L}{2(2L-1)(2L+1)}\left[\frac{4(N-1)^2+L}{2N-1}\right]^2$$
$$\times \frac{(2N-L)(2N-L+2)}{N(2N-3)[8(N-1)^2-L(L+1)]},$$

$$B(\mathrm{E2};\beta,L\to g,L)=\theta^2\frac{N}{9}\frac{L(L+1)}{(2L-1)(2L+3)}\left[\frac{4(N-1)^2+2L(L+1)-3}{2N-1}\right]^2$$
$$\times\frac{(2N-L)(2N+L+1)}{N(2N-3)[8(N-1)^2-L(L+1)]},$$

$$B(\mathrm{E2};\beta,L\to g,L+2)=\theta^2\frac{N}{9}\frac{3(L+1)(L+2)}{2(2L+1)(2L+3)}\left[\frac{4(N-1)^2-L}{2N-1}\right]^2$$
$$\times\frac{(2N+L+1)(2N+L+3)}{N(2N-3)[8(N-1)^2-L(L+1)]},$$

(b) $\gamma\to g$

$$B(\mathrm{E2};\gamma,L\to g,L-2)=\theta^2\frac{2}{3}N\frac{(L+1)(L+2)}{4(2L-1)(2L+1)}$$
$$\times\frac{2(N-1)(2N-L-2)(2N-L)(2N-L+2)(2N+L-1)}{N(2N-3)(2N-1)[8(N-1)^2-L(L+1)]},$$

$$B(\mathrm{E2};\gamma,L\to g,L-1)=\theta^2\frac{2}{3}N\frac{L+2}{2(2L+1)}$$
$$\times\frac{(2N-L-1)(2N-L+1)(2N+L)}{(2N)(2N-3)(2N-1)},$$

$$B(\mathrm{E2};\gamma,L\to g,L)=\theta^2\frac{2}{3}N\frac{3(L-1)(L+2)}{2(2L-1)(2L+3)}$$
$$\times\frac{2(N-1)(2N-L-2)(2N-L)(2N+L-1)(2N+L+1)}{N(2N-3)(2N-1)[8(N-1)^2-L(L+1)]},$$

$$B(\mathrm{E2};\gamma,L\to g,L+1)=\theta^2\frac{2}{3}N\frac{L-1}{2(2L+1)}$$
$$\times\frac{(2N-L-1)(2N+L)(2N+L+2)}{(2N)(2N-3)(2N-1)},$$

$$B(\mathrm{E2};\gamma,L\to g,L+2)=\theta^2\frac{2}{3}N\frac{(L-1)L}{4(2L+1)(2L+3)}$$
$$\times\frac{2(N-1)(2N-L-2)(2N+L-1)(2N+L+1)(2N+L+3)}{N(2N-3)(2N-1)[8(N-1)^2-L(L+1)]}.$$
(2.130)

Indeed, the technique outlined here, and used to derive Eqs. (2.128) and (2.130) is a general group-theoretic technique, which can be applied to evaluate the matrix elements of any operator for the group chain U(6) ⊃ SU(3) ⊃ O(3). It makes use of the concept of isoscalar factors (Wybourne, 1974) and of Racah's factorization lemma. It is particularly useful for complex algebraic structures. It will be discussed briefly in Sect. 5.5 and in detail in a second book, where it will be applied to the calculation of matrix elements of operators of interest in odd–even nuclei.

(iii) Chain III

In order to calculate matrix elements of the E2 operator in this case, it is convenient to return to (2.94). It turns out empirically that, in regions where the symmetry III applies, the second term in (2.94) is rather small. We therefore begin by considering matrix elements of the first term. These are easier to evaluate since the operator $[d^\dagger \times \tilde{s} + s^\dagger \times \tilde{d}]^{(2)}_\mu$ is a generator of O(6), Eq. (2.20). When taken between states $|[N], \sigma, \tau, \tilde{\nu}_\Delta, L, M_L\rangle$ it has selection rules

$$\Delta\sigma = 0, \quad \Delta\tau = \pm 1. \tag{2.131}$$

The first of these rules is a consequence of the fact that the operator is a generator of O(6), while the second is a consequence of the fact that this operator transforms as the tensor $\mathscr{R}^{[1,0,2]}$ under O(5) ⊃ O(3), $\tau = 1$, $\nu_\Delta = 0$, $L = 2$.

The evaluation of the matrix elements of the operator $[d^\dagger \times \tilde{s} + s^\dagger \times \tilde{d}]^{(2)}_\mu$ has been done by expanding the wave functions $|[N], \sigma, \tau, \tilde{\nu}_\Delta, L, M_L\rangle$ into those of chain I, as given by Eq. (2.57), and by using the results for chain I discussed under (i). One then obtains

$$\begin{aligned}
&\langle [N], \sigma, \tau' = \tau + 1, \tilde{\nu}'_\Delta = \tilde{\nu}_\Delta, L' \| [d^\dagger \times \tilde{s} + s^\dagger \times \tilde{d}]^{(2)} \| [N], \sigma, \tau, \tilde{\nu}_\Delta, L\rangle \\
&\quad = \sum_{n_d} \zeta^\sigma_{n_d+1,\tau+1} \zeta^\sigma_{n_d,\tau} (N - n_d)^{\frac{1}{2}} \left(\frac{n_d + \tau + 5}{2\tau + 5}\right)^{\frac{1}{2}} a_{LL'}(\tau) \\
&\qquad + \sum_{n_d} \zeta^\sigma_{n_d+1,\tau+1} \zeta^\sigma_{n_d+2,\tau} (N - n_d - 1)^{\frac{1}{2}} \left(\frac{n_d - \tau + 2}{2}\right)^{\frac{1}{2}} b_{L'L}(\tau),
\end{aligned} \tag{2.132}$$

where

$$\begin{aligned}
a_{L'L}(\tau) &= \langle n'_d = \tau + 1, v' = \tau + 1, \tilde{n}'_\Delta = \tilde{\nu}'_\Delta, L' \| d^\dagger \| n_d = \tau, v = \tau, \tilde{n}_\Delta = \tilde{\nu}_\Delta, L\rangle, \\
b_{L'L}(\tau) &= \langle n'_d = \tau + 1, v' = \tau + 1, \tilde{n}'_\Delta = \tilde{\nu}'_\Delta, L' \| \tilde{d} \| n_d = \tau + 2, v = \tau, \tilde{n}_\Delta = \tilde{\nu}_\Delta, L\rangle.
\end{aligned} \tag{2.133}$$

By inserting the appropriate expansion coefficients, one obtains the following B(E2) values (Arima and Iachello, 1979)

$$B(\mathrm{E2}; [N], \sigma=N, \tau+1, \tilde{\nu}'_\Delta=0, L'=2\tau+2 \rightarrow [N], \sigma=N, \tau, \tilde{\nu}_\Delta=0, L=2\tau)$$
$$=\alpha_2^2 \frac{\tau+1}{2\tau+5}(N-\tau)(N+\tau+4),$$

$$B(\mathrm{E2}; [N], \sigma=N, \tau+1, \tilde{\nu}'_\Delta=0, L'=2\tau \rightarrow [N], \sigma=N, \tau, \tilde{\nu}_\Delta=0, L=2\tau)$$
$$=\alpha_2^2 \frac{4\tau+2}{(2\tau+5)(4\tau-1)}(N-\tau)(N+\tau+4),$$

$$B(\mathrm{E2}; [N], \sigma=N, \tau+1, \tilde{\nu}'_\Delta=0, L'=2\tau-1 \rightarrow [N], \sigma=N, \tau, \tilde{\nu}_\Delta=0, L=2\tau)$$
$$=\alpha_2^2 \frac{(2\tau-2)(4\tau+1)}{(2\tau+5)(2\tau-1)(4\tau-1)}(N-\tau)(N+\tau+4),$$

$$B(\mathrm{E2}; [N], \sigma=N, \tau+1, \tilde{\nu}'_\Delta=0, L'=2\tau-1 \rightarrow [N], \sigma=N, \tau, \tilde{\nu}_\Delta=0, L=2\tau-2)$$
$$=\alpha_2^2 \frac{3(2\tau+1)}{(2\tau+1)(\tau-1)(4\tau-1)}(N-\tau)(N+\tau+4),$$

$$B(\mathrm{E2}; [N], \sigma=N, \tau+1, \tilde{\nu}'_\Delta=0, L'=2\tau \rightarrow [N], \sigma=N, \tau, \tilde{\nu}_\Delta=0, L=2\tau-2)$$
$$=\alpha_2^2 \frac{(\tau-1)(4\tau+3)}{(2\tau+5)(4\tau-1)}(N-\tau)(N+\tau+4),$$

$$B(\mathrm{E2}; [N], \sigma=N, \tau+1, \tilde{\nu}'_\Delta=0, L'=2\tau-1 \rightarrow [N], \sigma=N, \tau, \tilde{\nu}_\Delta=0, L=2\tau-3)$$
$$=\alpha_2^2 \frac{\tau(\tau-2)(2\tau+1)}{(\tau-1)(2\tau-1)(2\tau+5)}(N-\tau)(N+\tau+4). \quad (2.134)$$

Using $\tau=\frac{1}{2}L$ one can also rewrite the first of Eq. (2.134) as

$$B(\mathrm{E2}; [N], \sigma=N, \tau+1, \tilde{\nu}'_\Delta=0, L'=2\tau+2 \rightarrow [N], \sigma=N, \tau, \tilde{\nu}_\Delta=0, L=2\tau)$$
$$=\alpha_2^2 \frac{L+2}{2(L+5)} \frac{1}{4}(2N-L)(2N+L+8). \quad (2.135)$$

In particular one obtains

$$B(\mathrm{E2}; 2_1^+ \rightarrow 0_1^+)=\alpha_2^2 \frac{1}{5} N(N+4). \quad (2.136)$$

Because of the selection rules (2.131), all quadrupole moments are zero when only the first term in (2.94) is used,

$$Q_L=0. \quad (2.137)$$

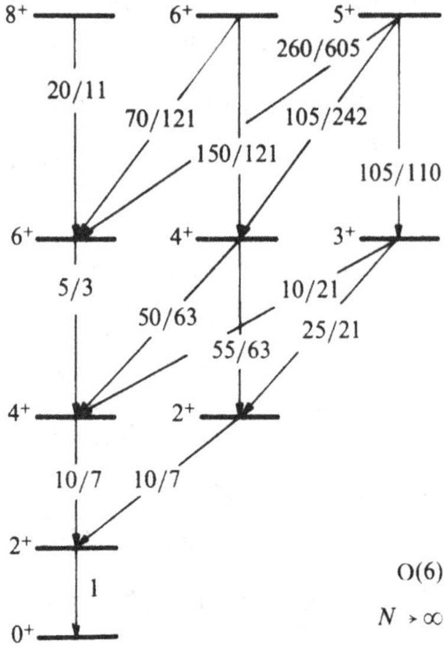

Fig. 2.12. $B(E2)$ values in units of $B(E2;\, 2_1^+ \rightarrow 0_1^+)$ for the dynamic symmetry III and $N \rightarrow \infty$, for some low-lying states, and assuming $\beta_2 = 0$ in (2.94).

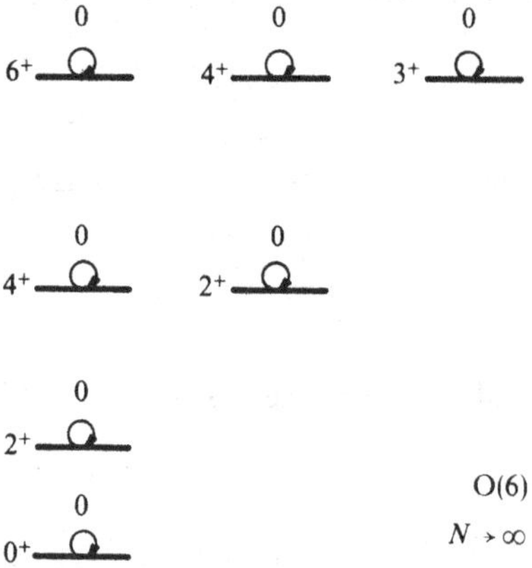

Fig. 2.13. Quadrupole moments in units of $Q_{2_1^+}$ for the dynamic symmetry III and $N \rightarrow \infty$. These moments are identically zero, if one assumes $\beta_2 = 0$ in (2.94).

In particular,

$$Q_{2_1^+} = 0. \tag{2.138}$$

The structure of the $B(E2)$ values and quadrupole moments in this case is shown in Figs. 2.12 and 2.13, in the limit of large N. Matrix elements of the operator $[d^\dagger \times \tilde{d}]^{(2)}_\mu$ have not been evaluated explicitly in this case.

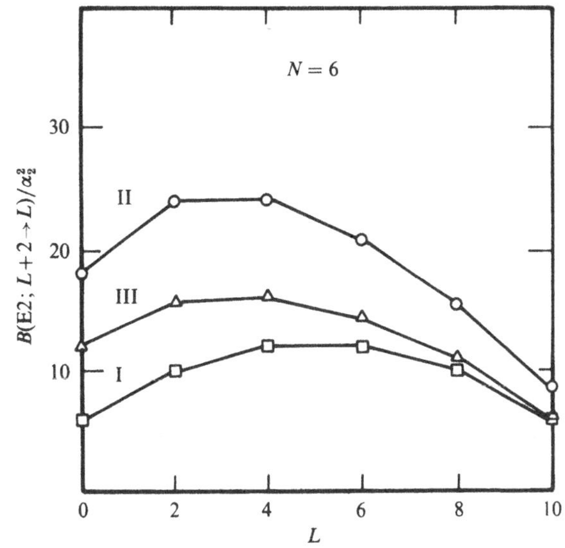

Fig. 2.14. B(E2) values along the g-band in the three limits, U(5) (I), SU(3) (II), O(6) (III) and $N=6$, in units α_2^2.

It is of interest to compare properties of electromagnetic E2 transition rates for the three limiting situations discussed above. In Fig. 2.14, the B(E2) values in the g-band are shown as a function of L. They are given by (2.109), (2.121) and (2.135) respectively. The B(E2) values are larger in II and III than in I, since in these cases they increase as N^2 rather than N. For finite N, they decrease as L approaches $2N$ (cut-off effect). This decrease is similar for II and III but rather different from I. From these simple considerations it is clear that the case I represents a situation intrinsically very different from II and III. Other important quantities that show the difference between the three cases are the ratios

$$R=\frac{B(\mathrm{E2};4_1^+\rightarrow 2_1^+)}{B(\mathrm{E2};2_1^+\rightarrow 0_1^+)},$$

$$R'=\frac{B(\mathrm{E2};2_2^+\rightarrow 2_1^+)}{B(\mathrm{E2};2_1^+\rightarrow 0_1^+)},$$

$$R''=\frac{B(\mathrm{E2};0_2^+\rightarrow 2_1^+)}{B(\mathrm{E2};2_1^+\rightarrow 0_1^+)}. \tag{2.139}$$

These are given by

$$R^{(\mathrm{I})}=2\frac{(N-1)}{N}\underset{N\to\infty}{\longrightarrow}2, \qquad \mathrm{U}(5);$$

$$R^{(\mathrm{II})}=\frac{10}{7}\frac{(N-1)(2N+5)}{N(2N+3)}\underset{N\to\infty}{\longrightarrow}\frac{10}{7}, \qquad \mathrm{SU}(3);$$

$$R^{(\mathrm{III})}=\frac{10}{7}\frac{(N-1)(N+5)}{N(N+4)}\underset{N\to\infty}{\longrightarrow}\frac{10}{7}, \qquad \mathrm{O}(6); \tag{2.140}$$

and

$$R'^{(\mathrm{I})}=2\frac{(N-1)}{N}\underset{N\to\infty}{\longrightarrow}2, \qquad \mathrm{U}(5);$$

$$R'^{(\mathrm{II})}=0, \qquad \mathrm{SU}(3);$$

$$R'^{(\mathrm{III})}=\frac{10}{7}\frac{(N-1)(N+5)}{N(N+4)}\underset{N\to\infty}{\longrightarrow}\frac{10}{7}, \qquad \mathrm{O}(6); \tag{2.141}$$

and

$$R''^{(\mathrm{I})}=2\frac{(N-1)}{N}\underset{N\to\infty}{\longrightarrow}2, \qquad \mathrm{U}(5);$$

$$R''^{(\mathrm{II})}=0, \qquad \mathrm{SU}(3);$$

$$R''^{(\mathrm{III})}=0, \qquad \mathrm{O}(6). \tag{2.142}$$

2.6.2 Examples of dynamic symmetries in E2 transitions

The formulas derived in Sect. 2.6.1 can be used to analyze experimental data for E2 transitions and moments. The best examples of dynamic symmetries in E2 transitions have been found for the symmetry III. We give here a brief survey of the present situation.

(i) Chain I

Table 2.11 shows a comparison between available experimental $B(\mathrm{E2})$ values in ^{110}Cd and those expected for chain I. All transitions forbidden by the selection rules appear to be small, $\sim 10^{-2}$, while all allowed transitions are in reasonable agreement with those predicted by the U(5) limit. The basic conditions for the observation of a U(5) symmetry in the electromagnetic transitions are:

(a) $\Delta n_{\mathrm{d}}=2$ transitions $\to 0$

(b) $R'=R''=R'''\to 2[(N-1)/N]\sim 1.6.$ (2.143)

Table 2.11. *Comparison between experimental* B(E2) *values in* $^{110}_{48}Cd_{62}$ ($N=7$) *and those expected for symmetry I*

B(E2) ratios	$^{110}_{48}Cd_{62}$	U(5)
$\frac{4_1 \to 2_1}{2_1 \to 0_1}$	(1.53 ± 0.19)	$2\left(\frac{N-1}{N}\right) = 1.71$
$\frac{2_2 \to 2_1}{2_1 \to 0_1}$	(1.08 ± 0.29)	$2\left(\frac{N-1}{N}\right) = 1.71$
$\frac{0_2 \to 2_1}{2_1 \to 0_1}$	—	$2\left(\frac{N-1}{N}\right) = 1.71$
$\frac{2_2 \to 0_1}{2_2 \to 2_1}$	$(5.5 \pm 0.9) \times 10^{-2}$	0
$\frac{3_1 \to 4_1}{3_1 \to 2_2}$	(0.47 ± 0.20)	$\frac{2}{5} = 0.40$
$\frac{3_1 \to 2_1}{3_1 \to 2_2}$	$(1.7 \pm 1.0) \times 10^{-2}$	0
$\frac{4_2 \to 4_1}{4_2 \to 2_2}$	$(0.23^{+0.38}_{0.23})$	$\frac{10}{11} = 0.91$
$\frac{4_2 \to 2_1}{4_2 \to 2_2}$	$(0.7 \pm 0.1) \times 10^{-2}$	0
$\frac{5_1 \to 6_1}{5_1 \to 3_1}$	(0.39 ± 0.23)	$\frac{104}{231} = 0.45$
$\frac{5_1 \to 4_2}{5_1 \to 3_1}$	(0.54 ± 0.10)	$\frac{5}{11} = 0.45$
$\frac{5_1 \to 4_1}{5_1 \to 3_1}$	$(0.7 \pm 0.6) \times 10^{-2}$	0

(*ii*) *Chain II*

The goodness of the SU(3) symmetry for electromagnetic E2 transitions has not been much investigated. This is due to the fact that most of the available data are on the decay properties $\gamma \to g$ and $\beta \to g$. Both transitions are forbidden in the limit of an SU(3) symmetry and of a quadrupole operator $\hat{Q}$. However, Casten, von Brentano and Haque (1985) have recently spelled out the conditions for an SU(3) symmetry when the condition that the transition operator is $\hat{Q}$ is slightly relaxed ($\alpha'_2 \ll \alpha_2$). These are:

(a) $B(\text{E2}; \gamma \to g) \to 0; \quad B(\text{E2}; \beta \to g) \to 0;$

(b) $\dfrac{B(\text{E2}; \beta \to g)}{B(\text{E2}; \gamma \to g)} = \dfrac{1}{6};$

Table 2.12. *Experimental examples of O(6) symmetry in electromagnetic transition rates. All ratios are in units of 10^{-2}*

	Nucleus				
B(E2) ratios	^{124}Xe	^{126}Xe	^{128}Xe	^{130}Xe	O(6)
$\frac{2_2 \to 0_1}{2_1 \to 2_1}$	3.9	1.4	1.2	0.6	0
$\frac{3_1 \to 4_1}{3_1 \to 2_2}$	46	47	37	25	40
$\frac{3_1 \to 2_1}{3_1 \to 2_2}$	1.6	1.1	1	1.4	0
$\frac{4_2 \to 3_1}{4_2 \to 2_2}$	—	—	—	—	0
$\frac{4_2 \to 4_1}{4_2 \to 2_2}$	91	42	133	107	91
$\frac{4_2 \to 2_1}{4_2 \to 2_2}$	0.4	1.0	1.7	3.2	0
$\frac{5_1 \to 4_2}{5_1 \to 3_1}$	106	127	88	—	46
$\frac{5_1 \to 6_1}{5_1 \to 3_1}$	—	—	204	—	45
$\frac{5_1 \to 4_1}{5_1 \to 3_1}$	3.8	4.9	3.7	—	0
$\frac{0_2 \to 2_1}{0_2 \to 2_2}$	1	9	14	26	0

(c) $B(\text{E2}; \gamma \to \beta) \to$ large;

(d) the mixing parameter Z_γ defined by Lipas (1962) and Riedinger, Johnson and Hamilton (1969), $\to 0$. (2.144)

On the basis of Casten's study, it appears that nuclei in the region of ^{178}Hf, ^{170}Er and ^{174}Yb are the best examples of SU(3) symmetry.

(iii) Chain III

Electromagnetic properties of nuclei that can be described by this chain have been extensively investigated. The first example was found by Cizewski *et al.* (1978) in $^{196}_{78}\text{Pt}_{118}$. Recently an extensive region of O(6) like nuclei has been found near $A \sim 130$ (Casten and von Brentano, 1985). The electromagnetic properties of these nuclei are summarized in Table 2.12.

Table 2.12. *continued*

	Nucleus					
B(E2) ratios	^{128}Ba	^{130}Ba	^{132}Ba	^{134}Ba	^{196}Pt	O(6)
$\frac{2_2\to 0_1}{2_2\to 2_1}$	9.2	5.7	0.2	0.6	0.0	0
$\frac{3_1\to 4_1}{3_1\to 2_2}$	—	30	73	40	95	40
$\frac{3_1\to 2_1}{3_1\to 2_2}$	—	1.5	0.2	1.0	0.1	0
$\frac{4_2\to 3_1}{4_2\to 2_2}$	—	—	—	14.5	—	0
$\frac{4_2\to 4_1}{4_2\to 2_2}$	42	89	75	77	109	91
$\frac{4_2\to 2_1}{4_2\to 2_2}$	1.7	3.9	2.2	2.5	1.7	0
$\frac{5_1\to 4_2}{5_1\to 3_1}$	$\leqslant$44	$\leqslant$57	—	—	—	46
$\frac{5_1\to 6_1}{5_1\to 3_1}$	$\leqslant$56	381	—	—	—	45
$\frac{5_1\to 4_1}{5_1\to 3_1}$	3.0	6.7	—	—	—	0
$\frac{0_2\to 2_1}{0_2\to 2_2}$	—	—	0	4	16	0

The basic features of the O(6) symmetry are:

(a) $\Delta\tau=2$ transitions $\to 0$;

(b) $R=R'\to\frac{10}{7}\frac{(N-1)(N+5)}{N(N+4)}\sim 1.4$;

(c) $R''\to 0$. (2.145)

In Table 2.12 the forbidden $\Delta\tau=2$ transitions are all of order 10^{-2}.

2.6.3 *Electric hexadecapole operators (E4)*

The electric hexadecapole operator was written in (1.24) as

$$T_\mu^{(E4)}=\beta_4(d^\dagger\times\tilde{d})_\mu^{(4)}. \tag{2.146}$$

Although experimental values of electromagnetic hexadecapole transitions are strongly influenced by contributions arising from $L=4$ pairs (g bosons) not included here, we give nonetheless for completeness the matrix elements of the E4 operator (2.146) in the cases I, II and III.

(i) Chain I

The operator (2.146) satisfies the selection rule $\Delta n_{\mathrm{d}}=0$. Thus it has vanishing matrix elements between states with different n_{d}. For example,

$$\langle [N], n_{\mathrm{d}}, v=n_{\mathrm{d}}, \tilde{n}_{\Delta}=0, L=2n_{\mathrm{d}} \| [d^{\dagger} \times \tilde{d}^{\dagger}]^{(4)} \| [N], n_{\mathrm{d}}+2, v=n_{\mathrm{d}}+2, \tilde{n}_{\Delta}=0, L=2n_{\mathrm{d}}+4 \rangle = 0. \quad (2.147)$$

As a result, one has for transitions between states in the g-band

$$B(\mathrm{E4}; L+4 \rightarrow L)=0, \quad (2.148)$$

and in particular

$$B(\mathrm{E4}; 4_1^+ \rightarrow 0_1^+)=0. \quad (2.149)$$

(ii) Chain II

Matrix elements of the E4 operator in this chain can be obtained using a technique similar to that used in Eqs. (2.126)–(2.128),

$$\begin{aligned}
&\langle [N], (2N,0), \tilde{\chi}=0, L \| [d^{\dagger} \times \tilde{d}]^{(4)} \| [N], (2N,0), \tilde{\chi}=0, L'=L+4 \rangle \\
&= \sqrt{9} \langle [N], (2N,0), \tilde{\chi}=0, L \| d^{\dagger} \| [N-1], (2N-2,0), \tilde{\chi}=0, L+2 \rangle \\
&\quad \times \langle [N-1], (2N-2,0), \tilde{\chi}=0, L+2 \| \tilde{d} \| [N], (2N,0), \tilde{\chi}=0, L+4 \rangle \\
&\quad \times \begin{Bmatrix} 2 & 2 & 4 \\ L+4 & L & L+2 \end{Bmatrix}. \quad (2.150)
\end{aligned}$$

This yields

$$\begin{aligned}
&\langle [N], (2N,0), \tilde{\chi}=0, L \| [d^{\dagger} \times \tilde{d}]^{(4)} \| [N], (2N,0), \tilde{\chi}=0, L'=L+4 \rangle \\
&= \sqrt{\left[\frac{(L+1)(L+2)(L+3)(L+4)}{(2L+3)(2L+5)(2L+7)}\right]} \frac{\sqrt{[(2N-L-2)(2N-L)(2N+L+3)(2N+L+5)]}}{2(2N-1)}. \quad (2.151)
\end{aligned}$$

and thus for the g-band,

$$\begin{aligned}
B(\mathrm{E2}; L+4 \rightarrow L) = \beta_4^2 \frac{(L+1)(L+2)(L+3)(L+4)}{(2L+3)(2L+5)(2L+7)(2L+9)} \times \\
\times (2N-L-2)(2N-L)(2N+L+3)(2N+L+5)/ \\
4(2N-1)^2. \quad (2.152)
\end{aligned}$$

In particular, one obtains

$$B(\text{E4};4_1^+\to 0_1^+)=\beta_4^2\frac{8}{315}\frac{(N-1)N(2N+3)(2N+5)}{(2N-1)^2}. \qquad (2.153)$$

(iii) Chain III

Matrix elements of the E4 operator have been evaluated only for

$$\langle[N],\sigma=N,\tau=0,\tilde{\nu}_\Delta=0,L=0\|[d^\dagger\times\tilde{d}]^{(4)}\|[N],\sigma=N,\tau=2,\tilde{\nu}_\Delta=0,L=4\rangle$$
$$=\frac{3}{\sqrt{70}}\frac{\sqrt{[(N-1)N(N+4)(N+5)]}}{(N+1)}, \qquad (2.154)$$

leading to

$$B(\text{E4};4_1^+\to 0_1^+)=\beta_4^2\frac{1}{70}\frac{(N-1)N(N+4)(N+5)}{(N+1)^2}. \qquad (2.155)$$

There are, at present, not enough systematic data on E4 transition rates to be able to test these formulas.

2.6.4 *Electric monopole operators (E0)*

The electric monopole operator was written in Eq. (1.40) as

$$T_0^{(\text{E0})}=\gamma_0'+\beta_0'[d^\dagger\times\tilde{d}]_0^{(0)}, \qquad (2.156)$$

where

$$\gamma_0'=\gamma_0+\alpha_0 N. \qquad (2.157)$$

The constant term γ_0' does not give rise to electromagnetic transitions. Thus one needs only to evaluate the matrix elements of the last operator in (2.156).

(i) Chain I

Since the operator $[d^\dagger\times\tilde{d}]_0^{(0)}=\hat{n}_d/\sqrt{5}$ is diagonal here, there are no E0 transitions in this limit. In particular,

$$\langle[N],n_d,v=n_d,\tilde{n}_\Delta=0,L=2n_d|T^{(\text{E0})}|[N],n_d'=n_d+2,v'=n_d,\tilde{n}_\Delta'=0,L'=2n_d\rangle=0. \qquad (2.158)$$

Matrix elements of the E0 operator are usually denoted by ρ. Thus here

$$\rho(n_d,L\to n_d-2,L)=0. \qquad (2.159)$$

(ii) Chain II

In order to evaluate matrix elements of the E0 operator in this chain, it is more convenient to rewrite the operator as

$$T^{(E0)}=\tilde{\gamma}_0+\tilde{\beta}_0\hat{n}_s, \tag{2.160}$$

where

$$\tilde{\gamma}_0=\gamma_0+\frac{\beta_0}{\sqrt{5}}N,$$

$$\tilde{\beta}_0=\alpha_0-\frac{\beta_0}{\sqrt{5}}. \tag{2.161}$$

Matrix elements of the operator $\hat{n}_s$ have been evaluated for transitions between the g-band and the β-band

$$\langle[N],(2N,0),\chi=0,L|\hat{n}_s|[N],(2N-4,2),\chi=0,L\rangle$$
$$=\sqrt{N}\sqrt{\left[\frac{(2N-L)(2N+L+1)}{3(2N-1)(2N)}\right]}\sqrt{\left[\frac{2(2N-1)^2-L(L+1)}{3(2N-2)(2N-1)}\right]}\sqrt{\left[\frac{2(N-1)}{(2N-3)}\right]}. \tag{2.162}$$

One thus has

$$\rho(\chi=0_2,L\to\chi=0_1,L)$$
$$=\tilde{\beta}_0\sqrt{N}\sqrt{\left[\frac{(2N-L)(2N+L+1)}{3(2N-1)(2N)}\right]}\sqrt{\left[\frac{2(2N-1)^2-L(L+1)}{3(2N-2)(2N-1)}\right]}\sqrt{\left[\frac{2(N-1)}{(2N-3)}\right]}, \tag{2.163}$$

For large N,

$$\rho(L\to L)\underset{N\to\infty}{\longrightarrow}\tilde{\beta}_0\sqrt{N}\frac{\sqrt{2}}{3}. \tag{2.164}$$

(iii) Chain III

Here it is also convenient to use Eq. (2.160). Matrix elements of the operator $\hat{n}_s$ have been evaluated explicitly between states with $\sigma=N$ and $\sigma=N-2$. They are given by

$$\langle[N],\sigma=N,\tau,\tilde{v}_\Delta=0,L=2\tau|\hat{n}_s|[N],\sigma=N-2,\tau,\tilde{v}_\Delta=0,L=2\tau\rangle$$
$$=\sqrt{N}\sqrt{\left[\frac{N(N+3)-\tau(\tau+3)}{2N(N+1)}\right]}\sqrt{\left[\frac{(N-1)(N-2)-\tau(\tau+3)}{2N(N+1)}\right]}. \tag{2.165}$$

Table 2.13. *Comparison between experimental ρ-values in $^{174}_{72}Hf_{102}$ and those expected on the basis of the* SU(3) *symmetry*

L	$\rho_{\exp}$ (fm^2)	$\rho_{\mathrm{SU(3)}}$ (fm^2)
0	12.0 ± 1.5	12[a]
2	12.0 ± 1.0	11.9
4	10.5 ± 1.5	11.6
6	11.5 ± 1.5	11.2

[a] Used to fix $\tilde{\beta}_0$.

The ρ-matrix elements are then

$$\rho(\sigma=N-2, L \rightarrow \sigma=N, L)$$

$$=\tilde{\beta}_0 \sqrt{N} \sqrt{\left[\frac{N(N+3)-\tau(\tau+3)}{2N(N+1)}\right]} \sqrt{\left[\frac{(N-1)(N-2)-\tau(\tau+3)}{2N(N+1)}\right]},$$

$$\tau=L/2. \tag{2.166}$$

For large N they become independent of L as in (2.164),

$$\rho(L \rightarrow L) \underset{N \rightarrow \infty}{\longrightarrow} \tilde{\beta}_0 \sqrt{N} \tfrac{1}{2}. \tag{2.167}$$

2.6.5 *Examples of dynamic symmetries in E0 transitions*

There are several existing measurements of E0 matrix elements in rare-earth nuclei. One of these is shown in Table 2.13 where it is compared with the values expected on the basis of a SU(3) symmetry. Although the experimental values compare well with the theoretical values, the evidence here is not sufficiently strong to make definitive statements based only on it. On the other hand, when combined with the results on electromagnetic E2 transition rates and energies, it appears to indicate that nuclei in the mass region $A \sim 170$ are, to a good approximation, the best examples found so far of SU(3) symmetry.

2.6.6 *Magnetic dipole operators (M1)*

The magnetic dipole operator was written in (1.24) as

$$T_{\mu}^{(\mathrm{M1})}=\beta_1 [d^{\dagger} \times \tilde{d}]_{\mu}^{(1)}. \tag{2.168}$$

Introducing the angular momentum operator, $\hat{L}$, one can rewrite (2.168) as

$$T_\mu^{(M1)} = \sqrt{\left[\frac{3}{4\pi}\right]} g_B \hat{L}, \tag{2.169}$$

where

$$g_B = \frac{\beta_1}{\sqrt{10}} \sqrt{\left[\frac{4\pi}{3}\right]} \tag{2.170}$$

is the effective boson g-factor. The factor $\sqrt{[4\pi/3]}$ has been introduced to conform with standard notation. Since the operator $\hat{L}$ is diagonal in any basis, no M1 transition can occur in this approximation. The diagonal matrix elements depend only on L and are the same for all three cases. They are given by

$$\langle L \| T^{(M1)} \| L \rangle = \sqrt{\left[\frac{3}{4\pi}\right]} g_B \sqrt{[L(L+1)(2L+1)]}, \tag{2.171}$$

where we have deleted all quantum numbers except L. From these, one can calculate magnetic moments. These are defined as

$$\begin{aligned} \mu_L &= \sqrt{\left[\frac{4\pi}{3}\right]} \langle L, M_L = L | T_0^{(M1)} | L, M_L = L \rangle \\ &= \sqrt{\left[\frac{4\pi}{3}\right]} \frac{L}{\sqrt{[L(L+1)(2L+1)]}} \langle L \| T^{(M1)} \| L \rangle. \end{aligned} \tag{2.172}$$

Use of (2.171) gives

$$\mu_L = g_B L. \tag{2.173}$$

Thus, the g-factors of the states, defined as

$$g_L = \frac{\mu_L}{L}, \tag{2.174}$$

are all equal in this approximation and given by

$$g_L = g_B. \tag{2.175}$$

Since the off-diagonal matrix elements of $T^{(M1)}$ vanish when only one-body terms are retained, one may wish in this case to consider higher-order terms. These terms may be written as in (1.32). There are several other representations of the two-body terms that have been used in practice. One of these is

$$T_\mu^{(M1)} = \beta_1 [d^\dagger \times \tilde{d}]_\mu^{(1)} + \alpha_1' [[d^\dagger \times \tilde{s} + s^\dagger \times \tilde{d}]^{(2)} \times [d^\dagger \times \tilde{d}]^{(1)}]_\mu^{(1)}$$
$$+ \gamma_1' [[d^\dagger \times \tilde{d}]^{(0)} \times [d^\dagger \times \tilde{d}]^{(1)}]_\mu^{(1)} + \delta_1' [[d^\dagger \times \tilde{d}]^{(2)} \times [d^\dagger \times \tilde{d}]^{(1)}]_\mu^{(1)}$$
$$+ \eta_1' [[s^\dagger \times \tilde{s}]^{(0)} \times [d^\dagger \times \tilde{d}]^{(1)}]_\mu^{(1)}. \tag{2.176}$$

The coefficients α_1', γ_1', δ_1', η_1' are linear combinations of those in (1.32). Introducing the operators $\hat{L}$, $\hat{Q}$, $\hat{Q}'$ and $\hat{n}_d$, Eq. (2.176) can be rewritten as

$$T_\mu^{(M1)} = (\beta_1 + \eta_1' N) \frac{1}{\sqrt{10}} \hat{L} + \alpha_1' \frac{1}{\sqrt{10}} [\hat{Q} \times \hat{L}]^{(1)}$$
$$+ \left(\frac{\gamma_1'}{\sqrt{5}} - \eta_1' \right) \frac{1}{\sqrt{10}} \hat{n}_d \hat{L} + \left(\alpha_1' \frac{\sqrt{7}}{2} + \delta_1' \right) \frac{1}{\sqrt{10}} [\hat{Q}' \times \hat{L}]^{(1)}. \tag{2.177}$$

In the consistent-Q formalism of Sect. 1.4.8, Eq. (2.177) simplifies to

$$T_\mu^{(M1)} = \tilde{\beta}_1 \hat{L} + \tilde{\alpha}_1 [\hat{Q}^\chi \times \hat{L}]^{(1)} + \tilde{\gamma}_1 \hat{n}_d \hat{L}, \tag{2.178}$$

where $\tilde{\beta}_1, \tilde{\alpha}_1$ and $\tilde{\gamma}_1$ are obviously related to $\beta_1, \alpha_1', \gamma_1', \eta_1'$. It may look at first sight that the forms (2.176)–(2.178) are not hermitian. However, using the commutation relations of Chapter 1, one can show that they are. For example, one can easily show that

$$[\hat{n}_d, L_\mu] = 0, \quad \mu = 0, \pm 1, \tag{2.179}$$

i.e. the d boson number operator commutes with the angular momentum operator, $\hat{L}$.

Matrix elements of the additional terms have been evaluated in a variety of cases. Consider first those of the operator $\hat{Q}^\chi \times \hat{L}$. These are, in general, given by

$$\langle \phi, L \| [\hat{Q}^\chi \times \hat{L}]^{(1)} \| \phi', L' \rangle$$
$$= (-)^{L+1+L'} \sqrt{3} \langle \phi, L \| \hat{Q}^\chi \| \phi', L' \rangle \sqrt{[L'(L'+1)(2L'+1)]}$$
$$\times \begin{Bmatrix} 2 & 1 & 1 \\ L' & L & L' \end{Bmatrix} \tag{2.180}$$

where ϕ, ϕ' denote all additional quantum numbers. Since they can be measured directly, particularly important are the E2/M1 mixing ratios, defined by

$$\Delta^{(E2/M1)} = \frac{\langle \phi, L \| T^{(E2)} \| \phi', L' \rangle}{\langle \phi, L \| T^{(M1)} \| \phi', L' \rangle}. \tag{2.181}$$

For operators of the form

$$T^{(E2)}=\alpha_2\hat{Q}^{\chi},$$
$$T^{(M1)}=\tilde{\beta}_1\hat{L}+\tilde{\alpha}_1[\hat{Q}^{\chi}\times\hat{L}]^{(1)}, \tag{2.182}$$

the mixing ratios are independent of the quantum numbers ϕ and given by

$$\begin{aligned} L'&=L, & \Delta^{(E2/M1)}&=-A'/\surd[(2L-1)(2L+3)],\\ L'&=L+1, & \Delta^{(E2/M1)}&=-A'/\surd[3L(L+2)],\\ L'&=L-1, & \Delta^{(E2/M1)}&=-A'/\surd[3(L-1)(L+2)], \end{aligned} \tag{2.183}$$

where

$$A'=\surd 10\frac{\alpha_2}{\tilde{\alpha}_1}. \tag{2.184}$$

Matrix elements of the operator $\hat{n}_d\hat{L}$ can also be simply evaluated. They are given by

$$\langle\phi,L\|\hat{n}_d\hat{L}\|\phi',L'\rangle=\delta_{LL'}\langle\phi,L|\hat{n}_d|\phi',L'\rangle\surd[L(L+1)(2L+1)]. \tag{2.185}$$

These matrix elements depend on the quantum numbers ϕ.

(i) Chain I

Here the operator $\hat{n}_d\hat{L}$ contributes only to diagonal matrix elements

$$\langle[N],n_d,v,\tilde{n}_\Delta,L|\hat{n}_d|[N],n_d,v,\tilde{n}_\Delta,L\rangle=n_d. \tag{2.186}$$

(ii) Chain II

In this case, the operator $\hat{n}_d\hat{L}$ contributes to both diagonal and off-diagonal matrix elements. Particularly important are the off-diagonal matrix elements between the β-band and the g-band. Matrix elements of $\hat{n}_d$ can be obtained using (2.162) and $\hat{n}_d=\hat{N}-\hat{n}_s$,

$$\langle[N],(2N,0),\tilde{\chi}=0,L|\hat{n}_d|[N],(2N-4,2),\tilde{\chi}=0,L\rangle$$

$$=-\surd N\sqrt{\left[\frac{(2N-L)(2N+L+1)}{3(2N-1)(2N)}\right]}\sqrt{\left[\frac{2(2N-1)^2-L(L+1)}{3(2N-2)(2N-1)}\right]}\sqrt{\left[\frac{2(N-1)}{(2N-3)}\right]}. \tag{2.187}$$

(iii) Chain III

The operator $\hat{n}_d\hat{L}$ contributes here to both diagonal and off-diagonal matrix elements. For matrix elements between the representations $\sigma=N$

and $\sigma=N-2$, one can use (2.165) to obtain

$$\langle[N],\sigma=N,\tau,\tilde{\nu}_\Delta,L|\hat{n}_d|[N],\sigma=N-2,\tau,\tilde{\nu}_\Delta,L\rangle$$

$$=-\sqrt{N}\sqrt{\left[\frac{N(N+3)-\tau(\tau+3)}{2N(N+1)}\right]}\sqrt{\left[\frac{(N-1)(N-2)-\tau(\tau+3)}{2N(N+1)}\right]}. \quad (2.188)$$

Two-body terms also contribute to g-factors. Writing $T^{(M1)}$ as in (2.178) with

$$\tilde{\beta}_1=\sqrt{\left(\frac{3}{4\pi}\right)}(g_B+\tilde{\eta}_1 N), \quad (2.189)$$

one obtains:

(*i*) *Chain I*

$$g_{2_1^+}=g_B+\tilde{\eta}_1 N+\sqrt{\left(\frac{4\pi}{3}\right)}\tilde{\gamma}_1-\sqrt{\left(\frac{4\pi}{3}\right)}\frac{1}{5\sqrt{2}}\tilde{\alpha}_1\chi. \quad (2.190)$$

(*ii*) *Chain II*

$$g_{2_1^+}=g_B+\tilde{\eta}_1 N+\sqrt{\left(\frac{4\pi}{3}\right)}\frac{6-8N(N-1)}{6(2N-1)}\tilde{\gamma}_1$$
$$+\sqrt{\left(\frac{4\pi}{3}\right)}\frac{1}{7\sqrt{2}}(4N+3)\tilde{\alpha}_1. \quad (2.191)$$

(*iii*) *Chain III*

$$g_{2_1^+}=g_B+\tilde{\eta}_1 N+\sqrt{\left(\frac{4\pi}{3}\right)}\frac{4+N(N-1)}{2(N+1)}\tilde{\gamma}_1. \quad (2.192)$$

2.6.7 *Examples of dynamic symmetries in M1 transitions*

Because of the complex structure of the operator (2.176), tests of dynamic symmetries in M1 transitions are difficult to perform. Warner (1981) has analyzed the available data in the rare-earth region, by using the operator (2.178). From the discussion in the preceding section, it is clear that transitions from states with angular momentum $L+1$ to states with angular momentum L depend only on $\tilde{\alpha}_1$. In chain II these transitions will connect states $\gamma\to g, \gamma\to\gamma, \gamma\to\beta$. On the other side, transitions from states with angular momentum L depend both on $\tilde{\alpha}_1$ and $\tilde{\gamma}_1$. These will be particularly important for $\beta\to g$ transitions. Part of Warner's analysis is shown in Table 2.14. This table shows the large uncertainties in the data,

Table 2.14. *Comparison between predicted and experimental reduced* E2/M1 *mixing ratios for* $\gamma \to g$ *transitions* (*Warner, 1981*)

	$\Delta^{(E2/M1)}$			
	$^{166}_{68}Er_{98}$		$^{168}_{68}Er_{100}$	
$L_i^P \to L_f^P$	Exp	Th	Exp	Th
$2^+ \to 2^+$	$-27.2^{+8.5}_{-22}$	-30.0	$\lvert\Delta\rvert > 140$	-26.5
$3^+ \to 2^+$	-29.2^{+14}_{-287}	-28.3	$+24.7^{+3.7}_{-4.7}$	-25.0
$3^+ \to 4^+$	$-18.2^{+10}_{-\infty}$	-20.4	-9.3 ± 0.6	-18.1
$4^+ \to 4^+$	$-27.8^{+6.9}_{-47}$	-15.7	$-164^{+98}_{-\infty}$	-13.8
$5^+ \to 4^+$	$-32.0^{+3.1}_{-3.7}$	-16.1		-14.2
$5^+ \to 6^+$	-60.1^{+11}_{-18}	-13.5		-11.9

which is a typical feature of all M1 transitions in even–even nuclei, and the fact that the theoretical calculations account only qualitatively for the data.

Lipas (1984) has recently extended Warner's calculations to include all terms in (2.177) and found better agreement with experiment. However, it is important to note here that, contrary to E2 transitions, M1 transitions are greatly affected by proton–neutron degrees of freedom, and thus should be more properly treated using the formalism of Part II.

2.6.8 *Magnetic octupole operators* (*M3*)

The form of this operator is given by (1.24)

$$T_\mu^{(M3)} = \beta_3 [d^\dagger \times \tilde{d}]_\mu^{(3)}. \tag{2.193}$$

This operator has zero matrix elements between states with different n_d in chain I. Thus

(*i*) *Chain I*

$$B(M3; 0_1^+ \to 3_1^+) = 0. \tag{2.194}$$

The only other matrix element that has been explicitly evaluated is for chain II.

(*ii*) *Chain II*

Using the techniques of Eqs. (2.126)–(2.128) one can compute

$$\langle [N], (2N-4, 2), \tilde{\chi}=2, L=3 \| [d^\dagger \times \tilde{d}]^{(3)} \| [N], (2N, 0), \tilde{\chi}'=0, L'=0 \rangle$$

$$= N \sqrt{\left[\frac{2}{3}\left(\frac{N-2}{N-1}\right)\left(\frac{2N-2}{2N-1}\right)\right]}, \tag{2.195}$$

and thus

$$B(M3;0_1^+ \rightarrow 3_1^+)=\beta_3^2 N^2 \frac{2}{3}\left(\frac{N-2}{N-1}\right)\left(\frac{2N-2}{2N-1}\right). \tag{2.196}$$

2.7 Binding energies

For calculations of binding energies, one needs to introduce terms which depend on the number of particles. These, as well as the properties described in the next two sections, 2.8 and 2.9, depend explicitly on proton and neutron degrees of freedom and should, therefore, be more appropriately treated in the proton–neutron formalism (Part II). Here we limit ourselves to some general considerations.

The energy eigenvalues for any of the three symmetries are written in the form

$$E(N,\phi)=E_0+E_{\mathrm{D}}(N,\phi), \tag{2.197}$$

where E_0 is a quadratic form in N, Eq. (2.78). For use in this section, it is convenient to write E_0 as

$$E_0=\tilde{C}+\tilde{A}N+\tilde{B}\tfrac{1}{2}N(N-1). \tag{2.198}$$

The parameters $\tilde{C}$, $\tilde{A}$ and $\tilde{B}$ are related to e_0, e_1, e_2 in (2.78) by

$$\begin{aligned} \tilde{C}&=e_0, \\ \tilde{A}&=e_1+6e_2, \\ \tilde{B}&=2e_2. \end{aligned} \tag{2.199}$$

The quantity $E_{\mathrm{D}}(N,\phi)$ is the contribution to the energies coming from the terms containing explicitly s and d boson operators and depends on the quantum numbers ϕ of the state. Using (2.197) and the results of Sect. 2.5.2, one can evaluate the ground-state energies, E_{G}, of the three chains.

(*i*) *Chain I*

$$E_{\mathrm{G}}=E_0. \tag{2.200}$$

(*ii*) *Chain II*

$$E_{\mathrm{G}}=E_0+\tfrac{4}{3}\delta N(2N+3). \tag{2.201}$$

(*iii*) *Chain III*

$$E_{\mathrm{G}}=E_0+2\eta N(N+4). \tag{2.202}$$

The quantities usually measured are not the energies themselves but rather the two-nucleon separation energies. These are defined as the difference in binding energies between nuclei differing by two particles, i.e. one boson. The binding energies, E_B, are the negative of the ground-state energies, E_G,

$$E_B = -E_G, \tag{2.203}$$

and the two-nucleon separation energies are given by

$$S_2(N) = E_B(N+1) - E_B(N). \tag{2.204}$$

From these expressions one obtains,

$$\begin{aligned} S_2^{(I)}(N) &= -\tilde{A} - \tilde{B}N, \\ S_2^{(II)}(N) &= -\tilde{A} - \tilde{B}N - \tfrac{4}{3}\delta(4N+5), \\ S^{(III)}(N) &= -\tilde{A} - \tilde{B}N - 2\eta(2N+5). \end{aligned} \tag{2.205}$$

Thus, in all three limits, the separation energies are linear functions of N, although with different slopes.

A delicate problem arises when reaching the middle of a major shell. According to the microscopic theory of the interacting boson model in which the bosons represent fermion pairs, one counts the number N either as the number of valence pairs, from the beginning to the middle of the shell, or as the number of hole pairs, from the middle to the end of the shell. This means that in those regions, Eq. (2.204) should be replaced by

$$\bar{S}_2(N) = E_B(N-1) - E_B(N), \tag{2.206}$$

and the coefficients $\tilde{A}$ and $\tilde{B}$ by

$$\begin{aligned} \bar{A} &= -\tilde{A}, \\ \bar{B} &= -\tilde{B}, \end{aligned} \tag{2.207}$$

giving rise to

$$\begin{aligned} \bar{S}_2^{(I)}(N) &= -\tilde{A} - \tilde{B}N \\ \bar{S}^{(II)}(N) &= -\tilde{A} - \tilde{B}N + \tfrac{4}{3}\delta(4N+1), \\ \bar{S}^{(III)}(N) &= -\tilde{A} - \tilde{B}N + 2\eta(2N+3). \end{aligned} \tag{2.208}$$

The change in sign in (2.208) produces a discontinuity in $S_2(N)$ at the middle of major shells often observed experimentally.

2.8 Nuclear radii

Nuclear radii can also be analyzed using the interacting boson model. These quantities are related to matrix elements of the E0 operator. The operator representing the square radius r^2 can be written as

$$r^2 = {}_c^2 + \tilde{\gamma}_0 \hat{N} + \tilde{\beta}_0 \hat{n}_d . \tag{2.209}$$

This is the most general form of any scalar operator within the framework of the interacting boson model-1 (see (1.40)). Here r_c^2 has the meaning of the square radius of the closed shell. Square radii are then given by the expectation value of (2.209) in the state $|[N], \phi\rangle$, which we denote by

$$\langle r^2 \rangle_\phi^{(N)} = r_c^2 + \tilde{\gamma}_0 N + \tilde{\beta}_0 \langle \hat{n}_d \rangle_\phi^{(N)} . \tag{2.210}$$

Expectation values of $\hat{n}_d$ for the ground-state band in the three cases, I, II and III, have been evaluated and are given by

$$\langle \hat{n}_d \rangle_L^{(N)} = \frac{L}{2}, \qquad \text{(I)} \tag{2.211}$$

$$\langle \hat{n}_d \rangle_L^{(N)} = \frac{8N(N-1) + L(L+1)}{6(2N-1)}, \qquad \text{(II)} \tag{2.212}$$

$$\langle \hat{n}_d \rangle_L^{(N)} = \frac{4N(N-1) + L(L+6)}{8(N+1)}. \qquad \text{(III)} \tag{2.213}$$

2.8.1 *Isomer shifts*

Quantities that are experimentally accessible in many cases are the isomer, isotope and isotone shifts. The isomer shift, $\delta\langle r^2 \rangle$, is a measure in r^2 between the first 2^+ state and the ground state. Using (2.210) we find

$$\delta\langle r^2 \rangle^{(N)} = \tilde{\beta}_0 [\langle \hat{n}_d \rangle_{2_1^+}^{(N)} - \langle \tilde{n}_d \rangle_{0_1^+}^{(N)}]. \tag{2.214}$$

Equation (2.210) gives, in the three limits,

$$\delta\langle r^2 \rangle^{(N)} = \tilde{\beta}_0 \qquad \text{in (I)}, \tag{2.215}$$

$$\delta\langle r^2 \rangle^{(N)} = \tilde{\beta}_0 \frac{1}{(2N-1)} \qquad \text{in (II)}, \tag{2.216}$$

$$\delta\langle r^2 \rangle^{(N)} = \tilde{\beta}_0 \frac{2}{(N+1)} \qquad \text{in (III)}. \tag{2.217}$$

It is interesting to note that, in the limit $N \to \infty$, both isomer shifts II and III vanish, a property experimentally observed. (See, for example, Fig. 12 of Scholten, Iachello and Arima (1978).)

2.8.2 Isotope and isotone shifts

Isotope (or isotone) shifts, $\Delta\langle r^2\rangle$, are a measure of differences in radii between nuclei one neutron (or one proton) pair (one boson) away from each other,

$$\Delta\langle r^2\rangle^{(N)}=\langle r^2\rangle_{0_1^+}^{(N+1)}-\langle r^2\rangle_{0_1^+}^{(N)}. \tag{2.218}$$

Using (2.210), Eq. (2.218) can be rewritten as

$$\Delta\langle r^2\rangle^{(N)}=\tilde{\gamma}_0+\tilde{\beta}_0[\langle\hat{n}_d\rangle_{0_1^+}^{(N+1)}-\langle\hat{n}_d\rangle_{0_1^+}^{(N)}]. \tag{2.219}$$

Isotope shifts can be evaluated explicitly in the three cases, I, II and III, with the result

$$\Delta\langle r^2\rangle^{(N)}=\tilde{\gamma}_0 \qquad \text{in (I)}, \tag{2.220}$$

$$\Delta\langle r^2\rangle^{(N)}=\tilde{\gamma}_0+\tilde{\beta}_0\frac{4}{3}\left[\frac{N(N+1)}{2N+1}-\frac{N(N-1)}{2N-1}\right] \qquad \text{in (II)}, \tag{2.221}$$

$$\Delta\langle r^2\rangle^{(N)}=\tilde{\gamma}_0+\tilde{\beta}_0\left[\frac{N(N+1)}{2N+4}-\frac{N(N-1)}{2N+2}\right] \qquad \text{in (III)}. \tag{2.222}$$

Here also a delicate problem arises when crossing the middle of the shell. The isotope shifts (2.218) should be replaced by

$$\bar{\Delta}\langle r^2\rangle^{(N)}=\langle r^2\rangle_{0_1^+}^{(N-1)}-\langle r^2\rangle_{0_1^+}^{(N)}, \tag{2.223}$$

and the coefficient $\tilde{\gamma}_0$ by

$$\bar{\tilde{\gamma}}_0=-\tilde{\gamma}_0, \tag{2.224}$$

giving rise to

$$\bar{\Delta}\langle r^2\rangle^{(N)}=\tilde{\gamma}_0 \qquad \text{in (I)}, \tag{2.225}$$

$$\bar{\Delta}\langle r^2\rangle^{(N)}=\tilde{\gamma}_0-\tilde{\beta}_0\frac{4}{3}\left[\frac{N(N-1)}{2N-1}-\frac{(N-1)(N-2)}{2N-3}\right] \qquad \text{in (II)}, \tag{2.226}$$

$$\bar{\Delta}\langle r^2\rangle^{(N)}=\tilde{\gamma}_0-\tilde{\beta}_0\left[\frac{N(N-1)}{2N+2}-\frac{(N-1)(N-2)}{2N}\right] \qquad \text{in (III)}. \tag{2.227}$$

Again, this produces a discontinuity in $\Delta\langle r^2\rangle^{(N)}$ when crossing the middle of the shell.

2.9 Two-nucleon transfer reactions

Operators for calculating two-nucleon transfer reactions have been discussed in Sects. 1.4.9–1.4.13. Here, even more than in the case of binding energies and nuclear radii, proton–neutron effects are of primary importance. Two-nucleon transfer reactions have thus been treated within the framework of the interacting boson model-1 by a formalism which takes somewhat into account these effects. Introducing the index (π, ν) to indicate protons (π) and neutrons (ν), the $L=0$ transfer operators have been written as

$$
\begin{aligned}
P^{(0)}_{+,\pi,0} &= \alpha_\pi s^\dagger \left(\frac{N_\pi+1}{N+1}\right)^{\frac{1}{2}} \left(\Omega_\pi - N_\pi - \frac{N_\pi}{N}\hat{n}_d\right)^{\frac{1}{2}}, \\
P^{(0)}_{+,\nu,0} &= \alpha_\nu s^\dagger \left(\frac{N_\nu+1}{N+1}\right)^{\frac{1}{2}} \left(\Omega_\nu - N_\nu - \frac{N_\nu}{N}\hat{n}_d\right)^{\frac{1}{2}}, \\
P^{(0)}_{-,\pi,0} &= \alpha_\pi \left(\Omega_\pi - N_\pi - \frac{N_\pi}{N}\hat{n}_d\right)^{\frac{1}{2}} \left(\frac{N_\pi+1}{N+1}\right)^{\frac{1}{2}} s, \\
P^{(0)}_{-,\nu,0} &= \alpha_\nu \left(\Omega_\nu - N_\nu - \frac{N_\nu}{N}\hat{n}_d\right)^{\frac{1}{2}} \left(\frac{N_\nu+1}{N+1}\right)^{\frac{1}{2}} s.
\end{aligned}
\tag{2.228}
$$

These forms are similar to those in (1.63), except that an explicit dependence on proton and neutron numbers, N_π and N_ν, has been introduced. Furthermore, a microscopic input has been added since Ω_π and Ω_ν represent the proton and neutron pair degeneracies, i.e. the available number of pair states, as will be discussed in Part II.

When computing matrix elements of the operators (2.228), the operator $\hat{n}_d$ in the square root has usually been replaced by its expectation value in the ground state $\langle \hat{n}_d \rangle^{(N)}_{0_1^+}$. With this approximation, an explicit evaluation of the matrix elements of the transfer operators involves only an evaluation of matrix elements of $s^\dagger$ (or s). This can be done analytically. Once the matrix elements have been evaluated, one can compute the transfer intensities I using

$$
I(N, \phi \rightarrow N+1, \phi') = |\langle [N+1], \phi' | P^{(0)}_{+,0} | [N], \phi \rangle|^2. \tag{2.229}
$$

The matrix elements of $s^\dagger$ between ground states in the various cases are given by

$$
\begin{aligned}
&\langle [N+1], n_d=0, v=0, \tilde{n}_\Delta=0, L=0 | s^\dagger | [N], n_d=0, v=0, \tilde{n}_\Delta=0, L=0 \rangle \\
&\qquad = \sqrt{(N+1)} \quad \text{in (I)},
\end{aligned}
\tag{2.230}
$$

$$
\begin{aligned}
&\langle [N+1], (2N+2,0), \tilde{\chi}=0, L=0 | s^\dagger | [N], (2N,0), \tilde{\chi}=0, L=0 \rangle \\
&\qquad = \sqrt{(N+1)} \sqrt{\left[\frac{(2N+3)}{3(2N+1)}\right]} \quad \text{in (II)},
\end{aligned}
\tag{2.231}
$$

$$\langle [N+1], \sigma=N+1, \tau=0, \tilde{\nu}_\Delta=0, L=0|s^\dagger|[N], \sigma=N, \tau=0, \tilde{\nu}_\Delta=0, L=0\rangle$$

$$=\sqrt{(N+1)}\sqrt{\left[\frac{(N+1)(N+4)}{2(N+2)}\right]} \quad \text{in (III).} \qquad (2.232)$$

Using these expressions and (2.228)–(2.229) one obtains,

$$I^{(\mathrm{I})}(N_\nu, 0_1^+ \rightarrow N_\nu+1, 0_1^+)=\alpha_\nu^2(N_\nu+1)(\Omega_\nu-N_\nu),$$

$$I^{(\mathrm{II})}(N_\nu, 0_1^+ \rightarrow N_\nu+1, 0_1^+)=\alpha_\nu^2(N_\nu+1)\frac{2N+3}{3(2N+1)}\left(\Omega_\nu-N_\nu-\frac{4}{3}\frac{(N-1)}{(2N-1)}N_\nu\right),$$

$$I^{(\mathrm{III})}(N_\nu, 0_1^+ \rightarrow N_\nu+1, 0_1^+)=\alpha_\nu^2(N_\nu+1)\frac{N+4}{2(N+2)}\left(\Omega_\nu-N_\nu-\frac{(N-1)}{2(N+1)}N_\nu\right), \qquad (2.233)$$

and similar expressions with ν interchanged with π for two-proton transfer reactions. The behavior of $I(N_\nu, 0_1^+ \rightarrow N_\nu+1, 0_1^+)$ with N_ν is shown in Fig. 2.15. When crossing the middle of the shell the roles of the operators P_+ and P_- are interchanged, in accordance with the fact that pairs are counted as holes from the middle of the shell on.

Similar techniques allow one to evaluate matrix elements to excited states. One obtains

$$I^{(\mathrm{I})}(N_\nu, 0_1^+ \rightarrow N_\nu+1, 0_{n_d=2}^+)=0,$$

$$I^{(\mathrm{II})}(N_\nu, 0_1^+ \rightarrow N_\nu+1, 0_{\lambda=2N-2, \mu=2}^+)$$

$$=\alpha_\nu^2(N_\nu+1)\frac{8N^2}{3(4N^2-1)(N+1)}\left(\Omega_\mu-N_\nu-\frac{4}{3}\frac{N-1}{2N-1}N_\nu\right),$$

$$I^{(\mathrm{III})}(N_\nu, 0_1^+ \rightarrow N_\nu+1, 0_{\sigma=N-2}^+)$$

$$=\alpha_\nu^2(N_\nu+1)\frac{N(N+3)}{2(N+1)^2(N+2)}\left(\Omega_\nu-N_\nu-\frac{N-1}{2(N+1)}N_\nu\right). \qquad (2.234)$$

The ratios

$$R^t=\frac{I(N_\nu, 0_1^+ \rightarrow N_\nu+1, 0_x^+)}{I(N_\nu, 0_1^+ \rightarrow N_\nu+1, 0_1^+)} \qquad (2.235)$$

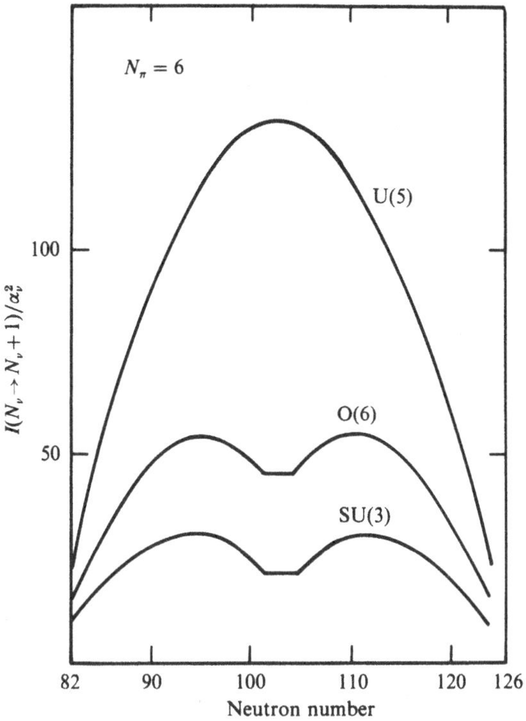

Fig. 2.15. Ground-state to ground-state two-neutron transfer reaction intensities, $I(N_\nu \to N_\nu + 1)$, in units of α_ν^2, in the three limits I, II and III of the interacting boson model-1.

are easily accessible experimentally. One has

$$
\begin{aligned}
R^{t(\mathrm{I})} &= 0, \\
R^{t(\mathrm{II})} &= \frac{8N^2(2N-1)}{(4N^2-1)(N+1)(N+3)} \underset{N\to\infty}{\longrightarrow} \frac{2}{N}, \\
R^{t(\mathrm{III})} &= \frac{N(N+3)}{(N+1)^2(N+4)} \underset{N\to\infty}{\longrightarrow} \frac{1}{N}.
\end{aligned}
\tag{2.236}
$$

It is interesting to note that, for the inverse reaction, $N_\nu + 1 \to N_\nu$, all matrix elements to excited states vanish, while those to the ground states are identical to (2.233).

Two-nucleon transfer reactions with $L = 2$ have not been much analyzed using explicit analytic expressions. In lowest order, the transfer operators have been written as

$$P^{(2)}_{+,\pi,\mu}=\beta_\pi d^\dagger_\mu\left(\frac{N_\pi+1}{N+1}\right)^{\frac{1}{2}}\left(\Omega_\pi-N_\pi-\frac{N_\pi}{N}\hat{n}_\mathrm{d}\right)^{\frac{1}{2}},$$

$$P^{(2)}_{+,\nu,\mu}=\beta_\pi d^+_\mu\left(\frac{N_\nu+1}{N+1}\right)^{\frac{1}{2}}\left(\Omega_\nu-N_\nu-\frac{N_\nu}{N}\hat{n}_\mathrm{d}\right)^{\frac{1}{2}},$$

$$P^{(2)}_{-,\pi,\mu}=\beta_\pi\left(\Omega_\pi-N_\pi-\frac{N_\pi}{N}\hat{n}_\mathrm{d}\right)^{\frac{1}{2}}\left(\frac{N_\pi+1}{N+1}\right)^{\frac{1}{2}}d_\mu, \tag{2.237}$$

$$P^{(2)}_{-,\nu,\mu}=\beta_\nu\left(\Omega_\nu-N_\nu-\frac{N_\nu}{N}\hat{n}_{\mathrm{d}_\nu}\right)^{\frac{1}{2}}\left(\frac{N_\nu+1}{N+1}\right)^{\frac{1}{2}}d_\mu.$$

Thus, under the same assumption of replacing $\hat{n}_\mathrm{d}$ by its expectation value $\langle\hat{n}_\mathrm{d}\rangle^{(N)}_{0_1^+}$, the evaluation of the matrix elements of the operators P_+, P_- is reduced to that of $d^\dagger$ and d. One can then calculate intensities of transfer reactions using

$$I(N,\phi\rightarrow N+1,\phi')=\frac{1}{2L+1}|\langle[N+1],\phi'\|P^{(2)}_+\|[N],\phi\rangle|^2. \tag{2.238}$$

Matrix elements of the $d^\dagger$ operator have been evaluated explicitly in some cases,

$$\langle[N+1],n_\mathrm{d}+1,v=n_\mathrm{d}+1,\tilde{n}_\Delta=0,L=2n_\mathrm{d}+2\|d^\dagger\|[N],n_\mathrm{d},v=n_\mathrm{d},n_\Delta=0,L'=2n_\mathrm{d}\rangle$$
$$=\sqrt{(n_\mathrm{d}+1)}\sqrt{(2L+1)}, \tag{2.239}$$

and

$$\langle[N+1],(2N+2,0),\tilde{\chi}=0,L\|d^\dagger\|[N],(2N,0),\tilde{\chi}=0.L'=L-2\rangle$$
$$=\sqrt{(N+1)}\sqrt{\left[\frac{(2N+1+L)(2N+3+L)(L-1)L}{(2N+1)(2N+2)(2L-1)(2L+1)}\right]}\sqrt{(2L+1)}. \tag{2.240}$$

These yield

$$I^{(\mathrm{I})}(N_\nu,0_1^+\rightarrow N_\nu+1,2_1^+)=\beta_\nu^2(N_\nu+1)\left(\frac{5}{N+1}\right)(\Omega_\nu-N_\nu), \tag{2.241}$$

$$I^{(\mathrm{II})}(N_\nu,0_1^+\rightarrow N_\nu+1,2_1^+)=\beta_\nu^2(N_\nu+1)\frac{(2N+3)(2N+5)2}{(2N+1)(2N+2)3}$$
$$\times\left(\Omega_\nu-N_\nu-\frac{4}{3}\frac{N-1}{2N-1}N_\nu\right). \tag{2.242}$$

The ratios

$$R''=\frac{I(N_v,0_1^+\to N_v+1,2_1^+)}{I(N_v,0_1^+\to N_v+1,0_1^+)}\tag{2.243}$$

are given by

$$R''^{(I)}=\frac{\beta_v^2}{\alpha_v^2}\frac{5}{N+1}\underset{N\to\infty}{\longrightarrow}\left(5\frac{\beta_v^2}{\alpha_v^2}\right)\frac{1}{N},\tag{2.244}$$

$$R''^{(II)}=\frac{\beta_v^2}{\alpha_v^2}\left(\frac{2N+5}{2N+2}\right)2\underset{N\to\infty}{\longrightarrow}\left(5\frac{\beta_v^2}{\alpha_v^2}\right)\frac{2}{5}.\tag{2.245}$$

2.9.1 Examples of dynamic symmetries in transfer reactions

The selection rules for $L=0$ transfer in the three limits I, II and III are summarized in Fig. 2.16. These appear to be reasonably well satisfied by the experimental results in regions where the symmetries apply. Cizewski (1981) and Vergnes (1981) have analyzed in detail the situation in the Os–Pt region and compared it with the predictions of the O(6) symmetry. The results for ground-state transitions are shown in Fig. 2.17.

The situation appears to be more complex for $L=2$ transfer where large discrepancies ($\sim 30\%$) have been observed experimentally with the results of the O(6) symmetry.

2.10 Transitional classes

As discussed in Sect. 2.5, the eigenvalue problem for H can be solved in closed form when H can be written in terms only of Casimir invariants of chain of groups $G\supset G'\supset\cdots$. Most nuclei are not in this ideal situation and thus one needs to treat the full Hamiltonian H, Eq. (2.74). After that of the dynamic symmetries in which all properties can be calculated analytically, the next situation, in order of complexity, arises when the Hamiltonian H can be written in terms of operators of *two* chains. One can schematically depict the situation as in Fig. 2.18, called Casten's triangle (Casten, 1981). The three limits (dynamic symmetries) are placed at the vertices of the triangle. Situations in which H contains only Casimir invariants of *two* chains are placed along the sides of the triangle, and situations in which Casimir invariants of all chains appear are placed inside the triangle. One can thus divide nuclei into four transitional classes:

(i) Class A, nuclei with properties intermediate between I and II;
(ii) Class B, nuclei with properties intermediate between II and III;

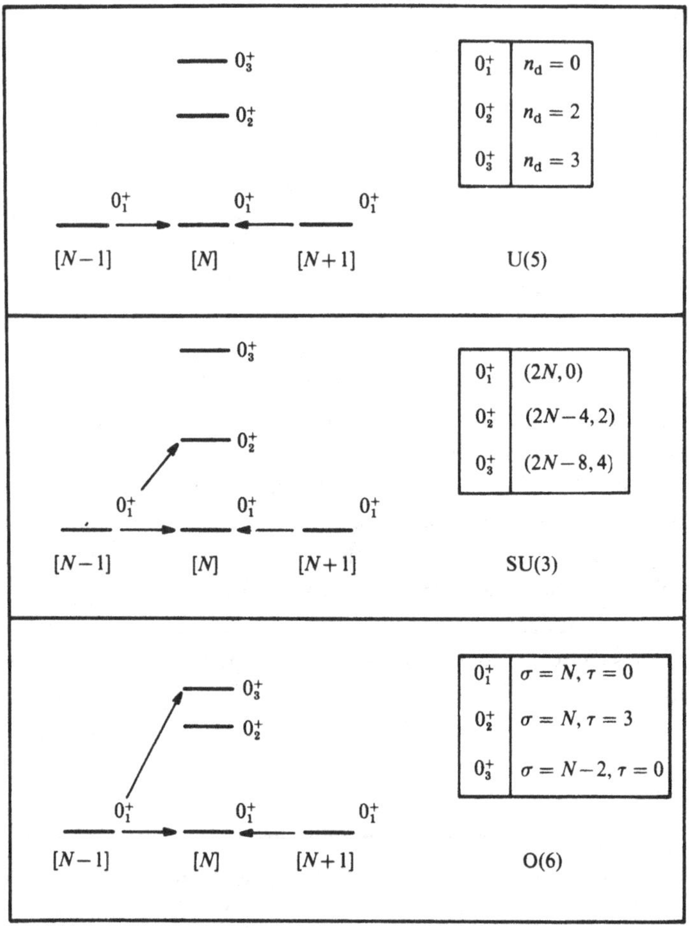

Fig. 2.16. Selection rules for two-particle $L=0$ transfer in the three limits of the interacting boson model-1.

(iii) Class C, nuclei with properties intermediate between III and I; and

(iv) Class D, nuclei with properties intermediate between all three limits.

The study of transitional classes can be made using any of the realizations discussed in Chapter 1. Here we discuss two of these.

2.10.1 *Transitional classes in the multipole realization*

(i) Transitional class A

This class is a mixture of limits I and II. One can study it by considering the

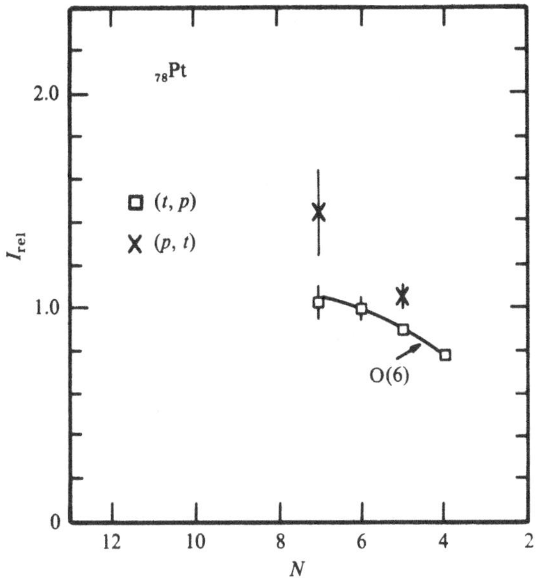

Fig. 2.17. Relative transfer intensities in Pt compared with the predictions of the O(6) symmetry.

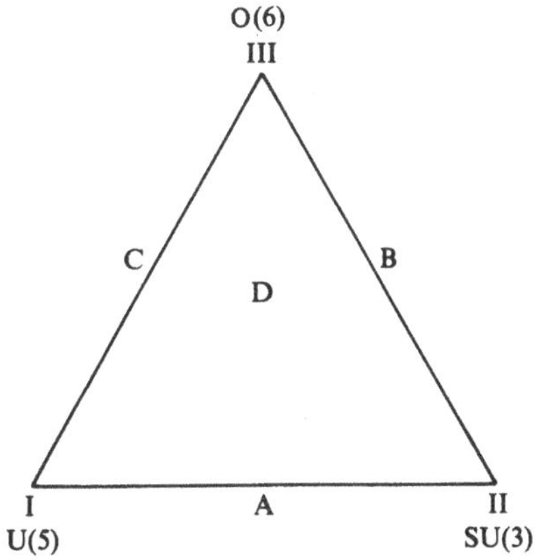

Fig. 2.18. Casten's triangle.

Hamiltonian

$$H^{(I)+(II)} = \varepsilon\mathscr{C}_1(\mathrm{U5}) + \alpha\mathscr{C}_2(\mathrm{U5}) + \beta\mathscr{C}_2(\mathrm{O5}) + \gamma\mathscr{C}_2(\mathrm{O3}) + \delta\mathscr{C}_2(\mathrm{SU3}), \quad (2.246)$$

where we have set $E_0 = 0$ for simplicity. In the multipole realization and neglecting the inessential terms α, β, Eq. (2.246) can be written as

$$H^{(I)+(II)} = \varepsilon\hat{n}_d - \kappa 2\hat{Q}\cdot\hat{Q} - \kappa'\hat{L}\cdot\hat{L}. \quad (2.247)$$

Since the term $\hat{L}\cdot\hat{L}$ is a Casimir invariant of both chains, I and II, the properties of the solutions of (2.247) depend only on the ratio ε/κ. This ratio takes the role of a control parameter for this transitional class. When ε/κ is large, the eigenfunctions of H are those appropriate to the limiting situation I, while, when ε/κ is small, the eigenfunctions of H are those appropriate to the limiting situation II. More on this control parameter will be said in connection with the classical limit of the interacting boson model (Chapter 3). Here, it suffices to say that a simple way to study the transitional class A is by expanding ε, κ and κ' as a function of the boson number N and retaining the first term in the expansion,

$$\begin{aligned}
\varepsilon(N) &= \varepsilon(N_0) + \left.\frac{\partial\varepsilon}{\partial N}\right|_{N=N_0}(N-N_0)+\cdots \\
\kappa(N) &= \kappa(N_0) + \left.\frac{\partial\kappa}{\partial N}\right|_{N=N_0}(N-N_0)+\cdots \\
\kappa'(N) &= \kappa'(N_0) + \left.\frac{\partial\kappa'}{\partial N}\right|_{N=N_0}(N-N_0)+\cdots.
\end{aligned} \tag{2.248}$$

In particular, calculations have been done by keeping κ and κ' constant and letting ε decrease with increasing N,

$$\begin{aligned}
\varepsilon &= \varepsilon^{(0)} - \varepsilon^{(1)}N, \\
\kappa &= \kappa^{(0)}, \\
\kappa' &= \kappa'^{(0)}.
\end{aligned} \tag{2.249}$$

This is a reasonable approximation if one investigates a small region of N. A similar approach has been followed for all operators. For example, if the E2 operator is written as in (1.24), it can be parametrized in terms of

$$\begin{aligned}
\alpha_2(N) &= \alpha_2(N_0) + \left.\frac{\partial\alpha_2}{\partial N}\right|_{N=N_0}(N-N_0)+\cdots, \\
\beta_2(N) &= \beta_2(N_0) + \left.\frac{\partial\beta_2}{\partial N}\right|_{N=N_0}(N-N_0)+\cdots.
\end{aligned} \tag{2.250}$$

A reasonable approximation here is to keep α_2 and β_2 constant,

$$\begin{aligned}
\alpha_2 &= \alpha_2^{(0)}, \\
\beta_2 &= \beta_2^{(0)},
\end{aligned} \tag{2.251}$$

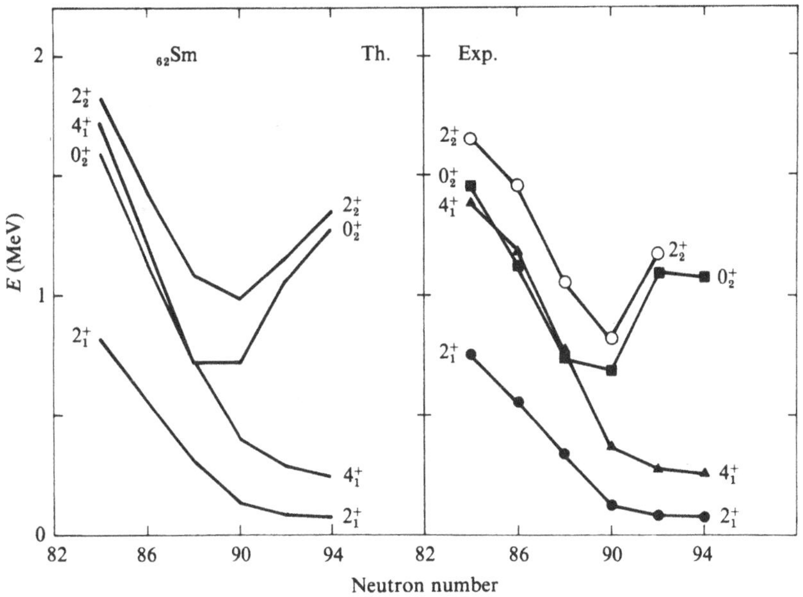

Fig. 2.19. Typical features of the transitional class A. Excitation energies.

in particular $\beta_2 = -\alpha_2\sqrt{7}/2$, in such a way that

$$T^{(\mathrm{E2})} = \alpha_2^{(0)}\hat{Q}. \tag{2.252}$$

The typical features of the spectra of this transitional class are shown in Fig. 2.19, which shows an application to the $_{62}$Sm isotopes. Certain electromagnetic properties are also very sensitive to the value of the control parameter ε/κ, for example the B(E2) ratios defined in (2.139). If one keeps $\beta_2 = -\alpha_2\sqrt{7}/2$ constant, these ratios do not depend on any parameter, since $\alpha_2^{(0)}$ drops out when taking ratios. They are then a distinctive signature of the transitional class. Another very distinctive ratio is

$$R''' = \frac{B(\mathrm{E2}; 2_2^+ \rightarrow 0_1^+)}{B(\mathrm{E2}; 2_2^+ \rightarrow 2_1^+)}, \tag{2.253}$$

which changes from

$$R''' = 0 \qquad \text{in (I)}, \tag{2.254}$$

to

$$R''' = 7/10 \quad \text{in (II)}. \tag{2.255}$$

Although both transitions in (2.253) vanish in the limit II, one can show

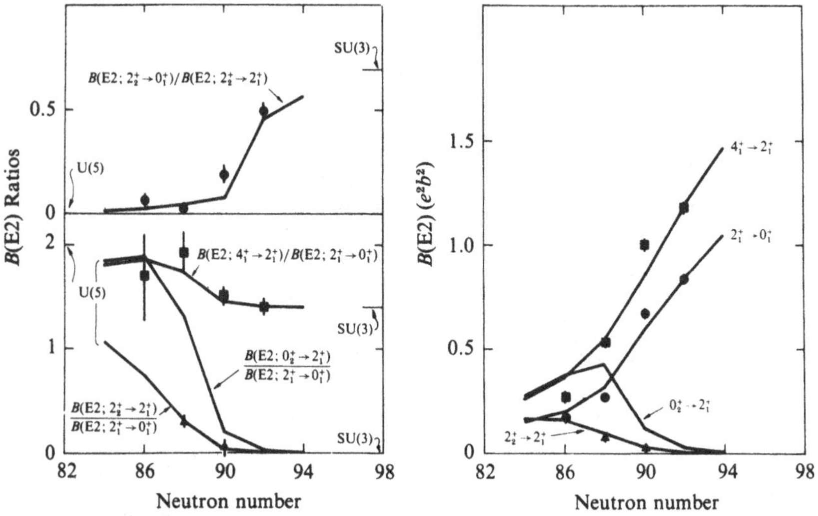

Fig. 2.20. Typical features of the transitional class A. Electromagnetic E2 transition rates.

that their ratio approaches the value 7/10. A comparison between theory and experiment is shown in Fig. 2.20. All calculations presented in this and subsequent figures have been performed using a computer program developed by Scholten (1976).

There are three more quantities that are extremely sensitive to the control parameter ε/κ, the isomer shift, $\delta\langle r^2\rangle$, the intensity of two-nucleon transfer reactions, $I(N \to N+1)$, and the two-neutron separation energies, $S_2(N)$. The behavior of these quantities in the transitional class A has been discussed by Scholten, Iachello and Arima (1978).

(ii) Transitional class B

This class is a mixture of limits (II) and (III). It can be studied by considering the Hamiltonian

$$H^{(\mathrm{II})+(\mathrm{III})} = \beta\mathscr{C}_2(\mathrm{O5}) + \gamma\mathscr{C}_2(\mathrm{O3}) + \delta\mathscr{C}_2(\mathrm{SU3}) + \eta\mathscr{C}_2(\mathrm{O6}), \qquad (2.256)$$

where again we have deleted E_0. In the multipole representation, and apart from terms that contribute to E_0, (2.256) can be rewritten as

$$H^{(\mathrm{II})+(\mathrm{III})} = A\,\hat{\mathrm{P}}_6 + B\,\hat{\mathrm{C}}_5 + C\,\hat{\mathrm{C}}_3 - \kappa 2\,\hat{\mathrm{Q}}\cdot\hat{\mathrm{Q}}, \qquad (2.257)$$

where the operators $\hat{\mathrm{P}}_6$, $\hat{\mathrm{C}}_5$ and $\hat{\mathrm{C}}_3$ are given by (2.91) and (2.73). Although when all terms are kept in (2.257) there is no single control parameter, it

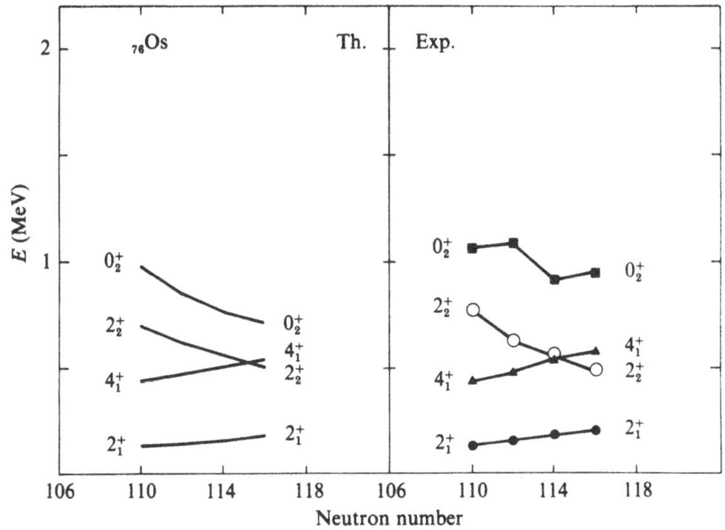

Fig. 2.21. Typical features of the transitional class B. Excitation energies.

appears that the situation in practice is controlled by the ratio A/κ. When this is large the eigenfunctions of H are those appropriate to symmetry III, while, when it is small, the eigenfunctions are those appropriated to symmetry II. In a limited region, one can again expand A, B, C and κ as

$$\begin{aligned} A(N) &= A(N_0) + \left.\frac{\partial A}{\partial N}\right|_{N=N_0} (N-N_0) + \cdots, \\ B(N) &= B(N_0) + \left.\frac{\partial B}{\partial N}\right|_{N=N_0} (N-N_0) + \cdots, \\ C(N) &= C(N_0) + \left.\frac{\partial C}{\partial N}\right|_{N=N_0} (N-N_0) + \cdots, \\ \kappa(N) &= \kappa(N_0) + \left.\frac{\partial \kappa}{\partial N}\right|_{N=N_0} (N-N_0) + \cdots. \end{aligned} \tag{2.258}$$

In particular, calculations have been done by keeping A, B and C constant and varying $\kappa(N)$ linearly with N (Iachello, 1980)

$$\kappa = \kappa^{(0)} + \kappa^{(1)} N. \tag{2.259}$$

These calculations are a slight variation of those performed by Casten and Cizewski (1978), who studied this transitional region for the first time.

The typical features of the spectra of this transitional class are shown in Fig. 2.21, which is an application to the $_{76}$Os isotopes. Similar changes

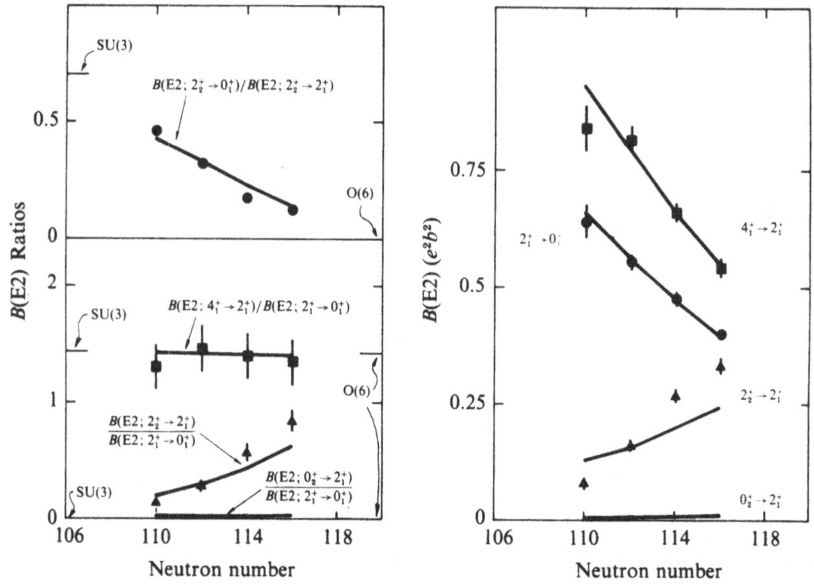

Fig. 2.22. Typical features of the transitional class B. Electromagnetic E2 transition rates.

occur in the electromagnetic transition rates. Studies have been done using the parametrization (1.24) and (2.251) with $\beta_2=0$. The corresponding results are shown in Fig. 2.22. The ratio R''' changes from

$$R'''=0, \qquad \text{in (III)}, \tag{2.260}$$

to

$$R'''=7/10, \quad \text{in (II)}. \tag{2.261}$$

Neither isomer shifts, nor separation energies and intensities of two-nucleon transfer reactions, are in this case very sensitive to the control parameter A/κ.

(iii) Transitional class C

This class is a mixture of limits (III) and (I). It can be studied by considering the Hamiltonian

$$H^{(\mathrm{III})+(\mathrm{I})}=\varepsilon\mathscr{C}_1(\mathrm{U5})+\alpha\mathscr{C}_2(\mathrm{U5})+\beta\mathscr{C}_2(\mathrm{O5})+\gamma\mathscr{C}_2(\mathrm{O3})+\eta\mathscr{C}_2(\mathrm{O6}). \tag{2.262}$$

In the multipole formalism, and neglecting the inessential term α, (2.262) can be rewritten as

$$H^{(\mathrm{III})+(\mathrm{I})}=\varepsilon\hat{n}_{\mathrm{d}}+A\,\hat{\mathrm{P}}_6+B\,\hat{\mathrm{C}}_5+C\,\hat{\mathrm{C}}_3. \tag{2.263}$$

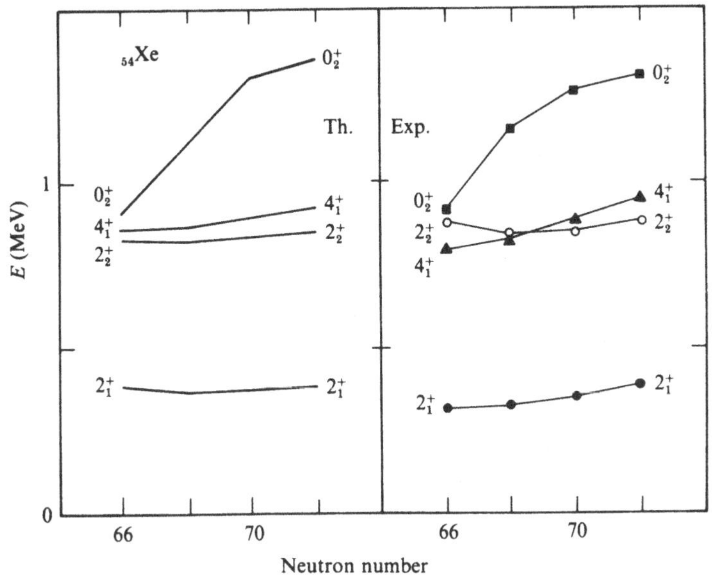

Fig. 2.23. Typical features of the transitional class C. Excitation energies.

Here again we have neglected terms which contribute to E_0. Since the terms $\hat{C}_5$ and $\hat{C}_3$ are common to both chains, the control parameter here is ε/A. When studying a limited region, the coefficients ε, A, B, C can be expanded as a function of N, as above:

$$\begin{aligned}
\varepsilon(N) &= \varepsilon(N_0) + \left.\frac{\partial \varepsilon}{\partial N}\right|_{N=N_0} (N-N_0) + \cdots \\
A(N) &= A(N_0) + \left.\frac{\partial A}{\partial N}\right|_{N=N_0} (N-N_0) + \cdots \\
B(N) &= B(N_0) + \left.\frac{\partial B}{\partial N}\right|_{N=N_0} (N-N_0) + \cdots \\
C(N) &= C(N_0) + \left.\frac{\partial C}{\partial N}\right|_{N=N_0} (N-N_0) + \cdots .
\end{aligned} \tag{2.264}$$

In particular, calculations have been done by keeping ε, B and C constant and varying A linearly with N (Arima and Iachello, 1984)

$$A = A^{(0)} + A^{(1)}N. \tag{2.265}$$

The resulting spectra have the properties shown in Fig. 2.23, which is an application to the $_{54}$Xe isotopes. Similarly, one can study electromagnetic

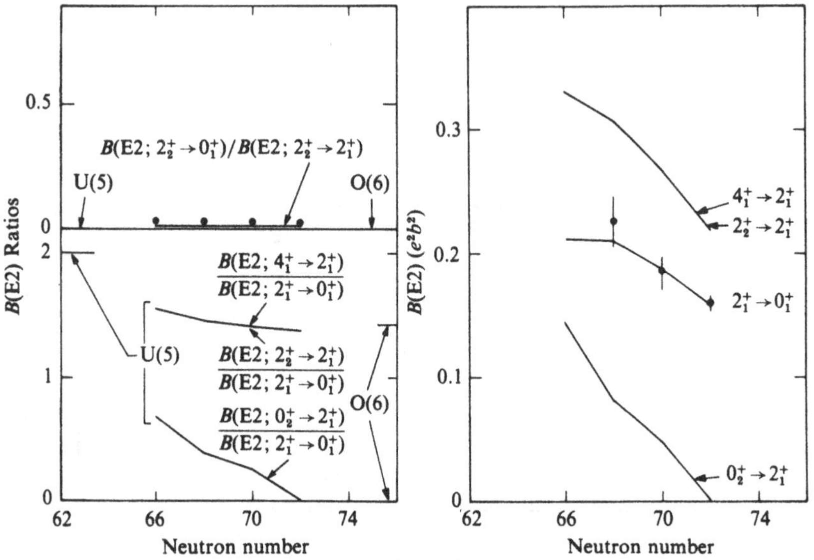

Fig. 2.24. Typical features of the transitional class C. Electromagnetic E2 transition rates.

transition rates. If the parametrization (1.24) and (2.250) is kept, with $\beta_2 = 0$, the results shown in Fig. 2.24 are obtained. The ratio R''', Eq. (2.253), remains in this transitional class identically zero,

$$R''' = 0 \quad \text{in (I)},$$
$$R''' = 0 \quad \text{in (III)}. \tag{2.266}$$

The presently available experimental data, Fig. 2.24, are not sufficient to clearly establish this transitional class. In fact, none of the distinguishing features, i.e. the ratio R'', Eq. (2.142), and the value $B(E2; 0_2^+ \to 2_1^+)$, have been measured. Another, slightly different, study of this transitional class has been done by Stachel, van Isacker and Heyde (1982).

(iv) Transitional class D

This class is intermediate between all three limits, I, II and III. Its properties must be studied by using the full Hamiltonian (2.74)

$$H^{(I)+(II)+(III)} = \varepsilon \mathscr{C}_1(U5) + \alpha \mathscr{C}_2(U5) + \beta \mathscr{C}_2(O5) + \gamma \mathscr{C}_2(O3) + \delta \mathscr{C}_2(SU3) + \eta \mathscr{C}_2(O6). \tag{2.267}$$

It is more convenient then to return to the form (1.35), using all parameters

ε', c'_L ($L=0, 2, 4$), v_L ($L=0, 2$). Although in principle this gives a better description of the data, there are a few calculations availabl، using the most general Hamiltonian, H. This is because the extra terms do not appear to introduce substantial improvements in the description of the data.

2.10.2 *Transitional classes in the consistent-Q formalism*

This formalism appears to provide the most economical approach to the study of nuclear spectra. In this formalism, it appears to be sufficient to use a parametrization of the Hamiltonian, obtained from (1.53) with $c_3 = c_4 = 0$,

$$H^{\chi} = E'_0 + \varepsilon\hat{n}_d + c_1(\hat{L}\cdot\hat{L}) + c_2(\hat{Q}^{\chi}\cdot\hat{Q}^{\chi}), \tag{2.268}$$

and of the transition operator, (1.55),

$$T^{(E2)} = \alpha_2\hat{Q}^{\chi}. \tag{2.269}$$

In this formalism, excitation energies depend only on *four* parameters $\varepsilon, c_1, c_2, \chi$ and E2 transitions on *one* more parameter. Furthermore, some of these parameters only set the scale of energy and E2 rates, while the term $\hat{L}\cdot\hat{L}$, being common to all chains, does not contribute to the structure. Thus all nuclear properties appear to depend only on two parameters ε/c_2 and χ. The Hamiltonian (2.268) is general enough that one can study all transitional classes.

(i) Transitional class A

This class can be studied by using (2.268), (2.269) with $\chi = -\frac{1}{2}\sqrt{7} = -1.323$. The Hamiltonian and transition operator E2 then reduce to (2.247) and (2.252), and the same results as in Sect. 2.10.1 are obtained.

(ii) Transitional class B

This is the most important transitional class since a large number of nuclei (for example, rare-earth nuclei) belong to it. In the consistent-Q formalism, this class can be studied by considering the Hamiltonian

$$H^{(\mathrm{II})+(\mathrm{III}),\chi} = c_1(\hat{L}\cdot\hat{L}) + c_2(\hat{Q}^{\chi}\cdot\hat{Q}^{\chi}), \tag{2.270}$$

for fixed c_1, c_2 and letting χ vary between $\chi = -\frac{1}{2}\sqrt{7} = -1.323$ (limit II) and $\chi = 0$ (limit III); χ plays here the role of a control parameter. Properties of this transitional class have been investigated in detail by Warner and Casten (1983). A schematic representation of the situation for energies and E2 electromagnetic transition rates is shown in Fig. 2.25. A detailed

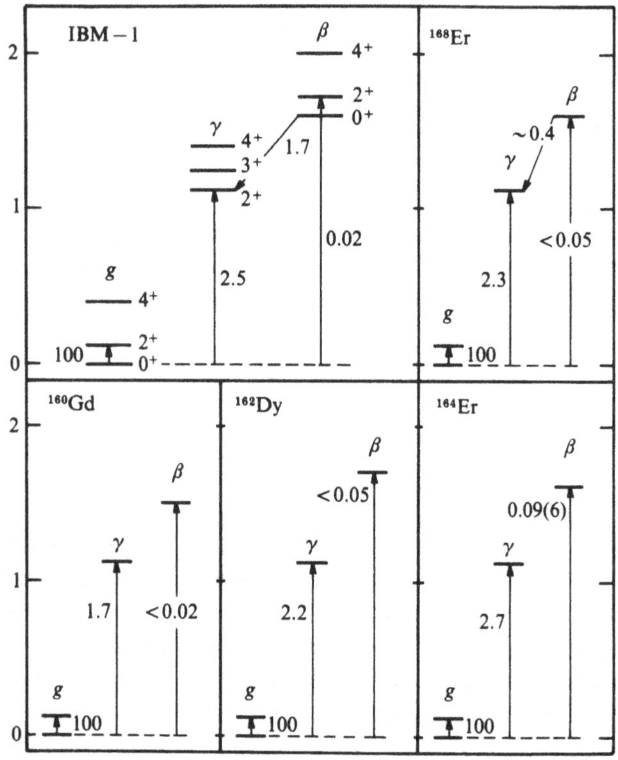

Fig. 2.25. Schematic representation of energies and E2 electromagnetic transition rates, in units such that $B(\mathrm{E2}; 0_1^+ \rightarrow 2_1^+) = 100$, compared with typical experimental situations in rare-earth nuclei. The excitation energies are in units of the intrinsic energy of the γ-band.

comparison (Warner and Casten, 1983) with experiments in ^{168}Er (Warner, Casten and Davidson, 1981) is shown in Table 2.15. In this table, the values appropriate to the case in which the nucleus behaves as a rigid rotor are also shown. The interacting boson model calculation appears to describe the situation very accurately. In Fig. 2.26 some results for the ratio $B(\mathrm{E2}; 2_\gamma^+ \rightarrow 0_g^+)/B(\mathrm{E2}; 2_\gamma^+ \rightarrow 2_g^+)$, and for the values of $Q_{2_1^+}$ are also shown. In the case of $Q_{2_1^+}$, since absolute values are plotted, one needs to specify the value of α_2, chosen to be $\alpha_2 = 0.13$ eb. Other results are given by Warner and Casten (1983). These authors also provide a chart for χ-values in several rare earth nuclei. (The value of $\chi = \chi_Q$ in their article is related to the value of χ used here by $\chi = \chi_Q/\sqrt{5}$.) The consistent-Q formalism for the transitional class B is the most economical one because it allows to calculate (apart from scales) all nuclear properties in terms of the single parameter χ.

Table 2.15. *Comparison between theory and experiment for relative* B(E2) *values for* $\gamma \rightarrow g$ *transitions in* ^{168}*Er (Warner and Casten, 1983),* $\chi = -0.492$

I_i	$I_f, \tilde{\chi}_f$	Exp	IBM$^{\chi}$	Rigid rotor
2	0, 0	54	54	70
	2, 0	100	100	100
	4, 0	7	8	5
3	2, 0	100	100	100
	4, 0	65	69	40
4	2, 0	20	18	34
	4, 0	100	100	100
	6, 0	14	16	9
5	4, 0	81	80	175
	6, 0	100	100	100
6	4, 0	12	9	27
	6, 0	100	100	100
	8, 0	37	20	11

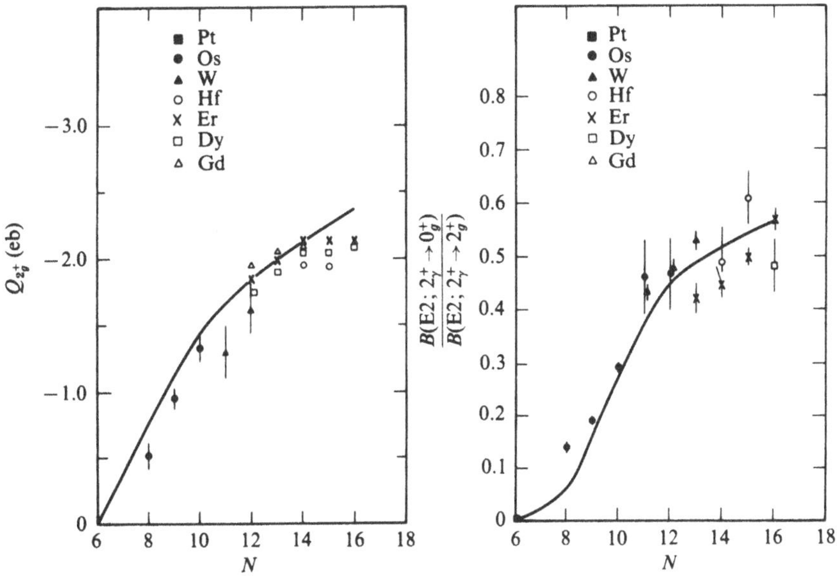

Fig. 2.26. Comparison between empirical and predicted trends of $Q_{2_1^+}$ and B(E2; $2_\gamma^+ \rightarrow 0_g^+$)/($B$(E2; $2_\gamma^+ \rightarrow 2_g^+$) in several rare-earth nuclei plotted as a function of boson number N (Warner and Casten, 1983).

In passing, it is interesting to note that the consistent-Q formalism when used in the O(6) limit, $\chi=0$, gives excitation energies

$$E^{(\mathrm{III}),\chi}(N,\sigma,\tau,\tilde{\nu}_\Delta,L,M_L)=2c_2[\sigma(\sigma+4)-\tau(\tau+3)]+c_1L(L+1). \tag{2.271}$$

Comparing with (2.92), one can see that (2.271) corresponds to a special case of (2.92) in which

$$A/4=B/6=-2c_2. \tag{2.272}$$

The reason why it is so can be found by inspecting Table 2.7, since

$$(\hat{Q}^\chi\cdot\hat{Q}^\chi)\underset{\chi\to 0}{\longrightarrow}\tfrac{1}{2}\mathscr{C}_2(\mathrm{O6})-\tfrac{1}{2}\mathscr{C}_2(\mathrm{O5}). \tag{2.273}$$

Empirically, it appears that (2.271) is closely satisfied. For example, in ^{196}Pt, $A/4=46.2$ KeV, $B/6=42$ KeV.

One further advantage of the consistent-Q formalism is that it is closely related to the microscopic understanding of the model, as discussed in Part II.

(iii) Transitional class C

This class can be studied by using (2.268) with $\chi=0$. The Hamiltonian then reduces to a special form of (2.263). The control parameter is ε/c_2 and the same results as in Sect. 2.10.1 are obtained.

(iv) Transitional class D

For this class, all terms in (2.268) must be retained and $\chi\neq 0$, -1.323. Properties of this class are intermediate between all three limits, (I), (II) and (III).

2.11 Transition densities

As mentioned in Sect. 1.5, electron-scattering experiments measure the spatial dependence of the matrix elements of the transition operators. For applications to these experiments the formulas derived in Sect. 2.6 should all be modified by replacing the numbers γ_0, α_0, α_2, β_L ($L=0,\ldots,4$) with the functions $\gamma_0(r)$, $\alpha_0(r)$, $\alpha_2(r)$, $\beta_L(r)$ ($L=0,\ldots,4$). The transition densities $\rho^{(L)}_{L_i\to L_f}(r)$ are then defined as the matrix elements of the corresponding operators.

2.11.1 Scalar densities

The transition operator is

$$T_0^{(\mathrm{E0})}(r)=\gamma_0(r)+\alpha_0(r)\hat{N}+\beta_0(r)\hat{n}_{\mathrm{d}}. \tag{2.274}$$

Particularly important here are the densities of 0^+ states. If we denote these states by $|N,0_i^+\rangle$, $(i=1,\ldots)$, we obtain

$$\rho^{(0)}_{0_i^+\to 0_i^+}(r)=\gamma_0(r)+\alpha_0(r)N+\beta_0(r)B_{ii}^{(0)}, \tag{2.275}$$

and

$$\rho^{(0)}_{0_i^+\to 0_j^+}(r)=\beta_0(r)B_{ij}^{(0)}, \quad i\neq j. \tag{2.276}$$

The coefficients

$$B_{ij}^{(0)}=\langle N,0_i^+|\hat{n}_{\mathrm{d}}|N,0_j^+\rangle \tag{2.277}$$

contain the nuclear-structure information. Their explicit expression in the case of a dynamic symmetry has been derived in Sect. 2.6. Knowing the diagonal densities, one can obtain mean-square radii. It is customary to normalize the diagonal densities as

$$\int_0^\infty 4\pi r^2\rho^{(0)}_{0_1^+\to 0_1^+}(r)\,\mathrm{d}r=Z. \tag{2.278}$$

Then,

$$\langle r^2\rangle_{0_1^+}=\frac{4\pi}{Z}\int_0^\infty r^4\rho^{(0)}_{0_1^+\to 0_1^+}(r)\,\mathrm{d}r. \tag{2.279}$$

2.11.2 Quadrupole densities

The transition operator here is

$$T_\mu^{(\mathrm{E2})}(r)=\alpha_2(r)[d^\dagger\times\tilde{s}+s^\dagger\times\tilde{d}]_\mu^{(2)}+\beta_2(r)[d^\dagger\times\tilde{d}]_\mu^{(2)}. \tag{2.280}$$

This operator yields transition densities

$$\rho^{(2)}_{0_1^+\to 2_i^+}(r)=\alpha_2(r)A_{1i}^{(2)}+\beta_2(r)B_{1i}^{(2)}, \tag{2.281}$$

where

$$A_{1i}^{(2)}=\langle N,0_1^+\parallel[d^\dagger\times\tilde{s}+s^\dagger\times\tilde{d}]^{(2)}\parallel N,2_i^+\rangle,$$

$$B_{1i}^{(2)}=\langle N,0_1^+\parallel[d^\dagger\times\tilde{d}]^{(2)}\parallel N,2_i^+\rangle. \tag{2.282}$$

The numbers $A_{1i}^{(2)}$ and $B_{1i}^{(2)}$ have been obtained in closed form in Sect. 2.6 in the case of a dynamic symmetry. Knowing the densities (2.281) one can obtain the $B(\mathrm{E2})$ values by

$$B(\mathrm{E2};0_1^+\to 2_1^+)=\left[\int_0^\infty r^4\rho^{(2)}_{0_1^+\to 2_i^+}(r)\,\mathrm{d}r\right]^2. \tag{2.283}$$

2.11.3 Hexadecapole densities

The transition operator here is

$$T_{\mu}^{(\mathrm{E4})}(r)=\beta_4(r)[d^{\dagger}\times\tilde{d}]_{\mu}^{(4)}, \tag{2.284}$$

with transition densities

$$\rho_{0_1^+\to 4_i^+}^{(4)}(r)=\beta_4(r)B_{1i}^{(4)}, \tag{2.285}$$

where

$$B_{1i}^{(4)}=\langle N,0_1^+\,\|\,[d^{\dagger}\times\tilde{d}]^{(4)}\,\|\,N,4_i^+\rangle. \tag{2.286}$$

The B(E4) values are given by

$$B(\mathrm{E4};0_1^+\to 4_i^+)=\left[\int_0^{\infty} r^6\rho_{0_1^+\to 4_i^+}^{(4)}(r)\,\mathrm{d}r\right]^2. \tag{2.287}$$

The contribution of $L=4$ pairs (g bosons), which is neglected in (2.284), makes a comparison with experiments here inappropriate.

2.11.4 Magnetic dipole densities

The lowest-order operator is

$$T_{\mu}^{(\mathrm{M1})}(r)=\beta_1(r)[d^{\dagger}\times\tilde{d}]_{\mu}^{(1)}. \tag{2.288}$$

All transition densities except the diagonal ones vanish here. Among the diagonal matrix elements of some importance are

$$\rho_{2_i^+\to 2_i^+}^{(1)}(r)=\beta_1(r)B_{ii}^{(1)}, \tag{2.289}$$

where

$$B_{ii}^{(1)}=\langle N,2_i^+\,\|\,[d^{\dagger}\times\tilde{d}]^{(1)}\,\|\,N,2_i^+\rangle. \tag{2.290}$$

From (2.289) one can obtain magnetic moments

$$\mu_{2_i^+}=\sqrt{\left(\frac{4\pi}{3}\right)}\sqrt{\left(\frac{2}{15}\right)}\int_0^{\infty} r^2\rho_{2_i^+\to 2_i^+}^{(1)}(r)\,\mathrm{d}r. \tag{2.291}$$

The form (2.288) is a poor approximation for magnetic dipole properties of nuclei. These are much influenced by proton–neutron degrees of freedom and should be treated more properly within the framework discussed in Part II.

2.11.5 Magnetic octupole densities

The appropriate operator is

$$T_{\mu}^{(\mathrm{M3})}(r)=\beta_3(r)[d^{\dagger}\times\tilde{d}]_{\mu}^{(3)}. \tag{2.292}$$

Table 2.16. *Structure coefficients* $A_{1i}^{(2)}$ *and* $B_{1i}^{(2)}$ *and* B(E2) *values for* $0_1^+ \rightarrow 2_1^+$ *transitions in* ^{150}Nd

Transition	$A_{1i}^{(2)}$	$B_{1i}^{(2)}$	$B(\mathrm{E2})_{\mathrm{calc}}$ (e^2 fm^4)	$B(\mathrm{E2})_{\mathrm{exp}}$ (e^2 fm^4)
$0_1^+ \rightarrow 2_1^+$	10.02	−2.26	27 200[a]	27 200
$0_1^+ \rightarrow 2_2^+$	1.68	1.30	134	76
$0_1^+ \rightarrow 2_3^+$	2.73	1.45	629	690

[a] Fitted.

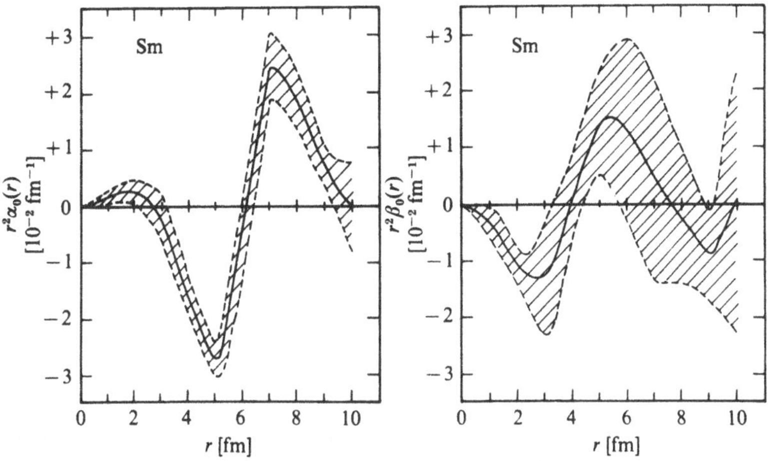

Fig. 2.27. The boson scalar densities $r^2\alpha_0(r)$ and $r^2\beta_0(r)$ extracted from fits to the ground state densities in $_{62}$Sm (Moinester *et al.*, 1982).

The transition densities which can be experimentally studied are

$$\rho_{0_1^+ \rightarrow 3_i^+}^{(3)}(r) = \beta_3(r) B_{1i}^{(3)}, \tag{2.293}$$

where

$$B_{1i}^{(3)} = \langle N, 0_1^+ \| [d^\dagger \times \tilde{d}]^{(3)} \| N, 3_i^+ \rangle. \tag{2.294}$$

From these one can obtain B(M3) values

$$B(\mathrm{M3}; 0_1^+ \rightarrow 3_i^+) = \left[\int_0^\infty r^4 \rho_{0_1^+ \rightarrow 3_i^+}^{(3)}(r)\, \mathrm{d}r \right]^2. \tag{2.295}$$

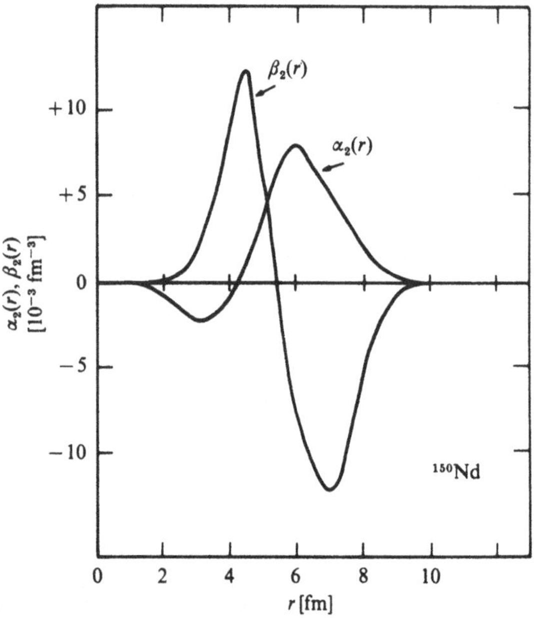

Fig. 2.28. Boson quadrupole densities $\alpha_2(r)$ and $\beta_2(r)$ extracted from a fit to the $0_1^+ \to 2_1^+$ and $0_1^+ \to 2_3^+$ transition densities in ^{150}Nd (Dieperink, 1979).

2.11.6 Experimental determination of transition densities in the transitional class A

Moinester *et al.* (1982) have extracted the scalar boson densities $\alpha_0(r)$ and $\beta_0(r)$ from fits to the ground-state scalar densities of the $_{62}$Sm isotopes (transitional class A). The functions $r^2\alpha_0(r)$ and $r^2\beta_0(r)$ are shown in Fig. 2.27. Dieperink (1979) has extracted the boson quadrupole densities $\alpha_2(r)$ and $\beta_2(r)$ from a fit to the $0_1^+ \to 2_1^+$ and $0_1^+ \to 2_3^+$ transition densities in $^{150}_{60}\mathrm{Nd}_{90}$. The structure coefficients were obtained by diagonalizing the Hamiltonian (2.247) and are reported in Table 2.16. With these values, Dieperink obtained the transition densities shown in Fig. 2.28. Similar results have been obtained by Moinester *et al.*

3
Geometry

3.1 Introduction

The interacting boson model has associated with it an intrinsic geometric structure. This is a property of all models with a definite group structure G, since each group has associated with it both an algebra $\mathscr{g}$ and a topology $\mathscr{G}$. The natural space for the geometric properties of a system with group structure G is the so-called coset space. The concept of a coset space is discussed in detail in the book by Gilmore (1974). Gilmore and Feng (1978) introduced it in nuclear physics through applications to the Lipkin model. Several authors have used it in connection with the interacting boson model, since the almost-simultaneous publication of three articles by Bohr and Mottelson (1980), Ginocchio and Kirson (1980) and Dieperink, Scholten and Iachello (1980).

The geometric properties of the interacting boson model are particularly important since they allow one to relate this model to the description of collective states in nuclei by shape variables introduced in the early 1950s by Bohr and Mottelson (Bohr, 1952; Bohr and Mottelson, 1953; Bohr and Mottelson, 1975).

The crucial ingredients in the study of geometric properties of the interacting boson model are the coherent (or intrinsic) states. There is a large variety of problems that can be attacked with these states, both of static and dynamic nature. Here, we confine ourselves to only a few of these.

3.2 Cosets and coset spaces

Given a group G, with algebraic structure $\mathscr{g}$, there is a straightforward mathematical procedure by means of which one can attach geometric spaces to it. This procedure is implicit, for example, in attaching the geometric variables θ, ϕ to the representations of the rotation group SO(3), thus using spherical harmonics to describe them. This procedure may or may not be unique, depending on the conditions imposed. If the procedure is not unique, one usually chooses in a particular application those geometric variables which have a direct physical meaning. The procedure consists in decomposing the algebra $\mathscr{g}$ into two parts,

$$\mathscr{g} = \mathscr{h} \oplus \mathscr{p}, \tag{3.1}$$

where $\mathscr{h}$ is a subalgebra of $\mathscr{g}$

$$\mathscr{g} \supset \mathscr{h} \tag{3.2}$$

and thus closed with respect to commutation, while $\mathscr{p}$, the remainder, is not closed with respect to commutation. Here $\mathscr{h}$ is called the stability algebra and $\mathscr{g}/\mathscr{h}$ the factor algebra. Geometric variables can then be associated with each element of $\mathscr{p}$. Their number, which we call the topological dimension of the space, is just the number of generators in $\mathscr{p}$.

There is a variety of ways in which the geometric variables corresponding to the decomposition (3.1) can be defined, all of them related by appropriate transformations. A commonly used definition, called the 'algebraic' definition, is as follows. The representations of $\mathscr{g}$ provide basis states, $|\Lambda\rangle$. Among these, there is a state, called the extremal state, $|\Lambda_{\mathrm{ext}}\rangle$, which has the property of being annihilated by the largest number of vectors in the algebra. Geometric variables, η_i, can be defined by introducing an 'algebraic' coherent state, through

$$|\eta_i\rangle = \exp\left[\sum_i \eta_i p_i\right]|\Lambda_{\mathrm{ext}}\rangle, \tag{3.3}$$

where p_i are the elements of $\mathscr{p}$. Two other definitions are the so-called 'group' and 'projective' definitions, which we shall discuss in the following sections and use throughout, since they have a direct physical meaning.

We remark here that, since the algebra $\mathscr{g}$ may have several possible subalgebras $\mathscr{h}$, there are in principle several possible geometric spaces. In particular applications, some possibilities are eliminated by some physical constraint. For example, if one insists that $\mathscr{h}$ contains the rotation algebra, O(3), one obtains limits on the possible subgroups of U(6), as in (2.25). Even within these constraints there may still be several possibilities left. One therefore usually defines as geometric space associated with $\mathscr{g}$ that which is connected with the maximum stability subalgebra $\mathscr{h} \subset \mathscr{g}$, where $\mathscr{h}$ is the largest possible algebra compatible with the physical conditions on $\mathscr{g}$. The procedure then uniquely determines the geometric variables associated with $\mathscr{g}$.

An important feature of this procedure when applied to the interacting boson model is that it naturally leads to the introduction of the shape variables of Bohr and Mottelson (Bohr, 1952) extensively used in the treatment of collective states in nuclei. This point will be discussed further below.

The maximum stability subalgebra of U(6) turns out to be U(5) ⊗ U(1), corresponding to the decomposition

$$\begin{aligned} \mathcal{g}: &\ [d^\dagger \times \tilde{d}]^{(k)}_\mu, \quad k=0,1,2,3,4; \\ &\ [d^\dagger \times \tilde{s}]^{(2)}_\mu, \\ &\ [s^\dagger \times \tilde{d}]^{(2)}_\mu, \\ &\ [s^\dagger \times \tilde{s}]^{(0)}_0, \end{aligned} \tag{3.4}$$

$$\begin{aligned} \mathcal{h}: &\ [d^\dagger \times \tilde{d}]^{(k)}_\mu, \quad k=0,1,2,3,4; \\ &\ [s^\dagger \times \tilde{s}]^{(0)}_0, \end{aligned} \tag{3.5}$$

$$\begin{aligned} \mathcal{p}: &\ [d^\dagger \times \tilde{s}]^{(2)}_\mu, \\ &\ [s^\dagger \times \tilde{d}]^{(2)}_\mu. \end{aligned} \tag{3.6}$$

Since $\mathcal{p}$ consists of five operators, $[d^\dagger \times \tilde{s}]^{(2)}_\mu$, and their hermitian conjugates, the geometric space of the interacting boson model-1 is a five-dimensional complex space. The subalgebra $\mathcal{h}$ corresponds to chain I in (2.25). The 'algebraic' definition of the geometric space is obtained through the extremal state

$$|\Lambda_{\text{ext}}\rangle: (1/\mathcal{N}_s)s^{\dagger N}|0\rangle \equiv |s^N\rangle, \tag{3.7}$$

annihilated by all $d^\dagger_\mu$ operators, and the coherent state

$$|N,\eta_\mu\rangle = \exp\left[\sum_\mu \eta_\mu [d^\dagger \times \tilde{s}]^{(2)}_\mu + \sum_\mu \eta^*_\mu [s^\dagger \times \tilde{d}]^{(2)}_\mu\right]|s^N\rangle. \tag{3.8}$$

It turns out, in general, that for boson systems in n dimensions defined by boson operators

$$b^\dagger_\alpha, \quad \alpha=1,\ldots,n; \tag{3.9}$$

with Lie algebra U(n),

$$\mathcal{g}: G_{\alpha\beta} = b^\dagger_\alpha b_\beta, \quad \alpha,\beta=1,\ldots,n, \tag{3.10}$$

the maximum stability group is U($n-1$) ⊗ U(1) (as long as there is no further restriction on the angular momenta of the bosons),

$$\begin{aligned} \mathcal{h}: &\ b^\dagger_1 b_1, b^\dagger_\alpha b_\beta, \quad \alpha,\beta=2,\ldots,n, \\ \mathcal{p}: &\ b^\dagger_1 b_\alpha, b^\dagger_\alpha b_1, \quad \alpha=2,\ldots,n. \end{aligned} \tag{3.11}$$

Table 3.1. *Coset spaces extensively investigated*

Space	Dimension	Variables
$\frac{\mathrm{U}(n)}{\mathrm{U}(n-1)\otimes\mathrm{U}(1)}$	$2(n-1)$	$(n-1)$ complex
$\frac{\mathrm{SO}(n)}{\mathrm{SO}(n-1)}$	$(n-1)$	$(n-1)$ real

Thus, the coset space is

$$\mathrm{U}(n)/\mathrm{U}(n)\otimes\mathrm{U}(1)\approx\mathrm{SU}(n)/\mathrm{U}(n-1), \tag{3.12}$$

with topological dimension

$$\dim=2(n-1). \tag{3.13}$$

These spaces have been extensively investigated (Table 3.1).

3.3 Coherent states

For practical purposes, it is more convenient to use in the discussion of the geometric properties of the interacting boson model another set of coherent states, the 'projective' states. These were introduced by Bohr and Mottelson (1980), Ginocchio and Kirson (1980) and Dieperink, Scholten and Iachello (1980). The corresponding variables, α_μ, have a straightforward connection with the shape variable of Bohr and Mottelson.

The projective coherent states, which we shall also call intrinsic states, are defined as

$$|N;\alpha_\mu\rangle=\left(s^\dagger+\sum_\mu\alpha_\mu d_\mu^\dagger\right)^N|0\rangle, \tag{3.14}$$

where α_μ are five complex variables, One can use the states (3.14) to study both static and dynamic problems. For static problems one can choose the α_μ real, while for dynamic problems the α_μ must be retained as complex numbers. We first discuss static problems.

Instead of using the five variables α_μ, one can use the three Euler angles $(\theta_1,\theta_2,\theta_3)$ defining the orientation in space of an intrinsic frame, and two intrinsic variables β,γ, Bohr variables (Bohr, 1952). The variables α_μ are

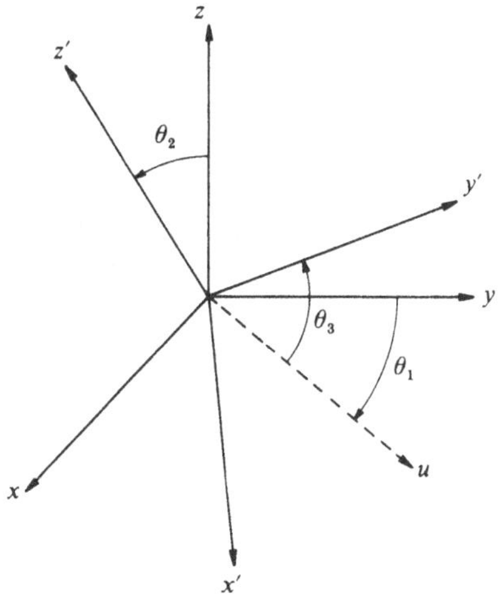

Fig. 3.1. The Euler angles $\theta_1, \theta_2, \theta_3$.

related to the intrinsic variables by

$$\alpha_\mu = \sum_\nu a_\nu \mathscr{D}^{(2)}_{\mu\nu}(\Omega), \tag{3.15}$$

where $\Omega \equiv (\theta_1, \theta_2, \theta_3)$ and $\mathscr{D}^{(2)}_{\mu\nu}$ represents a Wigner $\mathscr{D}$-function, and

$$\begin{aligned} a_0 &= \beta \cos\gamma \\ a_{\pm 2} &= 1/\sqrt{2}\, \beta \sin\gamma \\ a_{\pm 1} &= 0. \end{aligned} \tag{3.16}$$

The Euler angles $\theta_1, \theta_2, \theta_3$ are shown in Fig. 3.1. In terms of the intrinsic variables β, γ, the intrinsic state (3.14), can be written as

$$|N; \beta, \gamma\rangle = \{s^\dagger + \beta[\cos\gamma d^\dagger_0 + 1/\sqrt{2} \sin\gamma (d^\dagger_{+2} + d^\dagger_{-2})]\}^N |0\rangle. \tag{3.17}$$

Using the intrinsic state (3.17), one can evaluate, in a simple way and with considerable accuracy (order $1/N$), a large number of properties of nuclei. We begin by considering ground-state properties. These can be obtained by evaluating the quantity

$$E(N; \beta, \gamma) = \frac{\langle N; \beta, \gamma | H | N; \beta, \gamma\rangle}{\langle N; \beta, \gamma | N; \beta, \gamma\rangle}, \tag{3.18}$$

where H is the interacting boson model Hamiltonian, and by minimizing

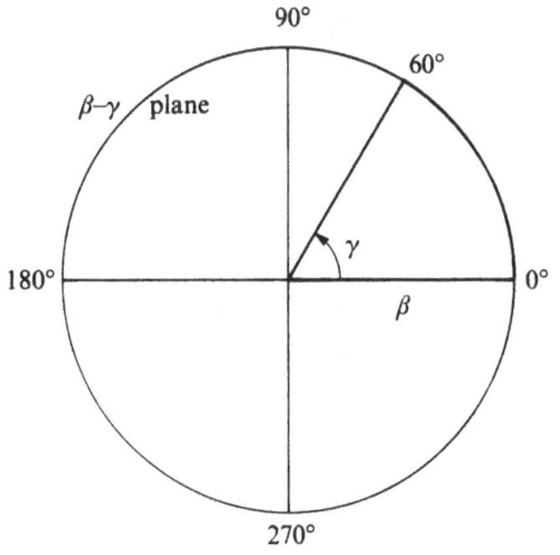

Fig. 3.2. The β–γ plane.

$E(N;\beta,\gamma)$ with respect to β and γ,

$$\frac{\partial E}{\partial \beta}=0; \quad \frac{\partial E}{\partial \gamma}=0. \tag{3.19}$$

This procedure (Hartree–Bose) gives the equilibrium 'shape' corresponding to the boson Hamiltonian H. For the most general boson Hamiltonian H, Eq. (1.21), van Isacker and Chen (1981) found

$$E(N;\beta,\gamma)=E_0+\frac{N}{(1+\beta^2)}(\varepsilon_s+\varepsilon_d\beta^2)$$
$$+\frac{N(N-1)}{(1+\beta^2)^2}(f_1\beta^4+f_2\beta^3\cos 3\gamma+f_3\beta^2+\tfrac{1}{2}u_0), \tag{3.20}$$

with

$$f_1=\tfrac{1}{10}c_0+\tfrac{1}{7}c_2+\tfrac{9}{35}c_4,$$
$$f_2=-2\left(\frac{1}{35}\right)^{\frac{1}{2}}\tilde{v}_2, \tag{3.21}$$
$$f_3=\left(\frac{1}{5}\right)^{\frac{1}{2}}(\tilde{v}_0+u_2).$$

It is convenient to plot in each case the energy surface (3.20) as a contour plot in the β–γ plane, Fig. 3.2. Since E in (3.20) depends only on $\cos 3\gamma$, it is sufficient to consider the portion of the β–γ plane with

$$0°\leqslant\gamma<60°. \tag{3.22}$$

The variables β and γ are related to those introduced by Bohr and Mottelson (1975), denoted here by $\hat{\beta}$ and $\hat{\gamma}$, by

$$\begin{aligned}\hat{\gamma}&=\gamma,\\ \hat{\beta}&=c\beta,\end{aligned} \tag{3.23}$$

where c is a scale variable which depends on the definition of $\hat{\beta}$. Two definitions of the Bohr variable $\hat{\beta}$ are often used. One is the distortion parameter $\hat{\delta}$ related to the intrinsic quadrupole moment eQ_0 by

$$eQ_0=eZ\frac{4}{3}\left\{\sum_{k=1}^{Z} r_k^2\right\}\hat{\delta}. \tag{3.24}$$

The other is the deformation parameter $\hat{\beta}_2$, which parametrizes the nuclear shape when it is considered as a liquid drop with radius

$$\begin{aligned}R(\theta,\phi)&=R_0\left(1+\sum_{\mu}\hat{\alpha}_{2\mu}Y^*_{2\mu}(\theta,\phi)\right),\\ \hat{\beta}_2^2&=\sum_{\mu}|\hat{\alpha}_{2\mu}|^2.\end{aligned} \tag{3.25}$$

The two parameters $\hat{\delta}$ and $\hat{\beta}_2$ are related by

$$\begin{aligned}\hat{\delta}&=\left(\frac{45}{16\pi}\right)^{\frac{1}{2}}\left(\frac{4}{5}\frac{\langle r\rangle R_0}{\langle r^2\rangle}\hat{\beta}_2+\frac{3}{5}\frac{R_0^2}{\langle r^2\rangle}\left(\frac{20}{49\pi}\right)^{\frac{1}{2}}\hat{\beta}_2^2+\cdots\right),\\ \hat{\delta}&\approx 0.95\,\hat{\beta}_2,\end{aligned} \tag{3.26}$$

where $\langle r\rangle$ and $\langle r^2\rangle$ are the average radius and square radius of the nucleus.

The value of the constant c can be obtained by equating (3.24) and the corresponding expression in the interacting boson model, to be derived below in Eq. (3.55),

$$\hat{\delta}=\frac{5}{4}\left(\frac{e_{\mathrm{B}}N}{eZR_0^2}\right)\frac{2\beta-\sqrt{\frac{2}{7}}\chi\beta^2}{1+\beta^2}, \tag{3.27}$$

where e_{B} is the boson effective charge. For typical values of the quantities in (3.27), one has, in rare-earth nuclei,

$$\hat{\delta}\approx 0.15\,\beta. \tag{3.28}$$

The parametrization (3.25) is very useful, since it allows one to visualize the nuclear surface.

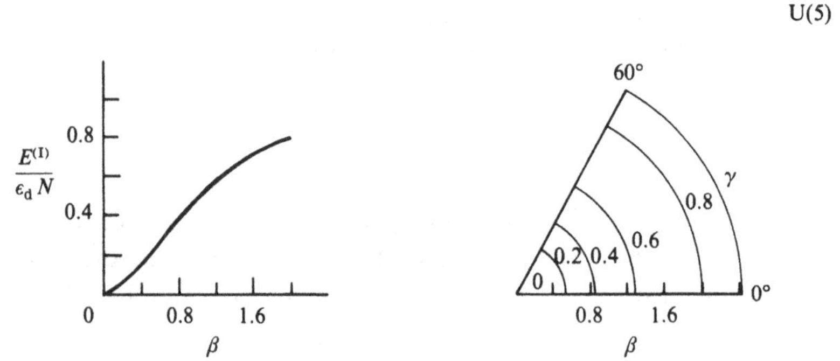

Fig. 3.3. (*a*) The energy functional $E^{(\mathrm{I})}(N;\beta,\gamma)$ as a function of β for $f_1=0$, $E_0=0$; (*b*) its β–γ plot.

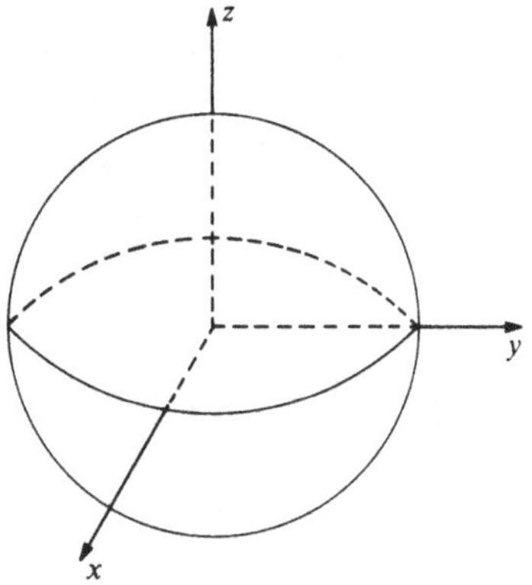

Fig. 3.4. The nuclear shape corresponding to chain I.

3.4 Shape structure of the dynamic symmetries

Equation (3.20) can be used to analyze the shape structure of the three dynamic symmetries, I, II and III.

(*i*) *Chain I*

Here it is convenient to start from the parametrization (2.81) of $H^{(\mathrm{I})}$. This yields

$$E^{(\mathrm{I})}(N;\beta,\gamma)=E_0+\varepsilon_{\mathrm{d}}N\frac{\beta^2}{(1+\beta^2)}+f_1N(N-1)\frac{\beta^4}{(1+\beta^2)^2}. \qquad (3.29)$$

This energy functional is γ-independent and has a minimum at $\beta_e=0$, Fig. 3.3. In the β–γ plane the lines of equal energy are circles. In the parametrization of Eq. (3.25), the drop is a sphere, Fig. 3.4.

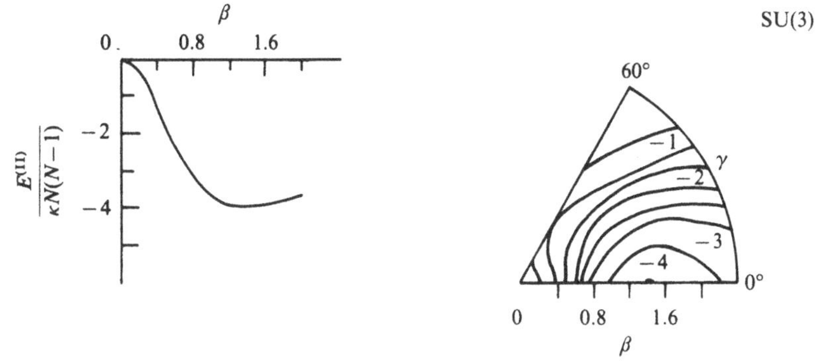

Fig. 3.5. (*a*) The energy functional $E^{(\mathrm{II})}(N;\beta,\gamma)$ as a function of β for $\gamma=0^\circ$, $\kappa'=3/4\kappa$, $E_0=10\kappa N$; (*b*) the corresponding β–γ plot.

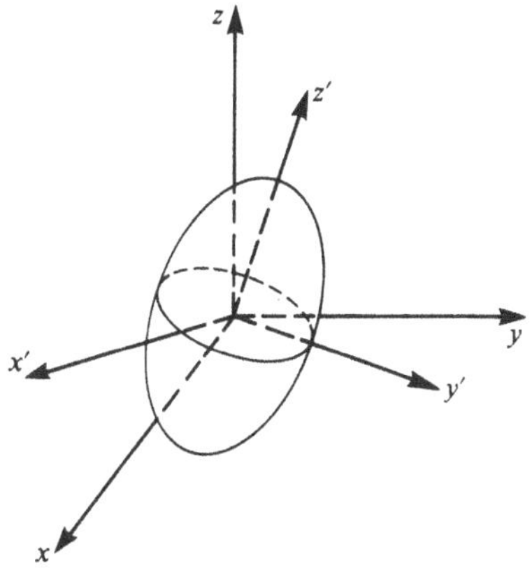

Fig. 3.6. The nuclear shape corresponding to chain II.

(*ii*) *Chain II*

We consider here the Hamiltonian $H^{(\mathrm{II})}$ in the form (2.85), obtaining

$$E^{(\mathrm{II})}(N;\beta,\gamma)=E_0-\kappa 2\left[\frac{N}{(1+\beta^2)}\left(5+\frac{11}{4}\beta^2\right)+\frac{N(N-1)}{(1+\beta^2)^2}\right.$$
$$\left.\times\left(\frac{\beta^4}{2}+2\sqrt{2}\,\beta^3\cos 3\gamma+4\beta^2\right)\right]-\kappa'\frac{6N\beta^2}{(1+\beta^2)}. \qquad (3.30)$$

This energy functional has a sharp minimum at $\gamma_e=0^\circ$ and at a value $\beta_e\neq 0$ (for $N\to\infty$, the minimum occurs at $\beta=\sqrt{2}$), Fig. 3.5. The lines of equal energy in the β–γ plane are as shown in the same figure. In the parametrization (3.25), the drop is a deformed body with axial symmetry and prolate shape. It is interesting to note that the same Hamiltonian (3.29)

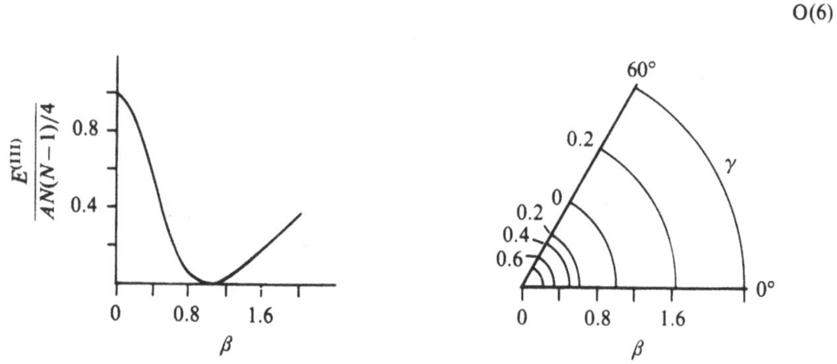

Fig. 3.7. (*a*) The energy functional $E^{(\mathrm{III})}(N;\beta,\gamma)$ as function of β for $B=C=0$, $E_0=0$; (*b*) the corresponding β–γ plot.

would give rise to an oblate shape (minimum at $\gamma_e=60°$) if the sign of the $[d^\dagger\times\tilde{d}]^{(2)}$ term in the operator $\hat{Q}$ is reversed from $-\sqrt{7}/2$ to $+\sqrt{7}/2$. The nuclear shape corresponding to Chain II is shown in Fig. 3.6.

(iii) Chain III

One can use here the parametrization of $H^{(\mathrm{III})}$ given by (2.90). One then obtains the energy functional

$$E^{(\mathrm{III})}(N;\beta,\gamma)=E_0'+(2B+6C)\frac{N\beta^2}{(1+\beta^2)}+\frac{A}{4}N(N-1)\left(\frac{1-\beta^2}{1+\beta^2}\right)^2. \quad (3.31)$$

This energy functional is γ-independent and has a minimum at a value $|\beta_e|\neq 0$ (for $N\to\infty$, the minimum is at $|\beta_e|=1$), Fig. 3.7. The lines of equal energy are shown in part (*b*) of Fig. 3.7. In the parametrization (3.25) the drop is a deformed body with γ-instability. This shape cannot be easily drawn.

3.5 Higher-order terms

The energy functional corresponding to some higher-order terms in H, Eq. (1.29), has been evaluated by van Isacker and Chen (1981). They analyze

Table 3.2. *Geometric limit of the cubic terms in the Hamiltonian*

r	0	2	3	4	6
l	2	0	2	2	4
A_r	0	$\frac{1}{5}$	$-\frac{1}{7}$	$\frac{3}{49}$	$\frac{14}{55}$
B_r	$\frac{2}{35}$	0	$\frac{1}{7}$	$\frac{3}{35}$	$-\frac{8}{385}$

the contribution arising from

$$H_3=\sum_r c_r[[d^\dagger \times d^\dagger]^{(l)} \times d^\dagger]^{(r)} \cdot [[\tilde{d} \times \tilde{d}]^{(l)} \times \tilde{d}]^{(r)}, \qquad (3.32)$$

and obtain

$$E_3(N;\beta,\gamma)=\sum_r N(N-1)(N-2)\frac{\beta^6}{(1+\beta^2)^3}[A_r+B_r\cos^2 3\gamma]. \qquad (3.33)$$

The values of r and l and the coefficients A_r and B_r are given in Table 3.2. It is interesting to note that, since some of the cubic terms contain $\cos^2 3\gamma$, a combination of these with the others may produce a minimum at values of γ different from 0° and 60° (triaxial shapes). The importance of these terms has been investigated numerically by Casten *et al.* (1985).

3.6 Shape-phase transitions

Coherent states can also be used to study the nature of the phase transitions between limiting solutions of the interacting boson model-1. Phase transitions are only defined for $N \rightarrow \infty$, where they appear as discontinuities in some quantities. For finite N, the discontinuities are smoothed out. The study of phase transitions can be done using an algorithm developed by Gilmore (1979). This algorithm consists in:

(i) taking Hamiltonians of the type discussed in Sect. 2.10, which are mixtures of Casimir invariants of two (or more) chains;

(ii) constructing the energy functional (3.18), $E(N;\beta,\gamma)$ and the energy per particle, $\mathscr{E}=E(N;\beta,\gamma)/N$;

(iii) minimizing $\mathscr{E}$ as a function of β and γ, thus determining $\mathscr{E}_{\min}$;

(iv) studying the behavior of $\mathscr{E}_{\min}$ and its derivatives as a function of the coupling constants which appear in front of the invariant operators. The phase transition is said to be of zero order if $\mathscr{E}_{\min}$ is discontinuous at the critical point, of first order if $\partial\mathscr{E}_{\min}/\partial\eta$ is discontinuous, of second order if $\partial^2\mathscr{E}_{\min}/\partial\eta^2$ is discontinuous, etc. Here η is a control parameter, to be defined below.

(i) Transitional class A

This transitional class has been studied by considering the Hamiltonian

$$H'^{(\mathrm{I})+(\mathrm{II})} = \varepsilon \hat{n}_d - \kappa[2\hat{Q}\cdot\hat{Q} + \tfrac{3}{4}\hat{L}\cdot\hat{L}] \tag{3.34}$$

which is a special form of (2.247), $\kappa' = \frac{3}{4}\kappa$. The energy functional corresponding to (3.34) can be easily obtained from (3.30)

$$E^{(\mathrm{I})+(\mathrm{II})}(N;\beta,\gamma) = \varepsilon N \frac{\beta^2}{(1+\beta^2)} - 2\kappa \frac{N(N-1)}{(1+\beta^2)^2}\left(\frac{\beta^4}{2} + 2\sqrt{2}\,\beta^3 \cos 3\gamma + 4\beta^2\right) - 10\kappa N. \tag{3.35}$$

Introducing the energy per particle, scaling it with ε and removing the constant term $-10\kappa N$, one can rewrite (3.35) as

$$\mathscr{E}^{(\mathrm{I})+(\mathrm{II})}(\beta,\gamma) = \frac{\beta^2}{(1+\beta^2)} - \eta \frac{\beta^4 + 4\sqrt{2}\,\beta^3 \cos 3\gamma + 8\beta^2}{(1+\beta^2)^2}, \tag{3.36}$$

where $\mathscr{E} = (E + 10\kappa N)/\varepsilon N$ and the control parameter η is given by

$$\eta = \frac{\kappa(N-1)}{\varepsilon}. \tag{3.37}$$

As a function of η, the values of β and γ for which $\mathscr{E}$ is a minimum, $\mathscr{E}_{\min}$, shift from $\beta_e = 0$ and $\gamma_e = 0°$ (spherical shape, $\eta < 1/9$) to $\beta_e \neq 0$ ($\beta_e \to \sqrt{2}$ for large η), $\gamma_e = 0°$ (axially deformed shape, $\eta > 1/9$). A study of $\mathscr{E}_{\min}$ and its derivatives at the critical point, $\eta = \eta_c = 1/9$, determines the nature of the phase transition. Since, in this case, $\partial\mathscr{E}_{\min}/\partial\eta$ changes discontinuously, the phase transition corresponding to the transitional class A is a first-order phase transition. This situation is illustrated in Fig. 3.8, where the energy for particle $\mathscr{E}$ is plotted as a function of β for $\gamma = 0°$ and several values of η.

(ii) Transitional class B

This class has been studied by considering a Hamiltonian which is a mixture of invariant operators of chains II and III,

$$H'^{(\mathrm{II})+(\mathrm{III})} = A\,\hat{\mathrm{P}}_6 - \kappa[2\hat{Q}\cdot\hat{Q} + \tfrac{3}{4}\hat{L}\cdot\hat{L}]. \tag{3.38}$$

This is a special case of (2.257) with $B = 0$, $C = -3\kappa/4$. The energy

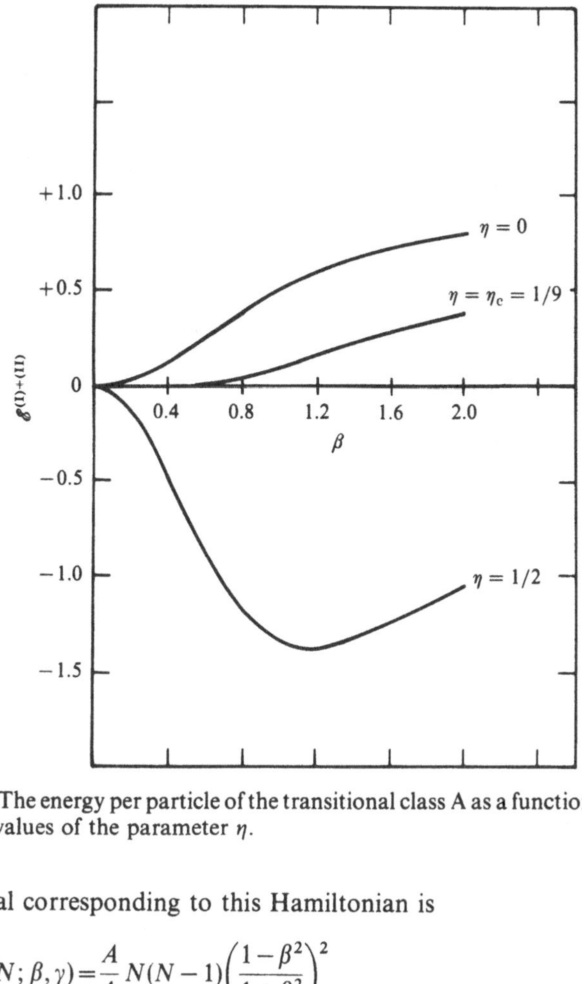

Fig. 3.8. The energy per particle of the transitional class A as a function of β and for different values of the parameter η.

functional corresponding to this Hamiltonian is

$$E^{(\mathrm{II})+(\mathrm{III})}(N;\beta,\gamma)=\frac{A}{4}N(N-1)\left(\frac{1-\beta^2}{1+\beta^2}\right)^2 - 2\kappa\frac{N(N-1)}{(1+\beta^2)^2}\left(\frac{\beta^4}{2}+2\sqrt{2}\,\beta^3\cos 3\gamma+4\beta^2\right)-10\kappa N. \tag{3.39}$$

Introducing again the energy for particle, scaling it with A and removing the constant term yields

$$\mathscr{E}^{(\mathrm{II})+(\mathrm{III})}(\beta,\gamma)=(N-1)\left[\frac{1}{4}\left(\frac{1-\beta^2}{1+\beta^2}\right)^2-\eta'\frac{(\beta^4+4\sqrt{2}\beta^3\cos 3\gamma+8\beta^2)}{(1+\beta^2)^2}\right], \tag{3.40}$$

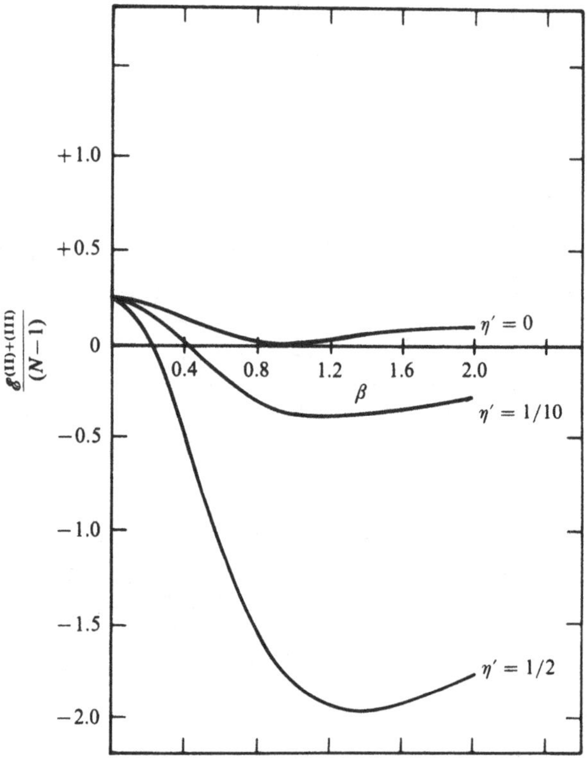

Fig. 3.9. The energy per particle of the transitional class B as a function of β and for different values of the parameter η'.

where $\mathscr{E} = (E + 10\kappa N)/AN$ and the control parameter η' is given by

$$\eta' = \kappa/A. \tag{3.41}$$

A study of $\mathscr{E}_{\min}$ and its derivatives shows that no phase transition occurs for physical values of the parameters $A, \kappa, \eta' \geqslant 0$. This situation is illustrated in Fig. 3.9, where the energy per particle is plotted as a function of β for $\gamma = 0°$ and various values of η'.

(iii) Transitional class C

This class has been studied by considering a Hamiltonian which is a mixture of invariant operators of chains I and III,

$$H'^{(III)+(I)} = \varepsilon \hat{n}_d + A\,\hat{P}_6. \tag{3.42}$$

This is a special form of (2.263) with $B = C = 0$. The energy functional corresponding to this Hamiltonian is

$$E^{(\mathrm{III})+(\mathrm{I})}(N;\beta,\gamma)=\varepsilon N\frac{\beta^2}{(1+\beta^2)}+\frac{A}{4}N(N-1)\left(\frac{1-\beta^2}{1+\beta^2}\right)^2. \tag{3.43}$$

Introducing the energy per particle and scaling it with ε, one obtains

$$\mathscr{E}^{(\mathrm{III})+(\mathrm{I})}=\frac{\beta^2}{(1+\beta^2)}+\eta''\left(\frac{1-\beta^2}{1+\beta^2}\right)^2, \tag{3.44}$$

where $\mathscr{E}=E/\varepsilon N$ and

$$\eta''=\frac{A(N-1)}{4\varepsilon}. \tag{3.45}$$

As a function of the control parameter η'', the value of β for which $\mathscr{E}$ is at a minimum, $\mathscr{E}_{\mathrm{min}}$, shifts from $\beta_e=0$ (spherical shape, $\eta''<1/4$) to $\beta_e=[(4\eta''-1)/(4\eta''+1)]^{\frac{1}{2}}$ (γ-unstable deformed shape, $\eta''>1/4$). A study of $\mathscr{E}_{\mathrm{min}}$ and its derivatives at the critical point, $\eta''=\eta''_c=1/4$, determines the nature of the phase transition. Since here $\partial^2\mathscr{E}_{\mathrm{min}}/\partial\eta''^2$ is discontinuous, the phase transition is a second-order phase transition. This situation is illustrated in Fig. 3.10, where the energy per particle $\mathscr{E}$ is plotted as a function of β for several values of η''.

(iv) Transitional class D

Feng, Gilmore and Deans (1981) have studied the phase structure of the most general Hamiltonian, (2.74). Their results can be most conveniently displayed by using Casten's triangle, and are shown in Fig. 3.11. There is a line of first-order phase transitions ending in a point of second-order phase transitions.

3.7 Experimental examples of shape-phase transitions

Shape-phase transitions in nuclei can be studied experimentally by considering the behavior of the ground-state energies of a series of isotopes, E_{G}, or, more conveniently, the behavior of two-nucleon separation energies $S_2(N)$, Eq. (2.204). These quantities can be written as

$$S_2(N)=E_{\mathrm{B}}(N+1)-E_{\mathrm{B}}(N)=\frac{\Delta E_{\mathrm{B}}}{\Delta N}. \tag{3.46}$$

Using the parametrization discussed in Sect. 2.10, it is possible to convert $\Delta E_{\mathrm{B}}/\Delta N$ into a variation with respect to the control parameters η,η',η''.

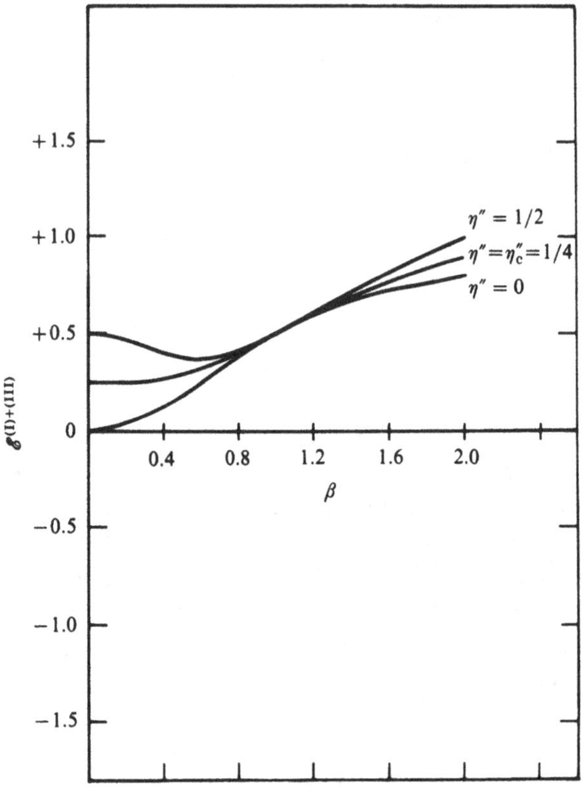

Fig. 3.10. The energy per particle of the transitional class C as a function of β and for different values of the parameter η''.

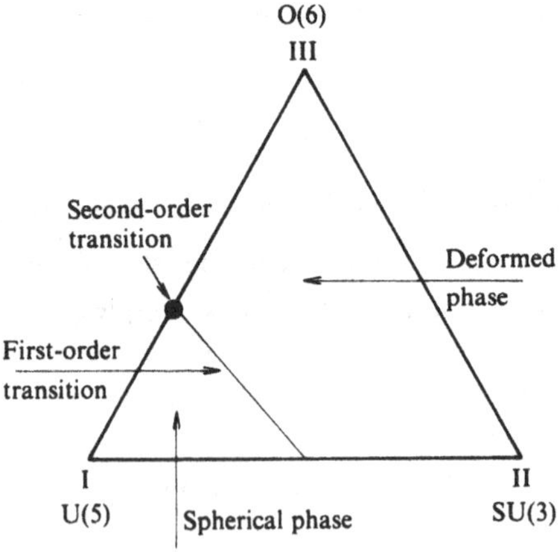

Fig. 3.11. Schematic representation of the phase structure of the interacting boson model-1 (Feng *et al.*, 1981).

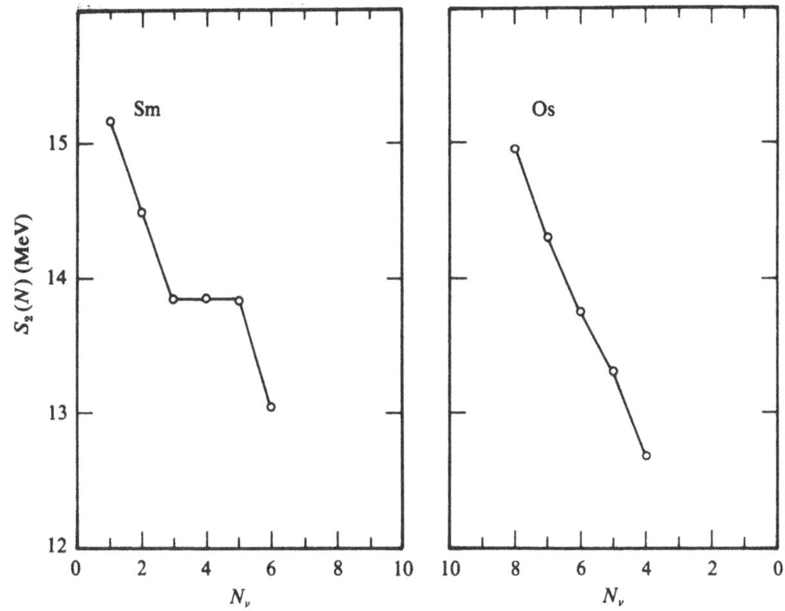

Fig. 3.12. Two-neutron separation energies in ${}_{62}$Sm and ${}_{76}$Os, as a function of neutron boson number N_ν.

For example, for transitional class A,

$$\frac{\Delta E_{\mathrm{B}}}{\Delta N}=\frac{\Delta E_{\mathrm{B}}}{\Delta \eta}\frac{\Delta \eta}{\Delta N}\approx\frac{\kappa^{(0)}}{\varepsilon^{(0)}}\frac{\Delta E_{\mathrm{B}}}{\Delta \eta}. \tag{3.47}$$

As a result, first-order phase transitions are expected to appear as discontinuities in $S_2(N)$ as a function of N, while second-order phase transitions are expected to appear as discontinuities in

$$\frac{\Delta S_2}{\Delta N}=\frac{\Delta^2 E_{\mathrm{B}}}{\Delta N^2}=S_2(N+1)-S_2(N). \tag{3.48}$$

The two-neutron separation energies in the Sm isotopes (class A) and in the Os isotopes (class B) are shown in Fig. 3.12. The experimental data show a distinct discontinuity in ${}_{62}$Sm (first-order phase transition) but no discontinuity in ${}_{76}$Os. This is consistent with the analysis of phase transitions discussed above.

3.8 Boson condensates

The intrinsic states (3.14) and their generalizations can be used to derive a variety of results, as first discussed by Bohr and Mottelson (1980). Consider, for example, the intrinsic state (3.17) for $\gamma=0°$ (axially-symmetric situation). This state describes the ground-state band of a deformed nucleus, and it can be written, including a normalization factor, as

$$|g\rangle=(N!)^{-\frac{1}{2}}(b_g^\dagger)^N|0\rangle, \tag{3.49}$$

where

$$b_g^\dagger=(1+\beta^2)^{-\frac{1}{2}}(s^\dagger+\beta d_0^\dagger). \tag{3.50}$$

In the pure SU(3) limit, $\beta\rightarrow\sqrt{2}$. Equation (3.49) represents a boson condensate and has been used in the previous sections to evaluate properties of the ground-state band. However, one can now consider excited bands as intrinsic excitations of the condensate (3.49). In order to construct β-vibrations ($K^P=0^+$), one can replace one of the bosons in the condensate by the orthogonal combination of $s^\dagger$ and $d_0^\dagger$,

$$|\beta_v\rangle=N^{-\frac{1}{2}}b_{\beta v}^\dagger b_g|g\rangle \tag{3.51}$$

with

$$b_{\beta v}^\dagger=(1+\beta^2)^{-\frac{1}{2}}(-\beta s^\dagger+d_0^\dagger). \tag{3.52}$$

The intrinsic γ-vibration (with $K^P=2^+$) is obtained by using $d_{+2}^\dagger$ and $d_{-2}^\dagger$,

$$|\gamma_v\rangle=N^{-\frac{1}{2}}b_{\gamma v}^\dagger b_g|g\rangle, \tag{3.53}$$

with

$$b_{\gamma v}^\dagger=\frac{1}{\sqrt{2}}(d_{+2}^\dagger+d_{-2}^\dagger). \tag{3.54}$$

The coherent states (3.51) and (3.53) can be used to derive properties of β- and γ-vibrations. An interesting application is the evaluation of the intrinsic matrix elements of the E2 operators (Bijker and Dieperink, 1982). If one uses an E2 operator of the form

$$T_\mu^{(2)}=e_B[[d^\dagger\times\tilde{s}+s^\dagger\times\tilde{d}]_\mu^{(2)}+\chi[d^\dagger\times\tilde{d}]_\mu^{(2)}], \tag{3.55}$$

where $e_B=\alpha_2$ is the boson effective charge, one obtains for the intrinsic matrix elements of the E2 operator the following results:

$$\begin{aligned}
\langle g|T_0^{(E2)}|g\rangle&=e_BN(1+\beta^2)^{-1}[2\beta-\sqrt{(\tfrac{2}{7})}\,\chi\beta^2],\\
\langle\beta_v|T_0^{(E2)}|\gamma_v\rangle&=e_BN^{\frac{1}{2}}(1+\beta^2)^{-1}[1-\sqrt{(\tfrac{2}{7})}\,\chi\beta-\beta^2],\\
\langle\gamma_v|T_{+2}^{(E2)}+T_{-2}^{(E2)}|g\rangle&=e_BN^{\frac{1}{2}}(1+\beta^2)^{-\frac{1}{2}}2^{-\frac{1}{2}}[2+2\sqrt{(\tfrac{2}{7})}\,\chi\beta],\\
\langle\gamma_v|T_{+2}^{(E2)}+T_{-2}^{(E2)}|\beta_v\rangle&=e_B(1+\beta^2)^{-\frac{1}{2}}2^{-\frac{1}{2}}[-2\beta+2\sqrt{(\tfrac{2}{7})}\,\chi].
\end{aligned} \tag{3.56}$$

The first of these equations has been used in relating the distortion parameter $\hat{\delta}$ to the parameter β, Eq. (3.27).

It is interesting to study what happens as a function of χ. For fixed (finite) N and $\chi<0$ the matrix elements for transitions $\beta_v, \gamma_v \rightarrow g$, which in the SU(3) limit correspond to different SU(3) representations $(2N-4,2)$ and $(2N,0)$ respectively, are sensitive to χ and vanish when $\chi \rightarrow -\sqrt{7}/2$. In addition, when $\beta=\sqrt{2}$, the ratio of intrinsic $\beta_v \rightarrow g$ and $\gamma_v \rightarrow g$ matrix elements does not depend on χ. In fact, for large N one can obtain a simple estimate for the ratio

$$B_{\beta\gamma}=B(\mathrm{E2}; 0_g \rightarrow 2_\beta)/B(\mathrm{E2}; 0_g \rightarrow 2_\gamma), \tag{3.57}$$

namely

$$B_{\beta\gamma} \approx \frac{|\langle\beta_v|T_0^{(\mathrm{E2})}|g\rangle|^2}{|\langle\gamma_v|T_{+2}^{(\mathrm{E2})}+T_{-2}^{(\mathrm{E2})}|g\rangle|^2}=\frac{1}{6}. \tag{3.58}$$

The behavior of the intrinsic $B(\mathrm{E2})$ values is shown in Fig. 3.13 as a function of χ. The coherent states (3.51) and (3.53) can also be used to evaluate the energies of the β- and γ-vibrations. Using the Hamiltonian

$$H=\kappa'''\hat{Q}\cdot\hat{Q}-\kappa'\hat{L}\cdot\hat{L}+\kappa''\,\hat{\mathrm{P}}_6, \tag{3.59}$$

Schaaser and Brink (1984) find an intrinsic excitation energy

$$\begin{aligned}
E_{\beta v,\,\mathrm{exc}} &= \mathrm{E}_{\beta v}-\mathrm{E}_g, \\
E_{\beta v,\,\mathrm{exc}} &= -\left(\frac{9}{4}\kappa'''+6\kappa'\right)\left(\frac{1-\beta^2}{1+\beta^2}\right)+\frac{(N-1)}{(1+\beta^2)^2}G_1, \\
G_1 &= (2\kappa'''-\kappa''/2)+4\sqrt{2}\,\kappa'''\beta+(5\kappa''-18\kappa'')\beta^2 \\
&\quad -8\sqrt{2}\,\kappa'''\beta^3+(\kappa'''-\kappa''/2)\beta^4.
\end{aligned} \tag{3.60}$$

Similarly, they find an intrinsic excitation energy of the γ-vibration

$$\begin{aligned}
E_{\gamma v,\,\mathrm{exc}} &= E_{\gamma v}-E_g, \\
E_{\gamma v,\,\mathrm{exc}} &= -\left(\frac{9}{4}\kappa'''+6\kappa'\right)\left(\frac{1}{1+\beta^2}\right)+\frac{(N-1)}{(1+\beta^2)^2}G_2. \\
G_2 &= (2\kappa'''-\kappa''/2)-4\sqrt{2}\,\kappa'''\beta+(\kappa''-6\kappa''')\beta^2 \\
&\quad -8\sqrt{2}\,\kappa'''\beta^3-(\kappa'''+\kappa''/2)\beta^4.
\end{aligned} \tag{3.61}$$

3.9 Cranking and moments of inertia

All standard many-body techniques can be used to obtain results within the framework of the interacting boson model. For example, one may wish to

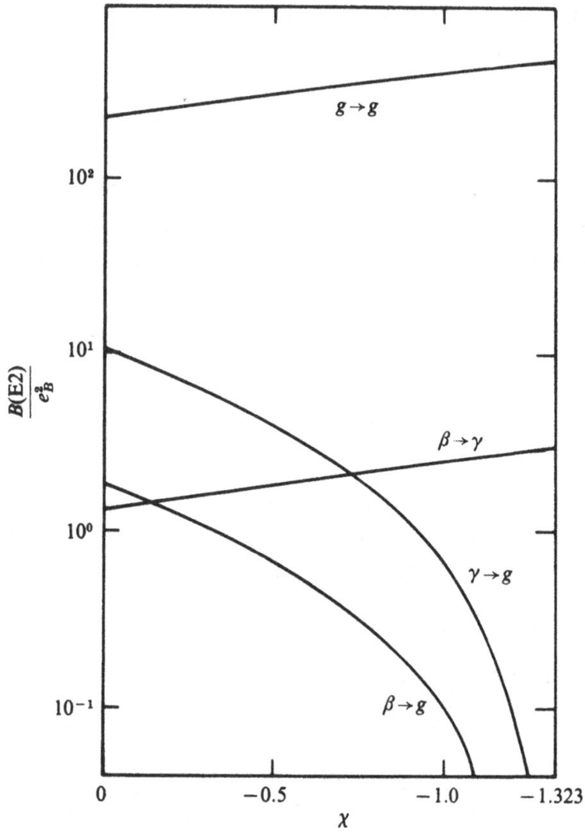

Fig. 3.13. Intrinsic $B(E2)$ values as a function of χ for $N=16$ and $\beta=\sqrt{2}$.

go one step beyond that of the evaluation of the ground-state energy, E_G. For deformed shapes, one wants to write an expression of the type

$$E_G(L)=E_G+(1/2\mathscr{I})L(L+1)+\cdots, \tag{3.62}$$

and determine explicitly the moment of inertia $\mathscr{I}$. This can be done by further generalizing the state (3.17) to the form

$$|N;a\rangle=(N!)^{-\frac{1}{2}}(1+a^2)^{-(N/2)}[s^\dagger+a_0d_0^\dagger+a_1(d_{+1}^\dagger+d_{-1}^\dagger)+a_2(d_{+2}^\dagger+d_{-2}^\dagger)]^N|0\rangle,$$
$$a^2=a_0^2+2a_1^2+2a_2^2, \tag{3.63}$$

where a_0 and a_2 are still given by (3.16), but now there is an explicit term containing a_1. The moment of inertia can be obtained by solving the self-consistent cranking problem (Schaaser and Brink, 1984)

$$\delta\langle N;a|H-\omega L_x|N;a\rangle=0; \qquad (3.64)$$

where ω is the cranking frequency and L_x is the x-component of the angular momentum operator,

$$L_x=-\frac{1}{\sqrt{2}}(L_{+1}-L_{-1}), \qquad (3.65)$$

which can be written in terms of d operators using (1.50). The terms containing a_1 in (3.63) are necessary, since the cranking Hamiltonian, $H-\omega L_x$, breaks rotational invariance. This procedure, when applied to the Hamiltonian (3.59) gives

$$\mathscr{I}=\beta^2\Big/\left[-\frac{1}{N}\left(\frac{3}{8}\kappa'''+\kappa'\right)+\frac{1}{6}\left(\frac{N-1}{N}\right)(1+\beta^2)^{-1}\tilde{F}\right], \qquad (3.66)$$

$$\tilde{F}=\sqrt{2}\,\kappa'''\beta-(15/2\,\kappa'''+12\kappa'-\kappa'')\beta^2$$
$$-3\sqrt{2}\,\kappa'''\beta^3-(\kappa''+\kappa'''/2+12\kappa')\beta^4.$$

3.10 Dynamic properties

Another way to study the dynamic properties of the system using a geometric framework, in addition to those described in Sects. 3.8 and 3.9, is to construct a differential equation in the variables β and γ and solving it. Again the coset spaces introduced in Sect. 3.2 offer a unique way to write differential equations associated with an algebra $\mathscr{g}$. The Casimir operators when realized on the coset space, become Laplace–Beltrami differential operators (Gilmore, 1974). The construction of the differential equations associated with the interacting boson model has been investigated by Ginocchio and Kirson (1980a, 1980b), Klein and Vallieres (1981), Weiguny (1981) and van Roosmalen (1982). The latter author considers 'group' coherent states, defined by

$$|N,\zeta_\mu\rangle=(N!)^{-\frac{1}{2}}\left(\sqrt{\left(1-\sum_\mu\zeta_\mu^*\zeta_\mu\right)}s^\dagger+\sum_\mu\zeta_\mu d_\mu^\dagger\right)^N|0\rangle, \qquad (3.67)$$

in terms of the five complex variables ζ_μ. Dynamic properties can be obtained by introducing coordinates q_μ and momenta p_μ through

$$\zeta_\mu=q_\mu+\mathrm{i}\,p_\mu, \qquad (3.68)$$

and evaluating then the expectation value of the Hamiltonian H in the intrinsic state (3.67). This gives the so-called classical limit, H_{cl}, of H

$$H_{cl}=\langle N,\zeta_\mu|H|N,\zeta_\mu\rangle. \tag{3.69}$$

The state $|N,\zeta_\mu\rangle$ is properly normalized. The rotational invariance of the problem can be exploited by transforming to an intrinsic frame as in Eq. (3.15),

$$\begin{aligned} q_\mu&=\sum_\nu u_\nu \mathscr{D}^{(2)}_{\mu\nu}(\Omega),\\ p_\mu&=\sum_\nu v_\nu \mathscr{D}^{(2)}_{\mu\nu}(\Omega), \end{aligned} \tag{3.70}$$

and introducing intrinsic variables through

$$\begin{aligned} u_0&=1/\sqrt{2}\,\tilde\beta\cos\gamma,\\ u_{\pm2}&=\tfrac{1}{2}\tilde\beta\sin\gamma,\\ u_{\pm1}&=0, \end{aligned} \tag{3.71}$$

where we have placed a tilde over β so as not to confuse it with the projective coordinate of the previous sections. The two variables are related by

$$\tilde\beta=\frac{\sqrt{2}\,\beta}{(1+\beta^2)^{\frac{1}{2}}}. \tag{3.72}$$

The five momenta v_ν can then be expressed in terms of the momenta conjugate to $\tilde\beta, \gamma, \theta_1, \theta_2, \theta_3$, denoted here by $p_{\tilde\beta}, p_\gamma, p_{\theta_1}, p_{\theta_2}$ and p_{θ_3}, not to be confused with the momenta, p_0, $p_{\pm1}$, $p_{\pm2}$ in the laboratory frame. One obtains

$$\begin{aligned} v_0&=p_{\tilde\beta}\cos\gamma-\frac{p_\gamma}{\tilde\beta}\sin\gamma,\\ v_{\pm1}&=\frac{-\mathrm{i}\,L_1}{2\sqrt{2}\,\tilde\beta\sin(\gamma-2\pi/3)}\pm\frac{L_2}{2\sqrt{2}\,\tilde\beta\sin[\gamma(4\pi/3)]},\\ v_{\pm2}&=\pm\frac{\mathrm{i}\,L_3}{2\sqrt{2}\,\tilde\beta\sin\gamma}+\frac{1}{\sqrt{2}}\frac{p_\gamma}{\tilde\beta}\cos\gamma+\frac{1}{\sqrt{2}}p_{\tilde\beta}\sin\gamma \end{aligned} \tag{3.73}$$

where L_k $(k=1,\ 2,\ 3)$ are the angular momentum components in the intrinsic frame

$$L_1=(p_{\theta_3}-p_{\theta_1}\cos\theta_2)\frac{\cos\theta_1}{\sin\theta_2}-p_{\theta_2}\sin\theta_1,$$

$$L_2=(p_{\theta_3}-p_{\theta_1}\cos\theta_2)\frac{\sin\theta_1}{\sin\theta_2}+p_{\theta_2}\cos\theta_1, \qquad (3.74)$$

$$L_3=p_{\theta_1}.$$

Using this technique, the classical limit of the most general Hamiltonian of the interacting boson model-1, parametrized in the form

$$H=\varepsilon\hat{n}_{\mathrm{d}}+a_0\hat{P}_6+a_1(\hat{L}\cdot\hat{L})+a_2(\hat{Q}\cdot\hat{Q})+a_5\hat{C}_5+a_6\hat{n}_{\mathrm{d}}^2 \qquad (3.75)$$

has been evaluated.

(i) Chain I

In order to study the classical limit for this case, it is sufficient to consider the Hamiltonian

$$H'^{(\mathrm{I})}=\varepsilon\hat{n}_{\mathrm{d}}. \qquad (3.76)$$

One obtains,

$$H_{\mathrm{cl}}^{(\mathrm{I})}=(\varepsilon N)\frac{1}{2}\left(p_{\hat{\beta}}^2+\hat{\beta}^2+\frac{\hat{T}^2}{\hat{\beta}^2}\right), \qquad (3.77)$$

where

$$\hat{T}=p_\gamma^2+\frac{1}{4}\sum_{k=1}^{3}\frac{L_k^2}{\sin^2(\gamma-2\pi k/3)}. \qquad (3.78)$$

Equation (3.77) should be compared with the Bohr Hamiltonian (Bohr, 1952)

$$H_{\mathrm{coll}}=\frac{1}{2B_2}\left[p_{\hat{\beta}}^2+\frac{p_\gamma^2}{\hat{\beta}^2}+\sum_{k=1}^{3}\frac{L_k^2}{4\hat{\beta}^2\sin^2(\gamma-2\pi k/3)}\right]+V(\hat{\beta},\gamma). \qquad (3.79)$$

By comparing (3.77) and (3.79), one can see that (3.77) is identical to the Hamiltonian of a five-dimensional harmonic vibrator, for which

$$V^{(\mathrm{I})}(\hat{\beta},\gamma)=\tfrac{1}{2}C_2\hat{\beta}^2. \qquad (3.80)$$

There are, however, major differences. This is because the eigenvalues of the operator $\hat{n}_{\mathrm{d}}$ are bounded by N, $n_{\mathrm{d}}=0, 1, \ldots, N$. As a result,

$$\frac{1}{2}\left(p_{\hat{\beta}}^2+\hat{\beta}^2+\frac{1}{\hat{\beta}^2}\right)N\leqslant N. \qquad (3.81)$$

Thus the phase space for this problem is bounded, Fig. 3.14.

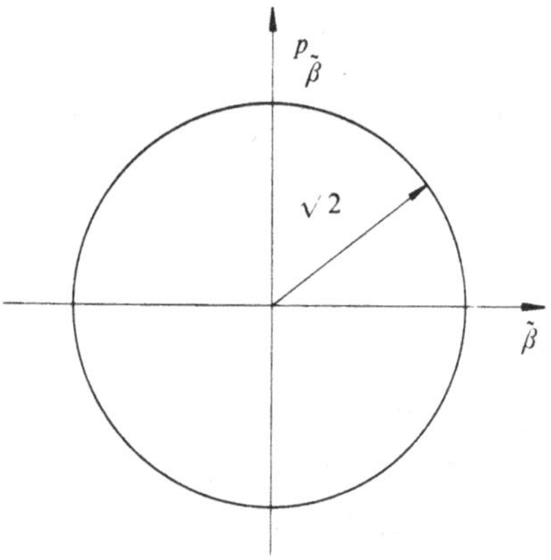

Fig. 3.14. Phase space in the $\tilde\beta$ variable when $p_\gamma = p_{\theta_i} = 0$.

The difference between the classical limit of the interacting boson model Hamiltonian and the Bohr Hamiltonian becomes even more apparent if one considers the classical limit of the full Hamiltonian with U(5) symmetry,

$$H^{(\mathrm{I})} = \varepsilon\hat{n}_{\mathrm{d}} + a_1(\hat{L}\cdot\hat{L}) + a_5\hat{C}_5 + a_6\hat{n}_{\mathrm{d}}^2. \tag{3.82}$$

This is

$$H_{\mathrm{cl}}^{(\mathrm{I})} = (\varepsilon + 6a_1 + \tfrac{4}{3}a_5 + a_6)N\frac{1}{2}\left(p_{\tilde\beta}^2 + \tilde\beta^2 + \frac{\hat{T}^2}{\tilde\beta^2}\right)$$

$$+ a_1N(N-1)\hat{L}^2 + a_5\frac{N(N-1)}{6}\hat{T}^2 + a_6N(N-1)\frac{1}{4}\left(p_{\tilde\beta}^2 + \tilde\beta^2 + \frac{\hat{T}^2}{\tilde\beta^2}\right)^2, \tag{3.83}$$

where

$$\hat{L}^2 = \sum_{k=1}^{3} L_k^2. \tag{3.84}$$

Apart from a more complex structure of the potential and kinetic energy terms, there appear in (3.83) higher-order terms in the momenta, $p_{\tilde\beta}^4$. In the Bohr Hamiltonian (3.79) it is usually assumed that one can have an arbitrarily complex function of the coordinates $V(\hat\beta, \gamma)$, but only a quadratic function of the momenta $p_{\tilde\beta}^2$, p_γ^2, L_k^2 (adiabatic condition).

Classical Hamiltonians of the type (3.83) are rather difficult to treat. Some insight into the structure of the solutions can be obtained by using a requantization procedure (van Roosmalen, 1982)

$$N\sum_\mu(-)^\mu \oint p_{-\mu}\,\mathrm{d}q_\mu = 2\pi t, \tag{3.85}$$

where q_μ and p_μ are the coordinates and momenta and t is an integer. This procedure is only straightforward in the case in which H_{cl} is given by (3.77). It gives a vibrational spectrum with frequency

$$\hbar\omega = \varepsilon, \qquad E_{cl}(n) = \hbar\omega n. \tag{3.86}$$

(ii) Chain II

The classical limit here can be studied by considering the Hamiltonian

$$H'^{(II)} = a_1(\hat{L}\cdot\hat{L}) + a_2(\hat{Q}\cdot\hat{Q}). \tag{3.87}$$

For the purposes of discussion in this section, it is sufficient to consider

$$H^{(II)} = -\kappa[2\hat{Q}\cdot\hat{Q} + \tfrac{3}{4}\hat{L}\cdot\hat{L}]. \tag{3.88}$$

The classical Hamiltonian that one obtains is

$$\begin{aligned} H_{cl}^{(II)} = {} & -\kappa 10N + \kappa 4N(N-1) - \frac{\tilde{T}^4}{16\tilde{\beta}^4} + \frac{\tilde{T}^2}{8\tilde{\beta}^2}(5\tilde{\beta}^2 - p_{\tilde{\beta}}^2) - \tilde{\beta}^2 \\ & + \frac{1}{2}\tilde{\beta}^2(\tilde{\beta}^2 + p_{\tilde{\beta}}^2) - \frac{1}{16}(\tilde{\beta}^2 + p_{\tilde{\beta}}^2) - \frac{1}{4}\sum_{k=1}^{3} L_k^2 \\ & + \frac{1}{2}\left(1 - \frac{\tilde{T}^2}{2\tilde{\beta}^2} - \frac{1}{2}p_{\tilde{\beta}}^2\right)^{\frac{1}{2}} \\ & \times \left\{\left(\frac{p_\gamma^2}{\tilde{\beta}} - \tilde{\beta}(\tilde{\beta}^2 + p_{\tilde{\beta}}^2)\right)\cos 3\gamma + 2p_\gamma p_{\tilde{\beta}}\sin 3\gamma \right. \\ & \left. + \sum_{k=1}^{3} \frac{L_k^2 \cos(\gamma - 2\pi k/3)}{4\tilde{\beta}^2 \sin^2(\gamma - 2\pi k/3)}\right\}. \end{aligned} \tag{3.89}$$

This Hamiltonian has a very complex form. van Roosmalen has studied its properties by expanding around the equilibrium values $\tilde{\beta}_e = 2/\sqrt{3}$ and $\gamma_e = 0°$, thus introducing variables

$$\Delta\tilde{\beta} = (\tilde{\beta} - \tilde{\beta}_e), \qquad \Delta\gamma = (\gamma - \gamma_e), \tag{3.90}$$

and by noting that for large N

$$p_{\Delta\tilde{\beta}}^2 \approx (\Delta\tilde{\beta})^2 \approx p_{\Delta\gamma}^2 \approx (\Delta\gamma)^2 \approx 1/N, \qquad \hat{L}^2 \approx 1/N^2. \tag{3.91}$$

He then obtains

$$H_{\rm cl}^{(\rm II)}=H_{\rm cl}^{(\rm II)}(\beta_{\rm e},\gamma_{\rm e})+4\kappa N^2\left[\frac{1}{2}p_{\Delta\beta}^2+\frac{9}{2}(\Delta\bar{\beta})^2+\frac{9}{8}p_{\Delta\gamma}^2+2(\Delta\gamma)^2+\frac{9}{32}\frac{L_3^2}{(\Delta\gamma)^2}\right]. \tag{3.92}$$

One thus has β and γ vibrations superimposed on the equilibrium value. In the limit $N\to\infty$ these vibrations are decoupled and harmonic, Eq. (3.92). This situation is identical to that obtained starting from (3.79) with (Bohr, 1952)

$$V(\hat{\beta},\gamma)=\tfrac{1}{2}C_{20}(\hat{\beta}-\hat{\beta}_{\rm e})^2+C_{22}(\gamma-\gamma_{\rm e})^2, \tag{3.93}$$

corresponding to a deformed body with axial deformation. The only difference is that in (3.92) the β and γ vibrations have the same frequency (SU(3) limit),

$$\hbar\omega=\hbar\omega_\beta=\hbar\omega_\gamma=12\kappa N. \tag{3.94}$$

They produce a spectrum of the type

$$E_{\rm cl}(n_\beta,n_\gamma)=E_{\rm cl}^{(\rm e)}+\hbar\omega(n_\beta+n_\gamma), \tag{3.95}$$

where

$$n_\beta=0,1,\ldots$$

$$n_\gamma=2m+\tfrac{1}{2}K;\quad m=0,1,2,\ldots;\quad K=0.2,4,\ldots. \tag{3.96}$$

The degeneracy of the β and γ vibrations is removed by adding other terms to $H^{(\rm II)}$ as shown explicitly in Eqs. (3.60) and (3.61) (i.e. moving away from the SU(3) limit).

(iii) Chain III

The classical limit here can be studied by considering the Hamiltonian

$$H'^{(\rm III)}=a_0\hat{P}_6+a_1(\hat{L}\cdot\hat{L})+a_5\hat{C}_5. \tag{3.97}$$

For purposes of discussion here, it is sufficient to consider

$$H^{(\rm III)}=a_0\hat{P}_6, \tag{3.98}$$

with classical limit

$$H_{\rm cl}^{(\rm III)}=a_0\frac{N(N-1)}{4}(\bar{\beta}^2p_{\bar{\beta}}^2+(1-\bar{\beta}^2)^2). \tag{3.99}$$

The structure of the excitations of (3.99) can be obtained by expanding

(3.93) again around the equilibrium value, $\tilde{\beta}_e = 1$. Equation (3.99) is independent of γ. One obtains, in the limit of large N,

$$H_{\mathrm{cl}}^{(\mathrm{III})} = a_0 \tfrac{1}{4} N^2 [p_{\Delta\tilde{\beta}}^2 + 4(\Delta\tilde{\beta})^2]. \tag{3.100}$$

One has thus β vibrations, with frequency

$$\hbar\omega = \hbar\omega_\beta = a_0 N, \tag{3.101}$$

and

$$E_{\mathrm{cl}}(n_\beta) = \hbar\omega_\beta n_\beta, \quad n_\beta = 0, 1, \ldots. \tag{3.102}$$

This spectrum corresponds to that of the Bohr Hamiltonian with

$$V(\hat{\beta}, \gamma) = \tfrac{1}{2} C_2 (\hat{\beta} - \hat{\beta}_e)^2, \tag{3.103}$$

discussed by Wilets and Jean (1956). Thus, in conclusion, a classical analysis of the interacting boson model-1 shows that it produces results identical to those obtained starting from the Bohr Hamiltonian. The only difference lies in the large amplitude motion, which is bounded in the interacting boson model and unbounded in the Bohr Hamiltonian (finite-N effects).

3.11 Non-spherical bosons and Goldstone bosons

The techniques described in Sects. 3.8 and 3.9 have been recently cast into a very general form by Leviatan (1985). Instead of introducing as building blocks of the interacting boson model the six spherical bosons of Eq. (1.1),

$$s^\dagger, d_{+2}^\dagger, d_{+1}^\dagger, d_0^\dagger, d_{-1}^\dagger, d_{-2}^\dagger, \tag{3.104}$$

one can introduce a set of six boson operators, which depend continuously on two parameters β and γ, as

$$b_c^\dagger = (1+\beta^2)^{-\frac{1}{2}} \left[\beta \cos\gamma \cdot d_0^\dagger + \frac{1}{\sqrt{2}} \beta \sin\gamma (d_{+2}^\dagger + d_{-2}^\dagger) + s^\dagger \right],$$

$$b_\beta^\dagger = (1+\beta^2)^{-\frac{1}{2}} \left[\cos\gamma \cdot d_0^\dagger + \frac{1}{\sqrt{2}} \sin\gamma (d_{+2}^\dagger + d_{-2}^\dagger) - \beta s^\dagger \right],$$

$$b_\gamma^\dagger = \frac{1}{\sqrt{2}} \cos\gamma (d_{+2}^\dagger + d_{-2}^\dagger) - \sin\gamma \cdot d_0^\dagger,$$

$$b_x^\dagger = \frac{1}{\sqrt{2}} (d_{+1}^\dagger + d_{-1}^\dagger),$$

$$b_y^\dagger = \frac{1}{\sqrt{2}}(d_{+1}^\dagger - d_{-1}^\dagger),$$

$$b_z^\dagger = \frac{1}{\sqrt{2}}(d_{+2}^\dagger - d_{-2}^\dagger). \tag{3.105}$$

Any choice of the parameters β and γ in (3.105) defines a complete and orthonormal boson basis,

$$\mathscr{B}^D: \frac{1}{\mathcal{N}} b_i^\dagger(\beta,\gamma) b_{i'}^\dagger(\beta,\gamma) \cdots |0\rangle. \tag{3.106}$$

The number operator is diagonal in this basis and given by

$$\hat{N} = b_c^\dagger b_c + b_\beta^\dagger b_\beta + b_\gamma^\dagger b_\gamma + b_x^\dagger b_x + b_y^\dagger b_y + b_z^\dagger b_z. \tag{3.107}$$

When $\gamma = 0°$, the first three operators in (3.105) become identical to those discussed in Sect. 3.8. In this case, one can if one wishes rearrange the bosons to form a basis with well-defined angular momentum projection on the symmetry axis:

$$\begin{aligned}
b_c^\dagger &= b_c^\dagger(\beta, \gamma=0°) = (1+\beta^2)^{-\frac{1}{2}}(\beta d_0^\dagger + s^\dagger) \\
b_\beta^\dagger &= b_\beta^\dagger(\beta, \gamma=0°) = (1+\beta^2)^{-\frac{1}{2}}(d_0^\dagger - \beta s^\dagger) \\
b_{\gamma,\pm 2}^\dagger &= 1/\sqrt{2}\,[b_\gamma^\dagger(\beta,\gamma=0°) \pm b_z^\dagger] = d_{\pm 2}^\dagger \\
b_{\pm 1}^\dagger &= 1/\sqrt{2}\,[b_x^\dagger \pm b_y^\dagger] = d_{\pm 1}^\dagger.
\end{aligned} \tag{3.108}$$

The fact that $\hat{N}$ is diagonal in this basis allows one to retain the advantages of the N-conserving formalism.

Returning to (3.105), one can now construct the condensate as in (3.49),

$$|g\rangle = (N!)^{-\frac{1}{2}}[b_c^\dagger(\beta,\gamma)]^N|0\rangle. \tag{3.109}$$

The equilibrium values (β_e, γ_e) are determined by a variational calculation, as before, with $|g\rangle$ as a trial function. Inserting these values in (3.109) produces the intrinsic ground-state wave function for the Hamiltonian H. The other members of the basis $b_i^\dagger(\beta=\beta_e, \gamma=\gamma_e)$, $i \neq c$, represent excitations of the condensate. The intrinsic β and γ vibrations are represented by $b_\beta^\dagger$ and $b_\gamma^\dagger$ as in Sect. 3.8, while $b_x^\dagger$, $b_y^\dagger$, $b_z^\dagger$ represent 'spurious' modes, connected with rotations about the x, y, z axes respectively. They can thus be thought of as Goldstone bosons. Leviatan has explicitly constructed the effects of operating with the generators of O(3) and O(5) on the state $|g\rangle$ and shown that in the case in which the Hamiltonian, in addition to having O(3) invariance, also has O(5) invariance, one then has, as well as the three Goldstone bosons, $b_x^\dagger$, $b_y^\dagger$, $b_z^\dagger$, a fourth one, $b_\gamma^\dagger$. In this case the $b_\gamma^\dagger$ represents a spurious mode (not an intrinsic excitation) corresponding to the γ-unstable nature of this situation (Wilets and Jean, 1956). Leviatan's approach is a generalization of the approach of Chen and Arima (1983), Arima (1984) and Leviatan (1984) which applies to nuclei in the proximity of the SU(3) and O(6) limits respectively. It has the advantage that it can also be used to describe triaxial situations. These will arise from higher-order terms in the interacting boson Hamiltonian, H.

Part II

The interacting boson model-2

4
Operators

4.1 Introduction

The interacting boson model-2 originated from the necessity to give to the collective nuclear states, as described by the interacting boson model-1, a microscopic foundation rooted in the spherical shell model (Mayer, 1949; Haxel, Jensen and Suess, 1949). It was originally introduced by Otsuka, Talmi and ourselves (Arima *et al.*, 1977; Otsuka *et al.*, 1978) following earlier ideas of Talmi, who had previously introduced the concept of generalized fermion seniority on which the interacting boson model-2 is based (Talmi, 1971; Shlomo and Talmi, 1972; Talmi, 1973).

In the interacting boson model-2, the bosons have a direct physical interpretation as correlated pairs of particles with $J^P=0^+$ and $J^P=2^+$. This physical interpretation yields, on the one hand, the possibility of constructing a microscopic theory of collective states based on the spherical shell model (to be discussed in a subsequent book) and, on the other hand, a richer algebraic structure which can be exploited to study nuclear properties. In Part II of this book we describe this richer algebraic structure, which has produced a large number of interesting results, the most notable being the discovery of low-lying collective modes in which protons and neutrons move out of phase (Bohle *et al.*, 1984).

Since, contrary to the case of the interacting boson model-1, all operators here have a direct microscopic counterpart, we shall occasionally point out this connection and use it as input in the choice of parameters.

4.2 Bosons from fermions

The underlying physical picture one has in mind is the spherical shell model (De Shalit and Talmi, 1963). Here particles move in the average field due to all others. This average field produces single-particle levels with quantum numbers l,j where l is the orbital angular momentum and j the total angular momentum, $j=l\pm\frac{1}{2}$. The nature of the average potential is such that it gives rise to a shell structure. Major shell closures occur at certain nucleon numbers. These 'magic' numbers are usually taken as 2, 8, 20, 28,

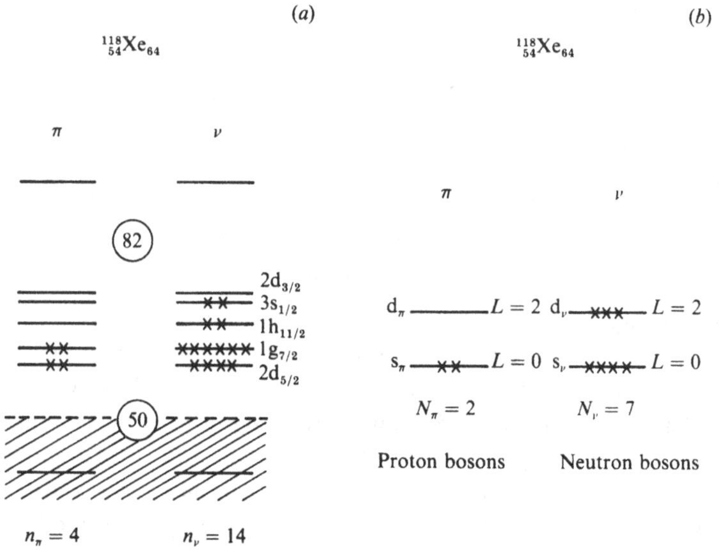

Fig. 4.1. (*a*) Schematic representation of the shell-model problem for $^{118}_{54}Xe_{64}$; (*b*) The boson problem which replaces the shell-model problem for $^{118}_{54}Xe_{64}$.

50, 82, 126. Sometimes, one also considers 'partially-magic' numbers where smaller but non-negligible gaps occur in the single-particle spectrum. Partially-magic numbers usually taken into account are 38, 40 and 64. A portion of the single-particle spectrum is shown in Fig. 4.1.

In order to calculate properties of a given even–even nucleus, one must then diagonalize the residual interaction in the space provided by the average field. This is an extremely difficult (and in some cases impossible) problem. One thus seeks truncations of the shell-model space. First, one usually assumes that the closed shells (corresponding to the magic numbers) are inert. Second, one assumes that the important particle configurations in even–even nuclei are those in which identical particles are paired together in states with total angular momentum and parity $J^P = 0^+$ and $J^P = 2^+$. Finally, one treats the pairs as bosons, much in the same way as Cooper pairs in the electron gas (Cooper, 1956). All these assumptions can be slightly relaxed and there exist already several calculations with active closed shells (Duval and Barrett, 1981), with $J^P = 4^+$ pair (van Isacker *et al.*, 1981; Heyde *et al.*, 1982; Wu and Zhou, 1984; Akiyama, 1985; Yoshinaga *et al.*, 1986), and with pairs explicitly treated as fermions (Yoshida, Arima and Otsuka, 1982).

If one retains all three approximations, one is led to consider a system of interacting bosons of two types, proton bosons and neutron bosons. The proton (neutron) bosons with $J^P=0^+$ will be denoted by $s_\pi(s_\nu)$. The proton (neutron) bosons with $J^P=2^+$ will be denoted by $d_\pi(d_\nu)$, Fig. 4.1. In order to take into account the particle–hole conjugation in particle space, the number of proton, N_π, and neutron, N_ν, bosons is counted from the nearest closed shell, i.e. if more than half of the shell is full, $N_{\pi(\nu)}$ is taken as the number of hole pairs. Thus, for example, for ${}^{118}_{54}Xe_{64}$ (Fig. 4.1), $N_\pi=(54-50)/2=2$ and $N_\nu=(64-50)/2=7$, while for ${}^{128}_{54}Xe_{74}$, $N_\pi=(54-50)/2$ and $N_\nu=(82-74)/2=\bar{4}$. A bar is placed sometimes over the number $\bar{N}_{\pi(\nu)}$ in order to denote the fact that these are hole states. The total number of bosons, N, which in the interacting boson model-1 could have been taken as a parameter, is now fixed by the microscopic interpretation of the bosons to be

$$N=N_\pi+N_\nu. \tag{4.1}$$

In this book, we discuss the consequences that the algebraic and geometric structure of the model has on nuclear properties of even–even nuclei. In subsequent publications, the consequences on nuclear properties of odd–even nuclei will be discussed, together with a full account of the microscopic structure of the bosons.

4.3 Boson operators

The building blocks here are boson operators with an extra label specifying proton (π) and neutron (ν) bosons. There are 12 creation and 12 annihilation operators:

$$\begin{cases} s_\pi^\dagger, d_{\pi,\mu}^\dagger, s_\nu^\dagger, d_{\nu,\mu}^\dagger, & (\mu=0, \pm 1, \pm 2), \\ s_\pi, d_{\pi,\mu}, s_\nu, d_{\nu,\mu}, & (\mu=0, \pm 1, \pm 2). \end{cases} \tag{4.2}$$

Proton operators commute with neutron operators, while each set $s_\pi^\dagger, d_{\pi,\mu}^\dagger, s_\pi, d_{\pi,\mu}$ and $s_\nu^\dagger, d_{\nu,\mu}^\dagger, s_\nu, d_{\nu,\mu}$ satisfies Bose commutation relations (1.2). A more compact notation is either

$$b_{\rho,l,m}^\dagger; b_{\rho,l,m}; \quad (\rho=\pi, \nu; \ l=0,2; \ -l \leqslant m \leqslant +l) \tag{4.3}$$

or

$$b_{\rho,\alpha}^\dagger; b_{\rho,\alpha}; \quad (\rho=\pi, \nu; \ \alpha=1, \ldots, 6), \tag{4.4}$$

with commutation relations

$$\begin{aligned} &[b_{\rho,l,m}, b_{\rho',l',m'}^\dagger]=\delta_{\rho\rho'}\,\delta_{ll'}\,\delta_{mm'}; \\ &[b_{\rho,l,m}, b_{\rho',l',m'}]=[b_{\rho,l,m}^\dagger, b_{\rho',l',m'}^\dagger]=0, \end{aligned} \tag{4.5}$$

or

$$[b_{\rho,\alpha}, b^{\dagger}_{\rho',\alpha'}] = \delta_{\rho\rho'}\,\delta_{\alpha\alpha'};$$
$$[b_{\rho,\alpha}, b_{\rho',\alpha'}] = [b^{\dagger}_{\rho,\alpha}, b^{\dagger}_{\rho',\alpha'}] = 0. \tag{4.6}$$

Once more, one needs to construct operators that transform as spherical tensors under rotations, Eq. (1.8). This is done by introducing

$$\tilde{s}_\pi = s_\pi, \quad \tilde{d}_{\pi,\mu} = (-)^\mu d_{\pi,-\mu},$$
$$\tilde{s}_\nu = s_\nu, \quad \tilde{d}_{\nu,\mu} = (-)^\mu d_{\nu,-\mu}. \tag{4.7}$$

In this part, too, we shall keep the tilde over the operators s, although not strictly necessary.

4.3.1 F-spin

Instead of writing explicitly the label π, ν, one can use a formalism similar to isotopic spin but applied now to bosons. Proton and neutron bosons can be considered as having an intrinsic quantity, called F-spin, of value $F=\frac{1}{2}$ (Otsuka *et al.*, 1978). Proton bosons then have a z-projection $F_z = +\frac{1}{2}$, while neutron bosons have $F_z = -\frac{1}{2}$, i.e.

$$|\pi\rangle = |\tfrac{1}{2}, +\tfrac{1}{2}\rangle,$$
$$|\nu\rangle = |\tfrac{1}{2}, -\tfrac{1}{2}\rangle. \tag{4.8}$$

The operators in (4.2) can be written as double tensors

$$b^{\dagger}_{\frac{1}{2},m_f,l,m},\ b_{\frac{1}{2},m_f,l,m}, \quad (m_f = \pm\tfrac{1}{2},\ l=0,2,\ -l \leqslant m \leqslant +l). \tag{4.9}$$

Operators that transform appropriately under rotations both in ordinary space and in F-spin space are

$$\tilde{b}_{\frac{1}{2},m_f,l,m} = (-)^{\frac{1}{2}+m_f+l+m} b_{\frac{1}{2},-m_f,l,-m}. \tag{4.10}$$

4.4 Basis states

Basis states are constructed by repeated application of boson operators on a boson vacuum $|0\rangle$,

$$\mathscr{B}:\ b^{\dagger}_{\pi,\alpha} b^{\dagger}_{\pi,\alpha'} \cdots b^{\dagger}_{\nu,\alpha} b^{\dagger}_{\nu,\alpha'} \cdots |0\rangle. \tag{4.11}$$

Here, too, it is convenient to couple operators to good angular momentum. This can be done by first coupling protons and neutrons separately and subsequently coupling them together,

$$\mathscr{B}:\ \{[b^{\dagger}_{\pi,l} \times b^{\dagger}_{\pi,l'} \times \cdots]^{(L_\pi)} \times [b^{\dagger}_{\nu,l} \times b^{\dagger}_{\nu,l'} \times \cdots]^{(L_\nu)}\}^{(L)}_{M_L} |0\rangle. \tag{4.12}$$

In general, other quantum numbers are needed in addition to L_π and L_ν. These depend on the basis used. If one uses the basis of chain I, Sect. 2.4.1, for both protons and neutrons, then

$$\mathscr{B}: \{s_\pi^{\dagger n_{s\pi}}[d_\pi^{\dagger n_{d\pi}}]_{v_\pi n_{\Delta\pi} L_\pi} s_\nu^{\dagger n_{s\nu}}[d_\nu^{\dagger n_{d\nu}}]_{v_\nu n_{\Delta\nu} L_\nu}\}_{LM_L}|0\rangle. \tag{4.13}$$

Conversely, one can use the concept of F-spin and characterize the states by the wave functions

$$|\psi\rangle = |LM_L FM_F\rangle. \tag{4.14}$$

4.5 Physical operators

4.5.1 The Hamiltonian operator

The Hamiltonian operator can be written as

$$H = H_\pi + H_\nu + V_{\pi\nu}. \tag{4.15}$$

Here, each piece H_π, H_ν is the same as in Eq. (1.19), which we rewrite now as

$$H_\rho = E_0^{(\rho)} + \sum_{\alpha\beta} \varepsilon_{\alpha\beta}^{(\rho)} b_{\rho,\alpha}^\dagger b_{\rho,\beta} + \sum_{\alpha\beta\gamma\delta} \tfrac{1}{2} u_{\alpha\beta\gamma\delta}^{(\rho)} b_{\rho,\alpha}^\dagger b_{\rho,\beta}^\dagger b_{\rho,\gamma} b_{\rho,\delta} + \cdots, (\rho = \pi, \nu), \tag{4.16}$$

while $V_{\pi\nu}$ can be written as

$$V_{\pi\nu} = \sum_{\alpha\beta\gamma\delta} w_{\alpha\beta\gamma\delta} b_{\pi,\alpha}^\dagger b_{\pi,\beta} b_{\nu,\gamma}^\dagger b_{\nu,\delta} + \cdots. \tag{4.17}$$

This $V_{\pi\nu}$ conserves separately the number of proton and neutron bosons, as it does H_ρ. This fact, that within the framework of the interacting boson model-1 could be viewed as an assumption, is here a consequence of the microscopic structure of the bosons and originates from the conservation of particle number.

In Eqs. (4.16) and (4.17), the fact that the Hamiltonian is a scalar under rotation is not evident. In order to make this apparent, it is convenient to rewrite H_ρ as

$$H_\rho = E_0^{(\rho)} + \sum_l \varepsilon_l^{(\rho)} (b_{\rho,l}^\dagger \cdot \tilde{b}_{\rho,l}) + \sum_{L,ll'l''l'''} \tfrac{1}{2} u_{ll'l''l'''}^{(\rho),(L)} [[b_{\rho,l}^\dagger \times b_{\rho,l'}^\dagger]^{(L)}$$

$$\times [\tilde{b}_{\rho,l''} \times \tilde{b}_{\rho,l'''}]^{(L)}]_0^{(0)} + \cdots, \tag{4.18}$$

and

$$V_{\pi\nu} = \sum_{L,ll'l''l'''} w_{ll'l''l'''}^{(L)} [[b_{\pi,l}^\dagger \times \tilde{b}_{\pi,l'}] \times [b_{\nu,l''}^\dagger \times \tilde{b}_{\nu,l'''}]^{(L)}]_0^{(0)} + \cdots. \tag{4.19}$$

In most calculations, only up to two-body terms are retained. When written explicitly in terms of s and d bosons, Eqs. (4.17) and (4.18) read

$$H_\rho = E_0^{(\rho)} + \varepsilon_s^{(\rho)}(s_\rho^\dagger \cdot \tilde{s}_\rho) + \varepsilon_d^{(\rho)}(d_\rho^\dagger \cdot \tilde{d}_\rho)$$
$$+ \sum_{L=0,2,4} \tfrac{1}{2}(2L+1)^{\frac{1}{2}} c_L^{(\rho)} [[d_\rho^\dagger \times d_\rho^\dagger]^{(L)} \times [\tilde{d}_\rho \times \tilde{d}_\rho]^{(L)}]_0^{(0)}$$
$$+ \frac{1}{\sqrt{2}} v_2^{(\rho)} [[d_\rho^\dagger \times d_\rho^\dagger]^{(2)} \times [\tilde{d}_\rho \times \tilde{s}_\rho]^{(2)} + [d_\rho^\dagger \times s_\rho^\dagger]^{(2)} \times [\tilde{d}_\rho \times \tilde{d}_\rho]^{(2)}]_0^{(0)}$$
$$+ \tfrac{1}{2} v_0^{(\rho)} [[d_\rho^\dagger \times d_\rho^\dagger]^{(0)} \times [\tilde{s}_\rho \times \tilde{s}_\rho]^{(0)} + [s_\rho^\dagger \times s_\rho^\dagger]^{(0)} \times [\tilde{d}_\rho \times \tilde{d}_\rho]^{(0)}]_0^{(0)}$$
$$+ u_2^{(\rho)} [[d_\rho^\dagger \times s_\rho^\dagger]^{(2)} \times [\tilde{d}_\rho \times \tilde{s}_\rho]^{(2)}]_0^{(0)} + \tfrac{1}{2} u_0^{(\rho)} [s_\rho^\dagger \times s_\rho^\dagger]^{(0)} \times [\tilde{s}_\rho \times \tilde{s}_\rho]^{(2)}]_0^{(0)}, \quad (4.20)$$

and

$$V_{\pi\nu} = \sum_{L=0,1,2,3,4} w_L [[d_\pi^\dagger \times \tilde{d}_\pi]^{(L)} \times [d_\nu^\dagger \times \tilde{d}_\nu]^{(L)}]_0^{(0)}$$
$$+ w_5 [[s_\pi^\dagger \times \tilde{s}_\pi]^{(0)} \times [s_\nu^\dagger \times \tilde{s}_\nu]^{(0)}]_0^{(0)} + w_6 [[s_\pi^\dagger \times \tilde{s}_\pi]^{(0)} \times [d_\nu^\dagger \times \tilde{d}_\nu]^{(0)}]_0^{(0)}$$
$$+ w_7 [[d_\pi^\dagger \times \tilde{d}_\pi]^{(0)} \times [s_\nu^\dagger \times \tilde{s}_\nu]^{(0)}]_0^{(0)}$$
$$+ w_8 [[d_\pi^\dagger \times \tilde{s}_\pi + s_\pi^\dagger \times \tilde{d}_\pi]^{(2)} \times [d_\nu^\dagger \times \tilde{s}_\nu + s_\nu^\dagger \times \tilde{d}_\nu]^{(2)}]_0^{(0)} \quad (4.21)$$
$$+ w_9 [[d_\pi^\dagger \times \tilde{s}_\pi - s_\pi^\dagger \times \tilde{d}_\pi]^{(2)} \times [d_\nu^\dagger \times \tilde{s}_\nu - s_\nu^\dagger \times \tilde{d}_\nu]^{(2)}]_0^{(0)}$$
$$+ w_{10} [[d_\pi^\dagger \times \tilde{d}_\pi]^{(2)} \times [d_\nu^\dagger \times \tilde{s}_\nu + s_\nu^\dagger \times \tilde{d}_\nu]^{(2)}]_0^{(0)}$$
$$+ w_{11} [[d_\pi^\dagger \times \tilde{s}_\pi + s_\pi^\dagger \times \tilde{d}_\pi]^{(2)} \times [d_\nu^\dagger \times \tilde{d}_\nu]^{(2)}]_0^{(0)}.$$

There are thus nine terms in H_π: $\varepsilon_L^\pi(L=0,\ 2)$, $c_L^\pi(L=0,2,4)$, $v_L^\pi(L=0,2)$, $u_L^\pi(L=0,2)$; nine terms in H_ν: $\varepsilon_L^\nu(L=0,2)$, $c_L^\nu(L=0,2,4)$, $v_L^\nu(L=0,2)$, $u_L^\nu(L=0,2)$; and 12 terms in $V_{\pi\nu}$: w_k $(k=0, \ldots, 11)$, giving a total of 30.

4.5.2 *Transition operators*

Transition operators can be written as

$$T^{(L)} = T_\pi^{(L)} + T_\nu^{(L)}. \quad (4.22)$$

Each term in (4.22) can be written as

$$T_\rho^{(L)} = t_{\rho,0}^{(0)} \delta_{L,0} + \sum_{\alpha\beta} t_{\rho,\alpha\beta}^{(L)} b_{\rho,\alpha}^\dagger b_{\rho,\beta} + \cdots. \quad (4.23)$$

Introducing explicitly s and d bosons and stopping at one-body terms yields,

$$
\begin{aligned}
T_0^{(\mathrm{E0})} &= \gamma_0 + \alpha_0^{(\pi)}[s_\pi^\dagger \times \tilde{s}_\pi]_0^{(0)} + \beta_0^{(\pi)}[d_\pi^\dagger \times \tilde{d}_\pi]_0^{(0)} \\
&\quad + \alpha_0^{(\nu)}[s_\nu^\dagger \times \tilde{s}_\nu]_0^{(0)} + \beta_0^{(\nu)}[d_\nu^\dagger \times \tilde{d}_\nu]_0^{(0)}, \\
T_\mu^{(\mathrm{M1})} &= \beta_1^{(\pi)}[d_\pi^\dagger \times \tilde{d}_\pi]_\mu^{(1)} + \beta_1^{(\nu)}[d_\nu^\dagger \times \tilde{d}_\nu]_\mu^{(1)}, \\
T_\mu^{(\mathrm{E2})} &= \alpha_2^{(\pi)}[d_\pi^\dagger \times \tilde{s}_\pi + s_\pi^\dagger \times \tilde{d}_\pi]_\mu^{(2)} + \beta_2^{(\pi)}[\mathrm{d}_\pi^\dagger \times \tilde{\mathrm{d}}_\pi]_\mu^{(2)} \\
&\quad + \alpha_2^{(\nu)}[d_\nu^\dagger \times \tilde{s}_\nu + s_\nu^\dagger \times \tilde{d}_\nu]_\mu^{(2)} + \beta_2^{(\nu)}[d_\nu^\dagger \times \tilde{d}_\nu]_\mu^{(2)}, \\
T_\mu^{(\mathrm{M3})} &= \beta_3^{(\pi)}[d_\pi^\dagger \times \tilde{d}_\pi]_\mu^{(3)} + \beta_3^{(\nu)}[d_\nu^\dagger \times \tilde{d}_\nu]_\mu^{(3)}, \\
T_\mu^{(\mathrm{E4})} &= \beta_4^{(\pi)}[d_\pi^\dagger \times \tilde{d}_\pi]_\mu^{(4)} + \beta_4^{(\nu)}[d_\nu^\dagger \times \tilde{d}_\nu]_\mu^{(4)}.
\end{aligned}
\tag{4.24}
$$

4.5.3 *Higher-order terms*

The phenomenological analysis of nuclear spectra appears to indicate that no higher-order terms are needed to describe their properties within the framework of the interacting boson model-2. In fact, it appears that the higher-order terms needed in some cases to describe nuclear properties within the framework of the interacting boson model-1, arise from a reduction of the space from two types to one type of bosons.

4.5.4 *Independent parameters*

For a fixed number of proton, N_π, and neutron, N_ν, bosons, not all terms in the Hamiltonian are independent. Using the relations

$$
\begin{aligned}
\hat{N}_\pi &= \hat{n}_{s\pi} + \hat{n}_{d\pi}, \\
\hat{N}_\nu &= \hat{n}_{s\nu} + \hat{n}_{d\nu},
\end{aligned}
\tag{4.25}
$$

one can rewrite H as

$$H = E_0' + h_\pi + h_\nu + v_{\pi\nu}, \tag{4.26}$$

where

$$
\begin{aligned}
h_\rho &= \varepsilon_\rho(d_\rho^\dagger \cdot \tilde{d}_\rho) + \sum_{L=0,2,4} \tfrac{1}{2}(2L+1)c_L'^{(\rho)}[[d_\rho^\dagger \times d_\rho^\dagger]^{(L)} \times [\tilde{d}_\rho \times \tilde{d}_\rho]^{(L)}]_0^{(0)} \\
&\quad + \frac{1}{\sqrt{2}} v_2^{(\rho)}[[d_\rho^\dagger \times d_\rho^\dagger]^{(2)} \times [\tilde{d}_\rho \times \tilde{s}_\rho]^{(2)} + [d_\rho^\dagger \times s_\rho^\dagger]^{(2)} \times [\tilde{d}_\rho \times \tilde{d}_\rho]^{(2)}]_0^{(0)} \\
&\quad + \tfrac{1}{2} v_0^{(\rho)}[[d_\rho^\dagger \times d_\rho^\dagger]^{(0)} \times [\tilde{s}_\rho \times \tilde{s}_\rho]^{(0)} + [s_\rho^\dagger \times s_\rho^\dagger]^{(0)} \times [\tilde{d}_\rho \times \tilde{d}_\rho]^{(0)}]_0^{(0)}, \\
v_{\pi\nu} &= w_0'(d_\pi^\dagger \cdot \tilde{d}_\pi)(d_\nu^\dagger \cdot \tilde{d}_\nu) + \sum_{L=1,2,3,4} w_L[[d_\pi^\dagger \times \tilde{d}_\pi]^{(L)} \times [d_\nu^\dagger \times \tilde{d}_\nu]^{(L)}]_0^{(0)}
\end{aligned}
$$

$$+w_8[[d_\pi^\dagger\times\tilde{s}_\pi+s_\pi^\dagger\times\tilde{d}_\pi]^{(2)}\times[d_\nu^\dagger\times\tilde{s}_\nu+s_\nu^\dagger\times\tilde{d}_\nu]^{(2)}]_0^{(0)}$$

$$+w_9[[d_\pi^\dagger\times\tilde{s}_\pi-s_\pi^\dagger\times\tilde{d}_\pi]^{(2)}\times[d_\nu^\dagger\times\tilde{s}_\nu-s_\nu^\dagger\times\tilde{d}_\nu]^{(2)}]_0^{(0)}$$

$$+w_{10}[[d_\pi^\dagger\times\tilde{d}_\pi]^{(2)}\times[d_\nu^\dagger\times\tilde{s}_\nu+s_\nu^\dagger\times\tilde{d}_\nu]^{(2)}]_0^{(0)}$$

$$+w_{11}[[d_\pi^\dagger\times\tilde{s}_\pi+s_\pi^\dagger\times\tilde{d}_\pi]^{(2)}\times[d_\nu^\dagger\times\tilde{d}_\nu]^{(2)}]_0^{(0)}. \tag{4.27}$$

In this expression,

$$E_0'=E_0+\varepsilon_s^{(\pi)}N_\pi+\tfrac{1}{2}u_0^{(\pi)}N_\pi(N_\pi-1)+\varepsilon_s^{(\nu)}N_\nu+\tfrac{1}{2}u_0^{(\nu)}N_\nu(N_\nu-1)+w_5N_\pi N_\nu,$$

$$\varepsilon_\pi=(\varepsilon_d^{(\pi)}-\varepsilon_s^{(\pi)})+\frac{1}{\sqrt{5}}u_2^{(\pi)}(N_\pi-1)-u_0^{(\pi)}(N_\pi-1)-w_5N_\nu+\frac{1}{\sqrt{5}}w_7N_\nu,$$

$$\varepsilon_\nu=(\varepsilon_d^{(\nu)}-\varepsilon_s^{(\nu)})+\frac{1}{\sqrt{5}}u_2^{(\nu)}(N_\nu-1)-u_0^{(\nu)}(N_\nu-1)-w_5N_\pi+\frac{1}{\sqrt{5}}w_6N_\pi,$$

$$c_L'^{(\rho)}=c_L^{(\rho)}+u_0^{(\rho)}-\frac{2}{\sqrt{5}}u_2^{(\rho)}, \tag{4.28}$$

$$w_0'=\tfrac{1}{5}w_0-\frac{1}{\sqrt{5}}(w_6+w_7)+w_5.$$

Since E_0 in itself is a quadratic function of N_π and N_ν

$$E_0=E_{00}+E_{01}^{(\pi)}N_\pi+\tfrac{1}{2}E_{02}^{(\pi)}N_\pi(N_\pi-1)+E_{01}^{(\nu)}N_\nu+\tfrac{1}{2}E_{02}^{(\nu)}N_\nu(N_\nu-1)$$
$$+E_{02}^{(\pi\nu)}N_\pi N_\nu, \tag{4.29}$$

one can write

$$E_0'=E_{00}+(E_{01}^{(\pi)}+\varepsilon_s^{(\pi)})N_\pi+\tfrac{1}{2}(E_{02}^{(\pi)}+u_0^{(\pi)})N_\pi(N_\pi-1)+(E_{01}^{(\nu)}+\varepsilon_s^{(\nu)})N_\nu$$
$$+\tfrac{1}{2}(E_{02}^{(\nu)}+u_0^{(\nu)})N_\nu(N_\nu-1)+(E_{02}^{(\pi\nu)}+w_5)N_\pi N_\nu. \tag{4.30}$$

Since E_0' does not contribute to excitation energies, there are $6+6+9=21$ parameters determining, in general, these energies. The other operator for which some terms can be eliminated is the E0 operator. This operator can be rewritten as

$$T_0^{(E0)}=\gamma_0'+\beta_0'^{(\pi)}[d_\pi^\dagger\times\tilde{d}_\pi]_0^{(0)}+\beta_0'^{(\nu)}[d_\nu^\dagger\times\tilde{d}_\nu]_0^{(0)}, \tag{4.31}$$

where

$$\gamma_0'=\gamma_0+\alpha_0^{(\pi)}N_\pi+\alpha_0^{(\nu)}N_\nu.$$

$$\beta_0'^{(\rho)}=\beta_0^{(\rho)}-\sqrt{5}\,\alpha_0^{(\rho)}. \tag{4.32}$$

Again, the constant term γ_0' does not contribute to transitions. Since γ_0 is itself a linear function of N_π and N_ν,

$$\gamma_0 = \gamma_{00} + \gamma_{01}^{(\pi)} N_\pi + \gamma_{01}^{(\nu)} N_\nu, \tag{4.33}$$

one can write

$$\gamma'_0 = \gamma_{00} + (\gamma_{01}^{(\pi)} + \alpha_0^{(\pi)}) N_\pi + (\gamma_{01}^{(\nu)} + \alpha_0^{(\nu)}) N_\nu. \tag{4.34}$$

4.5.5 *Multipole expansion*

The Hamiltonian (4.26) can be rewritten in terms of multipole operators for protons and neutrons. The two terms h_π and h_ν can be written in terms of the operator

$$\begin{aligned}
\hat{n}_{\mathrm{d}_\rho} &= (d_\rho^\dagger \cdot \tilde{d}_\rho), \\
\hat{\mathscr{P}}_\rho &= \tfrac{1}{2}(\tilde{d}_\rho \cdot \tilde{d}_\rho) - \tfrac{1}{2}(\tilde{s}_\rho \cdot \tilde{s}_\rho), \\
\hat{L}_\rho &= (10)^{\frac{1}{2}} [d_\rho^\dagger \times \tilde{d}_\rho]^{(1)}, \\
\hat{Q}_\rho &= [d_\rho^\dagger \times \tilde{s}_\rho + s_\rho^\dagger \times \tilde{d}_\rho]^{(2)} - \tfrac{1}{2}(7)^{\frac{1}{2}} [d_\rho^\dagger \times \tilde{d}_\rho]^{(2)}, \\
\hat{U}_\rho &= [d_\rho^\dagger \times \tilde{d}_\rho]^{(3)}, \\
\hat{V}_\rho &= [d_\rho^\dagger \times \tilde{d}_\rho]^{(4)},
\end{aligned} \tag{4.35}$$

while the proton–neutron interaction $v_{\pi\nu}$ requires in addition the operators

$$\begin{aligned}
\hat{Q}'_\rho &= [d_\rho^\dagger \times \tilde{d}_\rho]^{(2)}, \\
\hat{Q}''_\rho &= \mathrm{i}\, [d_\rho^\dagger \times \tilde{s}_\rho - s_\rho^\dagger \times \tilde{d}_\rho]^{(2)}.
\end{aligned} \tag{4.36}$$

The complete expressions are

$$\begin{aligned}
h_\rho = {} & \varepsilon_\rho \hat{n}_{\mathrm{d}\rho} + a_0^{(\rho)} (\hat{\mathscr{P}}_\rho^\dagger \cdot \hat{\mathscr{P}}_\rho) + a_1^{(\rho)} (\hat{L}_\rho \cdot \hat{L}_\rho) + a_2^{(\rho)} (\hat{Q}_\rho \cdot \hat{Q}_\rho) + a_3^{(\rho)} (\hat{U}_\rho \cdot \hat{U}_\rho) \\
& + a_4^{(\rho)} (\hat{V}_\rho \cdot \hat{V}_\rho), \\
v_{\pi\nu} = {} & c_0^{(\pi\nu)} (\hat{n}_{\mathrm{d}_\pi} \cdot \hat{n}_{\mathrm{d}_\nu}) + c_1^{(\pi\nu)} (\hat{L}_\pi \cdot \hat{L}_\nu) + c_2^{(\pi\nu)} (\hat{Q}_\pi \cdot \hat{Q}_\nu) + c_2^{\prime(\pi\nu)} (\hat{Q}_\pi \cdot \hat{Q}'_\nu) \\
& + c_2^{\prime\prime(\pi\nu)} (\hat{Q}'_\pi \cdot \hat{Q}_\nu) + c_2^{\prime\prime\prime(\pi\nu)} (\hat{Q}'_\pi \cdot \hat{Q}'_\pi) + c_2^{\mathrm{iv}(\pi\nu)} (\hat{Q}''_\pi \cdot \hat{Q}''_\nu) \\
& + c_3^{(\pi\nu)} (\hat{U}_\pi \cdot \hat{U}_\nu) + c_4^{(\pi\nu)} (\hat{V}_\pi \cdot \hat{V}_\nu).
\end{aligned} \tag{4.37}$$

4.5.6 *Effective charges and moments*

For transition operators, it is convenient to introduce parameters with a direct physical meaning. These are called effective boson charges and moments. They are defined in the following way. For E0 transitions,

$$T^{(\mathrm{E0})} = f_\pi \hat{n}_{\mathrm{d}_\pi} + f_\nu \hat{n}_{\mathrm{d}_\nu}. \tag{4.38}$$

For E2 transitions

$$T^{(E2)} = e_\pi \hat{Q}^\chi_\pi + e_\nu \hat{Q}^\chi_\nu,$$
$$\hat{Q}^\chi_\rho = [d^\dagger_\rho \times \tilde{s}_\rho + s^\dagger_\rho \times \tilde{d}_\rho]^{(2)} + \chi_\rho [d^\dagger_\rho \times \tilde{d}_\rho]^{(2)}. \tag{4.39}$$

For E4 transitions

$$T^{(E4)} = t_\pi \hat{V}_\pi + t_\nu \hat{V}_\nu. \tag{4.40}$$

For the magnetic transitions, one introduces boson effective g-factors. For M1 transitions these are defined by

$$T^{(M1)} = \sqrt{\left(\frac{3}{4\pi}\right)} (g_\pi \hat{L}_\pi + g_\nu \hat{L}_\nu). \tag{4.41}$$

while for M3 transitions they are defined by

$$T^{(M3)} = \sqrt{\left(\frac{7}{4\pi}\right)} (m_\pi \hat{U}_\pi + m_\nu \hat{U}_\nu). \tag{4.42}$$

The values of the effective charges and moments, f_ρ, e_ρ, t_ρ, g_ρ, m_ρ can be calculated microscopically.

4.5.7 *Consistent-Q formalism; Talmi Hamiltonian*

The Hamiltonian in (4.26) and (4.27) contains too many terms for a direct phenomenological study. However, the microscopic structure of the model suggests that only a few of the terms in (4.26) and (4.27) are important. This is due to the fact that the residual nucleon–nucleon interaction in the spherical shell model is dominated by a pairing term between identical nucleons. This produces in the boson Hamiltonian a term of the type $\varepsilon_\pi \hat{n}_{d\pi} + \varepsilon_\nu \hat{n}_{d\nu}$. In addition there is a quadrupole–quadrupole interaction between non-identical nucleons. This produces a term of the type $\kappa \hat{Q}^\chi_\pi \cdot \hat{Q}^\chi_\nu$, where $\hat{Q}^\chi_\pi$ and $\hat{Q}^\chi_\nu$ are given by (4.39). Finally, there is a symmetry energy, favoring states in which the protons and neutrons move in phase. This symmetry energy can be introduced, as will be discussed in more detail later on, by making use of an operator, called the Majorana operator, given by

$$\hat{M}_{\pi\nu} = [s^\dagger_\nu \times d^\dagger_\pi - s^\dagger_\pi \times d^\dagger_\nu]^{(2)} \cdot [\tilde{s}_\nu \times \tilde{d}_\pi - \tilde{s}_\pi \times \tilde{d}_\nu]^{(2)}$$
$$-2 \sum_{k=1,3} [d^\dagger_\nu \times d^\dagger_\pi]^{(k)} \cdot [\tilde{d}_\nu \times \tilde{d}_\pi]^{(k)}. \tag{4.43}$$

In conclusion, the adopted Hamiltonian is

$$H = E_0 + \varepsilon_\pi \hat{n}_{d_\pi} + \varepsilon_\nu \hat{n}_{d_\nu} + \kappa \hat{Q}^\chi_\pi \cdot \hat{Q}^\chi_\nu + \lambda' \hat{M}_{\pi\nu}, \tag{4.44}$$

Table 4.1. *Physical origin of the various terms in Talmi's Hamiltonian*

Nucleon–nucleon interaction	Boson counterpart
Pairing	$\varepsilon_\pi \hat{n}_{d_\pi} + \varepsilon_\nu \hat{n}_{d_\nu}$
Quadrupole	$\kappa Q_\pi^\chi \cdot Q_\nu^\chi$
Symmetry energy	$\lambda' \hat{M}_{\pi\nu}$

where, quite often, one puts $\varepsilon_\pi = \varepsilon_\nu = \varepsilon$. This Hamiltonian is specified by ε, κ, χ_π, χ_ν in addition to E_0 and λ'. The fact that only four parameters determine the excitation energies allows one to perform straightforward phenomenological studies. This Hamiltonian was suggested by Talmi (Otsuka *et al.*, 1978) and it will be referred to in the following section as the 'Talmi Hamiltonian'. Associated with it there is an electromagnetic quadrupole operator of the type

$$T^{(\mathrm{E2})} = e_\pi \hat{Q}_\pi^\chi + e_\nu \hat{Q}_\nu^\chi. \tag{4.45}$$

Since the same $\hat{Q}_\rho^\chi$ operators appear in (4.45) and (4.44), this formalism is also called 'consistent-Q formalism'. The complete situation is summarized in Table 4.1.

4.5.8 Binding energies

Ground-state energies can be calculated using either (4.26) or one of the various forms of the Hamiltonian H. Since the overall constant E_0' in H has the form (4.30), one can write, in general, for the ground-state energy, the expression

$$E_G(N_\pi, N_\nu) = E_c + A_\pi N_\pi + A_\nu N_\nu + \tfrac{1}{2} B_\pi N_\pi (N_\pi - 1) + \tfrac{1}{2} B_\nu N_\nu (N_\nu - 1) + C N_\pi N_\nu + E_\mathrm{D}(N_\pi, N_\nu). \tag{4.46}$$

Here E_c represents the energy of the closed shell and the coefficients A_π, A_ν, B_π, B_ν, C are linear combinations of those appearing in (4.26). The term $E_\mathrm{D}(N_\pi, N_\nu)$ represents the contribution coming from the remaining pieces h_π, h_ν and $v_{\pi\nu}$, and must be evaluated for each case. From (4.46) one can evaluate directly the two-particle separation energies as in (2.204). One must now distinguish between two-neutron and two-proton separation energies. Introducing the binding energy as in (2.203), one has

$$\begin{aligned} S_{2\pi}(N_\pi, N_\nu) &= E_\mathrm{B}(N_\pi + 1, N_\nu) - E_\mathrm{B}(N_\pi, N_\nu), \\ S_{2\nu}(N_\pi, N_\nu) &= E_\mathrm{B}(N_\pi, N_\nu + 1) - E_\mathrm{B}(N_\pi, N_\nu). \end{aligned} \tag{4.47}$$

Using (4.46), one has

$$S_{2\pi}(N_\pi, N_\nu) = -A_\pi - B_\pi N_\pi - CN_\nu - E_D(N_\pi+1, N_\nu) + E_D(N_\pi, N_\nu),$$
$$S_{2\nu}(N_\pi, N_\nu) = -A_\nu - B_\nu N_\nu - CN_\pi - E_D(N_\pi, N_\nu+1) + E_D(N_\pi, N_\nu). \tag{4.48}$$

As discussed in Sect. 2.7, a delicate problem arises when either the protons or the neutrons reach the middle of a major shell. As mentioned in Sect. 4.2, one shifts there from a description in terms of particles to one in terms of the holes. This means that the expressions in (4.48) should be replaced by

$$\bar{S}_{2\pi}(N_\pi, N_\nu) = E_B(N_\pi - 1, N_\nu) - E_B(N_\pi, N_\nu), \tag{4.49}$$

and the coefficients A_π, B_π and C by

$$\bar{A}_\pi = -A_\pi,$$
$$\bar{B}_\pi = -B_\pi, \tag{4.50}$$
$$\bar{C} = -C,$$

and

$$\bar{S}_{2\nu}(N_\pi, N_\nu) = E_B(N_\pi, N_\nu - 1) - E_B(N_\pi, N_\nu), \tag{4.51}$$

with

$$\bar{A}_\nu = -A_\nu,$$
$$\bar{B}_\nu = -B_\nu, \tag{4.52}$$
$$\bar{C} = -C.$$

4.5.9 Nuclear radii

Nuclear radii can also be analyzed within the framework of the interacting boson model-2. Using (4.24), they can be written as

$$r_\pi^2(N_\pi, N_\nu) = r_{c\pi}^2 + \alpha_{\pi\pi}\hat{N}_\pi + \alpha_{\pi\nu}\hat{N}_\nu + \beta_{\pi\pi}\hat{n}_{d\pi} + \beta_{\pi\nu}\hat{n}_{d\nu},$$
$$r_\nu^2(N_\pi, N_\nu) = r_{c\nu}^2 + \alpha_{\nu\pi}\hat{N}_\pi + \alpha_{\nu\nu}\hat{N}_\nu + \beta_{\nu\pi}\hat{n}_{d\pi} + \beta_{\nu\nu}\hat{n}_{d\nu}, \tag{4.53}$$

where $r_{c\pi}^2$ and $r_{c\nu}^2$ represent the square radii of the closed proton and neutron shells. From these expressions one can calculate isomer, isotope and isotone shifts.

(i) Isomer shifts

The isomer shift is a difference between the proton radius of the first excited 2_1^+ state and the ground state. Following the discussion in Sect. 2.8, it can be written as

$$\delta\langle r_\pi^2\rangle^{(N_\pi,N_\nu)} = \beta_{\pi\pi}[\langle \hat{n}_{d_\pi}\rangle_{2_1^+}^{(N_\pi,N_\nu)} - \langle \hat{n}_{d_\pi}\rangle_{0_1^+}^{(N_\pi,N_\nu)}]$$
$$+ \beta_{\pi\nu}[\langle \hat{n}_{d_\nu}\rangle_{2_1^+}^{(N_\pi,N_\nu)} - \langle \hat{n}_{d_\pi}\rangle_{0_1^+}^{(N_\pi,N_\nu)}]. \tag{4.54}$$

(ii) Isotope and isotone shifts

The isotope shifts can be written as

$$\begin{aligned}\Delta\langle r_\pi^2\rangle^{(N_\pi,N_\nu)} = \alpha_{\pi\nu} &+ \beta_{\pi\pi}[\langle\hat{n}_{d_\pi}\rangle_{0_1^+}^{(N_\pi,N_\nu+1)} - \langle\hat{n}_{d_\pi}\rangle_{0_1^+}^{(N_\pi,N_\nu)}] \\ &+ \beta_{\pi\nu}[\langle\hat{n}_{d_\nu}\rangle_{0_1^+}^{(N_\pi,N_\nu+1)} - \langle\hat{n}_{d_\nu}\rangle_{0_1^+}^{(N_\pi,N_\nu)}].\end{aligned} \tag{4.55}$$

Similarly, one can write the isotone shifts as

$$\begin{aligned}\Delta'\langle r_\pi^2\rangle^{(N_\pi,N_\nu)} = \alpha_{\pi\pi} &+ \beta_{\pi\pi}[\langle\hat{n}_{d_\pi}\rangle_{0_1^+}^{(N_\pi+1,N_\nu)} - \langle\hat{n}_{d_\pi}\rangle_{0_1^+}^{(N_\pi,N_\nu)}] \\ &+ \beta_{\pi\nu}[\langle\hat{n}_{d_\nu}\rangle_{0_1^+}^{(N_\pi+1,N_\nu)} - \langle\hat{n}_{d_\nu}\rangle_{0_1^+}^{(N_\pi,N_\nu)}].\end{aligned} \tag{4.56}$$

A situation similar to that encountered in the analysis of the binding energies arises here. When crossing the middle of the proton or neutron shells, the isotope and isotone shifts should be replaced by

$$\bar{\Delta}\langle r_\pi^2\rangle^{(N_\pi,N_\nu)} = \langle r_\pi^2\rangle_{0_1^+}^{(N_\pi,N_\nu-1)} - \langle r_\pi^2\rangle_{0_1^+}^{(N_\pi,N_\nu)}, \tag{4.57}$$

with

$$\bar{\alpha}_{\pi\nu} = -\alpha_{\pi\nu}, \tag{4.58}$$

and

$$\bar{\Delta}'\langle r_\pi^2\rangle^{(N_\pi,N_\nu)} = \langle r_\pi^2\rangle_{0_1^+}^{(N_\pi-1,N_\nu)} - \langle r_\pi^2\rangle_{0_1^+}^{(N_\pi,N_\nu)}, \tag{4.59}$$

with

$$\bar{\alpha}_{\pi\pi} = -\alpha_{\pi\pi}. \tag{4.60}$$

4.5.10 *Transfer operators*

Two-nucleon transfer operators can be directly expressed in terms of boson operators. Using the same notation as in Sect. 1.4.9, one can write

$$\begin{aligned}P_{+,\rho}^{(L)} &= \sum_\alpha p_{\rho,\alpha}^{(L)} b_{\rho,\alpha}^\dagger, \\ P_{-,\rho}^{(L)} &= \sum_\alpha p_{\rho,\alpha}^{(L)} b_{\rho,\alpha}.\end{aligned} \tag{4.61}$$

A more explicit form is

$$\begin{aligned}P_{+,\rho,m}^{(l)} &= p_{\rho,l} b_{\rho,l,m}^\dagger, \\ P_{-,\rho,m}^{(l)} &= p_{\rho,l} b_{\rho,l,m}.\end{aligned} \tag{4.62}$$

Introducing explicitly s and d bosons,

$$\begin{aligned}&P_{+,\pi,0}^{(0)} = p_{\pi,0} s_\pi^\dagger; \quad && P_{-,\pi,0}^{(0)} = p_{\pi,0} \tilde{s}_\pi; \\ &P_{+,\pi,\mu}^{(2)} = p_{\pi,2} d_{\pi,\mu}^\dagger; \quad && P_{-,\pi,\mu}^{(2)} = p_{\pi,2} \tilde{d}_{\pi,\mu}; \\ &P_{+,\nu,0}^{(0)} = p_{\nu,0} s_\nu^\dagger; \quad && P_{-,\nu,0}^{(0)} = p_{\nu,0} \tilde{s}_\nu; \\ &P_{+,\nu,\mu}^{(2)} = p_{\nu,2} d_{\nu,\mu}^\dagger; \quad && P_{-,\nu,\mu}^{(2)} = p_{\nu,2} \tilde{d}_{\nu,\mu}.\end{aligned} \tag{4.63}$$

For transfer operators, higher-order terms may be needed. A form often used is

$$
\begin{aligned}
P^{(0)}_{+,\rho,0} &= p_{\rho,0} s^{\dagger}_{\rho} \sqrt{(\Omega_{\rho} - N_{\rho} + \hat{n}_{d_{\rho}})},\\
P^{(0)}_{-,\rho,0} &= p_{\rho,0} \sqrt{(\Omega_{\rho} - N_{\rho} + \hat{n}_{d_{\rho}})}\tilde{s}_{\rho},\\
P^{(2)}_{+,\rho,\mu} &= p_{\rho,2} d^{\dagger}_{\rho,\mu} \sqrt{(\Omega_{\rho} - N_{\rho} + \hat{n}_{d_{\rho}})},\\
P^{(2)}_{-,\rho,\mu} &= p_{\rho,2} \sqrt{(\Omega_{\rho} - N_{\rho} + \hat{n}_{d_{\rho}})}\tilde{d}_{\rho,\mu}.
\end{aligned}
\tag{4.64}
$$

Because of the microscopic interpretation of the bosons, the coefficient Ω_{ρ} has now a precise physical interpretation. It represents the pair degeneracy of the shell. For example, for the shell between magic numbers 50 and 82, $\Omega = (82 - 50)/2 = 16$.

4.6 Transition densities

As discussed in Sect. 1.5, when analyzing electron scattering one must replace the coefficients α_L, β_L, γ_0 in Eq. (4.24) by appropriate functions of the radial variable r. This implies that the operators (4.24) must be rewritten as (Iachello, 1981)

$$
\begin{aligned}
T^{(E0)}_{0} &= \gamma_0(r) + \alpha^{(\pi)}_{0}(r)[s^{\dagger}_{\pi} \times \tilde{s}_{\pi}]^{(0)}_{0} + \beta^{(\pi)}_{0}(r)[d^{\dagger}_{\pi} \times \tilde{d}_{\pi}]^{(0)}_{0}\\
&\quad + \alpha^{(\nu)}_{0}(r)[s^{\dagger}_{\nu} \times \tilde{s}_{\nu}]^{(0)}_{0} + \beta^{(\nu)}_{0}(r)[d^{\dagger}_{\nu} \times \tilde{d}_{\nu}]^{(0)}_{0},\\
T^{(M1)}_{\mu} &= \beta^{(\pi)}_{1}(r)[d^{\dagger}_{\pi} \times \tilde{d}_{\pi}]^{(1)}_{\mu} + \beta^{(\nu)}_{1}(r)[d^{\dagger}_{\nu} \times \tilde{d}_{\nu}]^{(1)}_{\mu},\\
T^{(E2)}_{\mu} &= \alpha^{(\pi)}_{2}(r)[d^{\dagger}_{\pi} \times \tilde{s}_{\pi} + s^{\dagger}_{\pi} \times \tilde{d}_{\pi}]^{(2)}_{\mu} + \beta^{(\pi)}_{2}(r)[d^{\dagger}_{\pi} \times \tilde{d}_{\pi}]^{(2)}_{\mu}\\
&\quad + \alpha^{(\nu)}_{2}(r)[d^{\dagger}_{\nu} \times \tilde{s}_{\nu} + s^{\dagger}_{\nu} \times \tilde{d}_{\nu}]^{(2)}_{\mu} + \beta^{(\nu)}_{2}(r)[d^{\dagger}_{\nu} \times \tilde{d}_{\nu}]^{(2)}_{\mu},\\
T^{(M3)}_{\mu} &= \beta^{(\pi)}_{3}(r)[d^{\dagger}_{\pi} \times \tilde{d}_{\pi}]^{(3)}_{\mu} + \beta^{(\nu)}_{3}(r)[d^{\dagger}_{\nu} \times \tilde{d}_{\nu}]^{(3)}_{\mu},\\
T^{(E4)}_{\mu} &= \beta^{(\pi)}_{4}(r)[d^{\dagger}_{\pi} \times \tilde{d}_{\pi}]^{(4)}_{\mu} + \beta^{(\nu)}_{4}(r)[d^{\dagger}_{\nu} \times \tilde{d}_{\nu}]^{(4)}_{\mu}.
\end{aligned}
\tag{4.65}
$$

The transition densities are the matrix elements of these operators.

5
Algebras

5.1 Introduction

The algebraic structure of the interacting boson model-2 is at first sight a trivial extension of that of the interacting boson model-1. However, it turns out that if one wants to exploit the concept of dynamic symmetries introduced in Chapter 2, a much larger and richer variety occurs here.

Dynamic symmetries for coupled algebras have not been much used in the past. There are now three applications: (i) to coupled proton–neutron boson systems; (ii) to coupled bosons and fermion systems in odd–even nuclei (interacting boson–fermion model) (Iachello, 1980b); and (iii) to coupled rotovibrations in molecules (vibron model) (van Roosmalen *et al.*, 1983). Here we shall discuss the application to coupled proton–neutron boson systems. Since the technique for treating these cases is only now being developed, the results in this chapter will not be as complete as those in Chapter 2. An important concept in the treatment of dynamic symmetries for coupled systems is that of lattices of algebras. This will be also discussed briefly here. We shall assume in this chapter that the reader has already acquired a knowledge of algebraic structures of boson systems, as, for example, discussed in Chapter 2.

5.2 Boson algebras for coupled systems

Each set (π, ν) of 36 operators

$$\begin{aligned} G^{(\pi)}_{\alpha\beta} &= b^{\dagger}_{\pi,\alpha} b_{\pi,\beta}, \quad (\alpha, \beta = 1, \ldots, 6), \\ G^{(\nu)}_{\alpha\beta} &= b^{\dagger}_{\nu,\alpha} b_{\nu,\beta} \quad (\alpha, \beta = 1, \ldots, 6), \end{aligned} \tag{5.1}$$

generates the Lie algebra of U(6). The first obvious statement is that, taken together, the $36+36=72$ operators (5.1) generate the Lie algebra of the direct product $G = \mathrm{U}_\pi(6) \otimes \mathrm{U}_\nu(6)$. The main question then is how to reduce the algebra G to the rotation algebra, O(3), which we want always as a subalgebra, since nuclear states are characterized by a good value of the angular momentum. Since protons and neutrons are rotated simultaneously, the generators of O(3) are obtained by summing those of the two rotation algebras, $O_\pi(3)$ and $O_\nu(3)$

$$G_\mu^{(1)} = G_{\pi,\mu}^{(1)}(d,d) + G_{\nu,\mu}^{(1)}(d,d). \tag{5.2}$$

This corresponds to the familiar addition of angular momenta for protons and neutrons,

$$\hat{L} = \hat{L}_\pi + \hat{L}_\nu. \tag{5.3}$$

Since each of the U(6) algebras has a rich subalgebra structure, as discussed in Sect. 2, Eq. (2.25), there are a variety of ways in which the algebra $G = U_\pi(6) \otimes U_\nu(6)$ can be reduced to O(3). These are called lattices of algebras and we shall discuss here some in detail. We begin by considering the trivial case in which the only common algebra is that of O(3). This can be schematically written as

$$\begin{array}{ccccccc}
 & & U_\pi(5) \supset O_\pi(5) & & & & \\
 & \nearrow & & \searrow & & & \\
U_\pi(6) & \rightarrow & SU_\pi(3) & \rightarrow & O_\pi(3) & & \\
 & \searrow & & \nearrow & & \searrow & \\
 & & O_\pi(6) \supset O_\pi(5) & & & & \\
 & & & & & & O_{\pi+\nu}(3) \supset O_{\pi+\nu}(2), \\
 & & & & & \nearrow & \\
 & & U_\nu(5) \supset O_\nu(5) & & & & \\
 & \nearrow & & \searrow & & & \\
U_\nu(6) & \rightarrow & SU_\nu(3) & \rightarrow & O_\nu(3) & & \\
 & \searrow & & \nearrow & & & \\
 & & O_\nu(6) \supset O_\nu(5) & & & &
\end{array} \tag{5.4}$$

where we have added a subscript $\pi + \nu$ to O(3) in order to indicate that it is obtained by summing the generators of $O_\pi(3)$ and $O_\nu(3)$. From the practical point of view, this case is not particularly interesting since it does not produce anything new, although one of the chains in (5.4),

$$\begin{aligned}
U_\pi(6) \otimes U_\nu(6) &\supset U_\pi(5) \otimes U_\nu(5) \supset O_\pi(5) \otimes O_\nu(5) \\
&\supset O_\pi(3) \otimes O_\nu(3) \supset O_{\pi+\nu}(3) \supset O_{\pi+\nu}(2),
\end{aligned} \tag{5.5}$$

is used in the computer program NPBOS (Otsuka, 1977) which diagonalizes the Hamiltonian numerically.

There are two types of lattices of particular importance. The first is that in which the algebras of $U_\pi(6)$ and $U_\nu(6)$ are joined at the first step

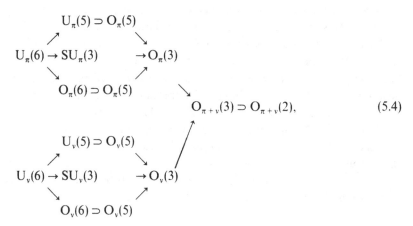

$$
\begin{array}{lllll}
\mathrm{U}_\pi(6) & & \mathrm{U}_{\pi+\nu}(5) \supset \mathrm{O}_{\pi+\nu}(5) & & (\mathrm{I}_1) \\
\searrow & \nearrow & & \searrow & \\
& \mathrm{U}_{\pi+\nu}(6) \rightarrow \mathrm{SU}_{\pi+\nu}(3) & & \rightarrow \mathrm{O}_{\pi+\nu}(3) \supset \mathrm{O}_{\pi+\nu}(2). & (\mathrm{II}_1) \\
\nearrow & \searrow & & \nearrow & \\
\mathrm{U}_\nu(6) & & \mathrm{O}_{\pi+\nu}(6) \supset \mathrm{O}_{\pi+\nu}(5) & & (\mathrm{III}_1)
\end{array}
\tag{5.6}
$$

The generators of the combined algebra $\mathrm{U}_{\pi+\nu}(6)$ are the sum of the generators of $\mathrm{U}_\pi(6)$ and $\mathrm{U}_\nu(6)$,

$$
G^{(k)}_{\pi+\nu}(l, l') = G^{(k)}_\pi(l, l') + G^{(k)}_\nu(l, l'), \quad (l, l' = 0, 2). \tag{5.7}
$$

Those of the subalgebras are obtained from (5.7) using the same procedure as described in Sect. 2.3.

A second type is obtained by joining the algebras at the second step,

$$
\begin{array}{lll}
\mathrm{U}_\pi(6) \supset \mathrm{U}_\pi(5) & & \\
\searrow & & \\
& \mathrm{U}_{\pi+\nu}(5) \supset \mathrm{O}_{\pi+\nu}(5) \supset \mathrm{O}_{\pi+\nu}(3) \supset \mathrm{O}_{\pi+\nu}(2), & (\mathrm{I}_2) \\
\nearrow & & \\
\mathrm{U}_\nu(6) \supset \mathrm{U}_\nu(5) & & \\
\mathrm{U}_\pi(6) \supset \mathrm{SU}_\pi(3) & & \\
\searrow & & \\
& \mathrm{SU}_{\pi+\nu}(3) \supset \mathrm{O}_{\pi+\nu}(3) \supset \mathrm{O}_{\pi+\nu}(2), & (\mathrm{II}_2) \\
\nearrow & & \\
\mathrm{U}_\nu(6) \supset \mathrm{SU}_\nu(3) & & \\
\mathrm{U}_\pi(6) \supset \mathrm{O}_\pi(6) & & \\
\searrow & & \\
& \mathrm{O}_{\pi+\nu}(6) \supset \mathrm{O}_{\pi+\nu}(5) \supset \mathrm{O}_{\pi+\nu}(3) \supset \mathrm{O}_{\pi+\nu}(2). & (\mathrm{III}_2) \\
\nearrow & & \\
\mathrm{U}_\nu(6) \supset \mathrm{O}_\nu(6) & &
\end{array}
\tag{5.8}
$$

Here an interesting new possibility arises, since the SU(3) algebra is generated by (2.15)–(2.16),

$$
G^{(1)}_\mu(d, d) = [d^\dagger \times \tilde{d}]^{(1)}_\mu,
$$

$$
\begin{aligned}
G^{(2)}_\mu(d, s) + G^{(2)}_\mu(s, d) \pm \tfrac{1}{2}\sqrt{7}\, G^{(2)}_\mu(d, d) & \\
= [d^\dagger \times \tilde{s} + s^\dagger \times \tilde{d}]^{(2)}_\mu \pm \tfrac{1}{2}\sqrt{7}\, [d^\dagger \times \tilde{d}]^{(2)}_\mu, &
\end{aligned}
\tag{5.9}
$$

with two possible signs ($\pm$) corresponding to positive ($+$) and negative

($-$) quadrupole moments. We shall denote the algebra with a negative sign SU(3), and that with a positive sign, $\overline{\mathrm{SU}(3)}$. When combining the two algebras $\mathrm{SU}_\pi(3)$ and $\mathrm{SU}_\nu(3)$, we can either have the same sign for protons and neutrons, or opposite signs. We call the combined algebra $\mathrm{SU}_{\pi+\nu}(3)$ in the former case and $\mathrm{SU}_{\pi+\nu}(3)^*$ in the latter case. Physically, these two situations correspond either to protons and neutrons both with prolate ($-$) or oblate ($+$) deformations, or to protons and neutrons with deformations of opposite sign. We shall therefore explicitly consider the chain

$$\begin{array}{lll} \mathrm{U}_\pi(6) \supset \mathrm{SU}_\pi(3) & & \\ & \searrow & \\ & & \mathrm{SU}_{\pi+\nu}(3)^* \supset \mathrm{O}_{\pi+\nu}(3) \supset \mathrm{O}_{\pi+\nu}(2). \quad \mathrm{(IV)} \\ & \nearrow & \\ \mathrm{U}_\nu(6) \supset \overline{\mathrm{SU}_\nu(3)} & & \end{array} \tag{5.10}$$

This peculiar property of the coupled algebras was first noted by Dieperink (Dieperink and Bijker, 1982).

The generators of the combined groups are obtained by simple addition as in (5.7), with the exception of $\mathrm{SU}_{\pi+\nu}(3)^*$ where particular care must be taken in choosing the appropriate signs.

5.3 Basis states for coupled systems

When one considers coupled systems with algebraic structure

$$G = G_1 \otimes G_2, \tag{5.11}$$

i.e. the direct product of two identical algebraic structures for systems 1 and 2, one has a straightforward way of constructing basis states. This is done through the Kronecker product of the representations $[\lambda_1]$ of G_1 and $[\lambda_2]$ of G_2, schematically indicated by

$$[\lambda] = [\lambda_1] \otimes [\lambda_2]. \tag{5.12}$$

Rules for taking this product are given in books on group theory (Hamermesh, 1962) and we shall make use of them extensively.

We begin with the simple example given by (5.5). The basis states in this chain are characterized by

$$|[N_\pi], n_{d\pi}, v_\pi, \tilde{n}_{\Delta\pi}, L_\pi; [N_\nu], n_{d\nu}, v_\nu, \tilde{n}_{\Delta\nu}, L_\nu; L, M_L\rangle. \tag{5.13}$$

Here, we have used the results of Sect. 2.4, to characterize the representations of the groups $\mathrm{U}_\pi(6) \supset \mathrm{U}_\pi(5) \supset \mathrm{O}_\pi(5) \supset \mathrm{O}_\pi(3)$ and $\mathrm{U}_\nu(6) \supset$

$U_\nu(5) \supset O_\nu(5) \supset O_\nu(3)$, where N_π and N_ν are the number of proton and neutron bosons. In order to find the values of the quantum number L we must use (5.12) as applied to

$$O_\pi(3) \otimes O_\nu(3) \supset O_{\pi+\nu}(3). \tag{5.14}$$

In this case, results for combined representations are trivial and are known from elementary methods. One has

$$|L_\pi - L_\nu| \leqslant L \leqslant |L_\pi + L_\nu|. \tag{5.15}$$

The values of M_L are then given as usual by

$$-L \leqslant M_L \leqslant +L. \tag{5.16}$$

As mentioned above, the basis states (5.13) are used in the numerical diagonalization of H.

5.3.1 F-spin basis

For studying dynamic symmetries, a more interesting basis is that provided by (5.6), which, for reasons which will become apparent below, we shall call the F-spin basis. In order to construct this basis, we need to construct the Kronecker product of two U(6) representations. This is relatively simple since both representations are totally symmetric, as they describe bosons. One finds

$$[N_\pi] \otimes [N_\nu] = \oplus \sum_{k=0}^{N_\nu} [N_\pi + N_\nu - k, k], \quad \text{if } N_\pi \geqslant N_\nu. \tag{5.17}$$

The sum extends to N_π if $N_\pi \leqslant N_\nu$. Equation (5.17) gives, for example,

$$[1] \otimes [1] = [2] \oplus [11]. \tag{5.18}$$

Thus, from the product of two totally-symmetric representations, one obtains Young tableaux, which in general are two-rowed,

$$\begin{array}{l} \overbrace{\square\,\square\ \cdots\ \square}^{N_1 = N_\pi + N_\nu - k} \\ \underbrace{\square\,\square}_{N_2 = k} \end{array} . \tag{5.19}$$

Instead of the two numbers $[N_1, N_2]$, one can use the quantities

$$N = N_1 + N_2,$$

$$F = \frac{N_1 - N_2}{2}, \tag{5.20}$$

Table 5.1. *Partial classification scheme for chain I_1*

$U_{\pi+\nu}(6)$	$U_{\pi+\nu}(5)$	$O_{\pi+\nu}(5)$	$O_{\pi+\nu}(3)$
$[N_1, N_2]$	(n_1, n_2)	(v_1, v_2)	L
[1, 1]	(1, 0)	(1, 0)	2
	(1, 1)	(1, 1)	3, 1
[2, 1]	(2, 0)	(2, 0)	4, 2
		(0, 0)	0
	(1, 0)	(1, 0)	2
	(2, 1)	(1, 0)	2
		(2, 1)	5, 4, 3, 2, 1
	(1, 1)	(1, 1)	3, 1
[3, 1]	(3, 0)	(3, 0)	6, 4, 3, 0
		(1, 0)	2
	(2, 0)	(2, 0)	4, 2
		(0, 0)	0
	(1, 0)	(1, 0)	2
	(3, 1)	(2, 0)	4, 2
		(3, 1)	$7, 6, 5^2, 4, 3^2, 2, 1$
		(1, 1)	3, 1
	(2, 1)	(1, 0)	2
		(2, 1)	5, 4, 3, 2, 1
	(1, 1)	(1, 1)	3, 1

to characterize the states, i.e. the total boson number $N = N_\pi + N_\nu$ and the value of the F-spin, hence the name given the basis.

The occurrence of states which are not totally symmetric is the new aspect brought in by the coupling of protons and neutrons. These states have been recently discovered (Bohle *et al.*, 1984), thus opening the way for a new type of nuclear spectroscopy.

(i) Chain I_1

In order to provide a complete classification scheme we now need to reduce representations of $U_{\pi+\nu}(6)$ to those of its subgroups. For totally-symmetric representations, which were the only ones occurring in the interacting boson model-1, the reduction was given in Sect. 2.4. Here, however, we need the reduction also for mixed-symmetry states. The rules to obtain this reduction are much more complex and are given in part by van Isacker, Frank and Sun (1984). Here we only list the quantum numbers and show in Table 5.1 a partial tabulation of the results. The quantum numbers needed to classify the states in this chain are

$$\left| \begin{array}{cccccccc} U_\pi(6)\otimes & U_\nu(6)\supset & U_{\pi+\nu}(6) & \supset U_{\pi+\nu}(5)\supset & O_{\pi+\nu}(5)\supset & & \\ \downarrow & \downarrow & \downarrow & \downarrow & \downarrow & & \\ [N_\pi], & [N_\nu], & [N_1,N_2], & (n_1,n_2), & (v_1,v_2), & \tilde{n}_{1\Delta}, & \tilde{n}_{2\Delta}, \\ O_{\pi+\nu}(3)\supset & O_{\pi+\nu}(2) & & & & & \\ \downarrow & \downarrow & & & & & \\ L, & M_L & & & & & \end{array} \right\rangle \tag{5.21}$$

where $\tilde{n}_{1\Delta}$, $\tilde{n}_{2\Delta}$ represents missing labels. Because of the complex structure of (5.21), we digress briefly here to discuss the question of how many labels are, in general, needed to classify uniquely basis states of a group G. This number can be found from Tables 2.2 and 2.3 as follows. Consider the number of integers needed to characterize the tensor representations of G, given in Table 2.3. This number is called the rank of the corresponding algebra $\mathscr{g}$, and denoted here by $\mathscr{r}$. Consider the number of generators of $\mathscr{g}$, given in Table 2.2, and denoted here by $\mathscr{n}$. The number of labels ℓ, needed to characterize uniquely the basis states of G is

$$\ell=\tfrac{1}{2}(\mathscr{n}+\mathscr{r}). \tag{5.22}$$

Equation (5.22) gives for SU(n),

$$\ell=(n^2+n-2)/2, \tag{5.23}$$

and for O(n), n = odd

$$\ell=(n^2-1)/4. \tag{5.24}$$

From (5.23) one has, for example, for SU(3), $\ell=5$, and for O(5), $\ell=6$. Thus while the reduction SU(3) ⊃ O(3) ⊃ O(2) requires the introduction of only one missing label, that of O(5) ⊃ O(3) ⊃ O(2) requires, in general, the introduction of two missing labels

$$\left| \begin{array}{ccc} O(5)\quad\supset & O(3)\supset & O(2) \\ \downarrow & \downarrow & \downarrow \\ (v_1,v_2)\tilde{n}_{1\Delta},\tilde{n}_{2\Delta}, & L, & M_L \end{array} \right\rangle. \tag{5.25}$$

For some special cases, the number of missing labels is reduced. For example, if the O(5) representations in (5.25) are totally symmetric, ($v_1=v$, $v_2=0$) as it is the case in the interacting boson model-1, only one missing label is needed, $(\tilde{n}_{1\Delta}=\tilde{n}_{\Delta}, \tilde{n}_{2\Delta}=0)$. For SU(6), Eq. (5.22) gives $\ell=20$. Thus 20 labels are needed, in general, to characterize the basis states of SU(6). However, one can show that, in the particular case of the interacting boson model-2 only 12 labels are actually needed to classify the basis states, as one can see directly from (5.13).

Returning now to (5.21), we note that the representations of $U_{\pi+\nu}(6)$, $U_{\pi+\nu}(5)$ and $O_{\pi+\nu}(5)$ are all two-rowed, i.e.

$$\begin{aligned} &[N_1, N_2] \equiv [N_1, N_2, 0, 0, 0, 0] \\ &(n_1, n_2) \equiv (n_1, n_2, 0, 0, 0) \\ &(v_1, v_2) \equiv (v_1, v_2). \end{aligned} \tag{5.26}$$

The values of (n_1, n_2) contained in the representation $[N, 0]$ of $U_{\pi+\nu}(6)$ are given by (2.29). For the representation $[N-1, 1]$ they are

$$\begin{aligned} (n_1, n_2) = (N-1, 0),\quad &(N-2, 0),\quad \ldots,\quad (1, 0); \\ (N-1, 1),\quad &(N-2, 1),\quad \ldots,\quad (1, 1). \end{aligned} \tag{5.27}$$

Similarly, the values of (v_1, v_2) contained in the representation $(n, 0)$ of $U_{\pi+\nu}(6)$ are given by (2.30). For the representation $(n-1, 1)$ they are

$$\begin{aligned} (v_1, v_2) = (n-2, 0),\quad (n-4, 0),\quad \ldots,\quad (2, 0) \text{ or } (1, 0);& \\ (n = \text{even or odd})& \\ (n-1, 1),\quad (n-3, 1),\quad \ldots,\quad (2, 1) \text{ or } (1, 1);& \\ (n = \text{odd or even}).& \end{aligned} \tag{5.28}$$

Finally, the values of L contained in a representation $(v, 0)$ are given by (2.34). Those contained in $(v-1, 1)$ are

$$\begin{aligned} L = 2v-1,\quad &2v-2,\quad \ldots,\quad 3; \\ v-1,\quad &v-2,\quad \ldots,\quad 1,\quad (v \geqslant 2) \\ v+1,\quad &v,\quad \ldots,\quad 5,\quad (v \geqslant 4) \\ v+2,\quad &v+1,\quad \ldots,\quad 7,\quad (v \geqslant 5) \\ \ldots.& \end{aligned} \tag{5.29}$$

These rules give the results shown in Table 5.1.

Table 5.2. *Partial classification scheme for chain* II_1

$U_{\pi+\nu}(6)$	$SU_{\pi+\nu}(3)$	$O_{\pi+\nu}(3)$
$[N_1, N_2]$	(λ, μ)	L
[1, 1]	(2, 1)	3, 2, 1
[2, 1]	(4, 1)	5, 4, 3, 2, 1
	(2, 2)	4, 3, 2^2, 0
	(1, 1)	2, 1
[3, 1]	(6, 1)	7, 6, 5, 4, 3, 2, 1
	(4, 2)	6, 5, 4^2, 3, 2^2, 0
	(3, 1)	4, 3, 2, 1
	(2, 3)	5, 4, 3^2, 2, 1
	(2, 0)	2, 0
	(1, 2)	3, 2, 1

(*ii*) *Chain* II_1

The quantum numbers needed to classify the states in this chain are

$$\left|\begin{array}{ccccc} U_\pi(6) & \otimes & U_\nu(6) & \supset U_{\pi+\nu}(6) & \supset SU_{\pi+\nu}(3) \supset \\ \downarrow & & \downarrow & \downarrow & \searrow \\ [N_\pi], & & [N_\nu], & [N_1, N_2], & \tilde{\chi}'_1, \tilde{\chi}'_2, \tilde{\chi}'_3, (\lambda,\mu), \tilde{\chi} \\ O_{\pi+\nu}(3) & \supset & O_{\pi+\nu}(2) & & \\ \downarrow & & \downarrow & & \\ L, & & M_L & & \end{array}\right\rangle, \tag{5.30}$$

where $\tilde{\chi}$ and $\tilde{\chi}'_1$, $\tilde{\chi}'_2$, $\tilde{\chi}_3$ are missing labels. The representations (λ, μ) contained in a representation $[N, 0]$ of U(6) are given by (2.41). Those contained in a representation $[N-1, 1]$ are

$$\begin{array}{llllll} (\lambda,\mu)=(\Gamma, 1), & (\Gamma-2,2), & (\Gamma-3,1), & (\Gamma-4,3), & (\Gamma-4,0), & (\Gamma-5,2), \\ \quad\Gamma\geq 1, & \Gamma\geq 4, & & \Gamma\geq 5, & & \\ \quad\ldots & & & & & \end{array} \tag{5.31}$$

with

$$\Gamma=2N-2, \quad 2N-8, \quad 2N-14, \quad \ldots. \tag{5.32}$$

The reduction from SU(3) to O(3) is the same as in (2.42). These rules give the results shown in Table 5.2.

(iii) Chain III_1

The quantum numbers needed to classify the states in this chain are

$$\left|\begin{array}{ccccccc} U_\pi(6) & \otimes & U_\nu(6) & \supset & U_{\pi+\nu}(6) & \supset & O_{\pi+\nu}(6) \supset \\ \downarrow & & \downarrow & & \downarrow & & \downarrow \\ [N_\pi], & & [N_\nu], & & [N_1, N_2], & & (\sigma_1, \sigma_2), \\ O_{\pi+\nu}(5) & \supset & O_{\pi+\nu}(3) & \supset & O_{\pi+\nu}(2) & & \\ \downarrow & & \searrow & & \downarrow & & \\ (\tau_1, \tau_2), & & \tilde{\nu}_{1\Delta}, \tilde{\nu}_{2\Delta}, L, & & M_L & & \end{array}\right\rangle \tag{5.33}$$

where $\tilde{\nu}_{1\Delta}, \tilde{\nu}_{2\Delta}$ represents missing labels. The representations of $O_{\pi+\nu}(6)$ and $O_{\pi+\nu}(5)$ are two-rowed

$$(\sigma_1, \sigma_2) \equiv (\sigma_1, \sigma_2, 0),$$
$$(\tau_1, \tau_2) \equiv (\tau_1, \tau_2). \tag{5.34}$$

The representations (σ_1, σ_2) contained in a representation $[N, 0]$ of $U_{\pi+\nu}(6)$ are given in (2.50). The representations (σ_1, σ_2) contained in $[N-1, 1]$ are given by

$$(\sigma_1, \sigma_2) = (N-2, 0), \quad (N-4, 0), \quad \ldots, \quad (2, 0) \quad \text{or} \quad (1, 0);$$
$$(N = \text{even or odd});$$
$$(N = 1, 1), \quad (N-3, 1), \quad \ldots, \quad (2, 1) \quad \text{or} \quad (1, 1);$$
$$(N = \text{odd or even}). \tag{5.35}$$

The representations (τ_1, τ_2) contained in a representation $(\sigma, 0)$ of $O_{\pi+\nu}(6)$ are given by (2.51). Those contained in $(\sigma-1, 1)$ are given by

$$(\tau_1, \tau_2) = (\sigma-1, 0), \quad (\sigma-2, 0), \quad \ldots, \quad (1, 0);$$
$$(\sigma-1, 1), \quad (\sigma-2, 1), \quad \ldots, \quad (1, 1). \tag{5.36}$$

The reduction from $O_{\pi+\nu}(5)$ to $O_{\pi+\nu}(3)$ is the same as in (5.29). These rules give the results shown in Table 5.3.

5.3.2 Coupled basis

If F-spin is not a good quantum number one may wish to couple protons and neutrons not at the level of U(6) but at a later stage. We shall consider here the chains (5.8) and (5.10), which we call the coupled basis.

Table 5.3. *Partial classification scheme for chain III_1*

$U_{\pi+\nu}(6)$	$O_{\pi+\nu}(6)$	$O_{\pi+\nu}(5)$	$O_{\pi+\nu}(3)$
$[N_1, N_2]$	(σ_1, σ_2)	(τ_1, τ_2)	L
[1, 1]	(1, 1)	(1, 0)	2
		(1, 1)	3, 1
[2, 1]	(1, 0)	(1, 0)	2
		(0, 0)	0
	(2, 1)	(2, 0)	4, 2
		(1, 0)	2
		(2, 1)	5, 4, 3, 2, 1
		(1, 1)	3, 1
[3, 1]	(2, 0)	(2, 0)	4, 2
		(1, 0)	2
		(0, 0)	0
	(3, 1)	(3, 0)	6, 4, 3, 0
		(2, 0)	4, 2
		(1, 0)	2
		(3, 1)	7, 6, 5^2, 4, 3^2, 2, 1
		(2, 1)	5, 4, 3, 2, 1
		(1, 1)	3, 1
	(1, 1)	(1, 0)	2
		(1, 1)	3, 1

(*i*) *Chain I_2*

For this basis we need to construct products of U(5) representations. Since both representations are totally symmetric, their product is given by the same rule (5.17), i.e.

$$(n_{d_\pi}) \otimes (n_{d_\nu}) = \oplus \sum_{k=0}^{n_{d_\nu}} [n_{d_\pi} + n_{d_\nu} - k, k], \quad \text{if } n_{d_\pi} \geqslant n_{d_\nu}, \tag{5.37}$$

with the sum extending to n_{d_π} if $n_{d_\pi} \leqslant n_{d_\nu}$. The representations of the combined $U_{\pi+\nu}(5)$ group are thus still two-rowed, (n_1, n_2). From there on, the reduction of the representations of U(5) into those of its subgroups proceeds as in Sect. 5.3.1. The complete classification scheme for this chain is thus

$$\left|\begin{array}{cccccccc} U_\pi(6) \otimes & U_\nu(6) \supset & U_\pi(5) \otimes & U_\nu(5) \supset & U_{\pi+\nu}(5) \supset & O_{\pi+\nu}(5) & & \\ \downarrow & \downarrow & \downarrow & \downarrow & \downarrow & \downarrow & & \\ [N_\pi], & [N_\nu], & (n_{d_\pi}), & (n_{d_\nu}), & (n_1, n_2), & (v_1, v_2), & \tilde{n}_{1\Delta}, \tilde{n}_{2\Delta}, & \\ O_{\pi+\nu}(3) \supset & O_{\pi+\nu}(2) & & & & & & \\ \downarrow & \downarrow & & & & & & \\ L, & M_L & & & & & & \end{array}\right\rangle. \tag{5.38}$$

(ii) Chains II_2 and IV

We treat these two chains together since from the formal point of view they are identical. Here we need to construct products of representations of SU(3)

$$(\lambda_\pi, \mu_\pi) \otimes (\lambda_\nu, \mu_\nu) = \oplus \sum (\lambda, \mu). \tag{5.39}$$

These are not so simple as the previous ones, since the representations of SU(3) are not, in general, totally symmetric. The most straightforward way to construct the products is to return to the quantum numbers (f_1, f_2) specifying the Young tableau, Eq. (2.40), and then using the multiplication rules of Young tableaux (Hamermesh, 1962).

In the same way in which there is a difference between the generators Q and $\bar{Q}$ of $\overline{\mathrm{SU}(3)}$, there appears here also a difference in the corresponding representations. This arises because there are in SU(3) two conjugate representations (2, 0) and (0, 2). The states in SU(3) are built out of (2, 0) while those of $\overline{\mathrm{SU}(3)}$ are built out of (0, 2), or vice-versa. When taking tensor products to form representations of $\mathrm{SU}_{\pi+\nu}(3)$, this causes major differences since the representations (2, 0) and (0, 2) have different Young tableaux. For example, the product (2, 0) ⊗ (2, 0) yields

$$\begin{array}{ccccccccc} \square\square & \otimes & \square\square & = & \square\square\square\square & \oplus & \begin{array}{l}\square\square\square\\ \square\end{array} & \oplus & \begin{array}{l}\square\square\\ \square\square\end{array} \\ \downarrow & & \downarrow & & \downarrow & & \downarrow & & \downarrow \\ (2,0) & \otimes & (2,0) & = & (4,0) & \oplus & (2,1) & \oplus & (0,2) \\ \downarrow & & \downarrow & & \downarrow & & \downarrow & & \downarrow \\ L=2,0 & & L=2,0 & & L=4,2,0 & & L=3,2,1 & & L=2,0, \end{array} \tag{5.40}$$

while the product (2, 0) ⊗ (0, 2) yields

$$\begin{array}{ccccccccc} \square\square & \otimes & \begin{array}{l}\square\square\\ \square\square\end{array} & = & \begin{array}{l}\square\square\square\square\\ \square\square\end{array} & \oplus & \begin{array}{l}\square\square\square\\ \square\square\\ \square\end{array} & \oplus & \begin{array}{l}\square\square\\ \square\square\\ \square\square\end{array} \\ \downarrow & & \downarrow & & \downarrow & & \downarrow & & \downarrow \\ (2,0) & \otimes & (0,2) & = & (2,2) & \oplus & (1,1) & \oplus & (0,0) \\ \downarrow & & \downarrow & & \downarrow & & \downarrow & & \downarrow \\ L=2,0 & & L=2,0 & & L=4,3,2^2,0 & & L=2,1 & & L=0. \end{array} \tag{5.41}$$

Thus the $SU_{\pi+\nu}(3)$ representations appearing in $SU_\pi(3) \otimes \overline{SU_\nu(3)}$ are different from those appearing in $SU_\pi(3) \otimes SU_\nu(3)$ and for this reason they are denoted by $SU_{\pi+\nu}(3)^*$. The further reduction from $SU_{\pi+\nu}(3)$ or $SU_{\pi+\nu}(3)^*$ to their subalgebras is as in Sect. 5.3.1. The quantum numbers characterizing the chains II_2 and IV are thus

$$\left| \begin{array}{ccccccccc} U_\pi(6) & \otimes & U_\nu(6) & \supset & SU_\pi(3) & \otimes & SU_\nu(3) & \supset & SU_{\pi+\nu}(3) \supset \\ \downarrow & & \downarrow & & \downarrow & & \downarrow & & \searrow \\ [N_\pi], & & [N_\nu], & & (\lambda_\pi, \mu_\pi), & & (\lambda_\nu, \mu_\nu), & & \tilde{\chi}'_1, \tilde{\chi}'_2, \tilde{\chi}'_3, \quad (\lambda, \mu), \tilde{\chi} \\ O_{\pi+\nu}(3) & \supset & O_{\pi+\nu}(2) & & & & & & \\ \downarrow & & \downarrow & & & & & & \\ L, & & M_L & & & & & & \end{array} \right\rangle \tag{5.42}$$

and

$$\left| \begin{array}{ccccccccc} U_\pi(6) & \otimes & U_\nu(6) & \supset & SU_\pi(3) & \otimes & \overline{SU_\nu(3)} & \supset & SU_{\pi+\nu}(3)^* \supset \\ \downarrow & & \downarrow & & \downarrow & & \downarrow & & \searrow \\ [N_\pi], & & [N_\nu], & & (\lambda_\pi, \mu_\pi), & & (\lambda_\nu, \mu_\nu), & & \tilde{\chi}'_1, \tilde{\chi}'_2, \tilde{\chi}'_3, \quad (\lambda, \mu), \tilde{\chi} \\ O_{\pi+\nu}(3) & \supset & O_{\pi+\nu}(2) & & & & & & \\ \downarrow & & \downarrow & & & & & & \\ L, & & M_L & & & & & & \end{array} \right\rangle . \tag{5.43}$$

The representations of $\overline{SU_\nu(3)}$ contained in the representation $[N_\nu]$ of $U_\nu(6)$ are given by (2.41) with λ and μ interchanged.

(iii) Chain III_2

For this chain one needs to construct products of O(6) representations. The multiplication rules for orthogonal groups are slightly more complex than those for unitary groups. For example, one has

$$(\sigma_\pi, 0, 0) \otimes (1, 0, 0) = (\sigma_\pi + 1, 0, 0) \oplus (\sigma_\pi, 1, 0) \oplus (\sigma_\pi - 1, 0, 0). \tag{5.44}$$

Using these rules, one can construct the classification scheme

$$\left|\begin{array}{ccccccc} \mathrm{U}_\pi(6)\otimes\mathrm{U}_\nu(6) & \supset & \mathrm{O}_\pi(6)\otimes\mathrm{O}_\nu(6) & \supset & \mathrm{O}_{\pi+\nu}(6) & \supset & \mathrm{O}_{\pi+\nu}(5) \supset \\ \downarrow\quad\downarrow & & \downarrow\quad\downarrow & & \downarrow & & \downarrow \\ [N_\pi],\ [N_\nu], & & (\sigma_\pi),\ (\sigma_\nu), & & (\sigma_1,\sigma_2), & & (\tau_1,\tau_2),\quad \tilde{\nu}_{1\Delta},\tilde{\nu}_{2\Delta} \\ \mathrm{O}_{\pi+\nu}(3) \supset \mathrm{O}_{\pi+\nu}(2) & & & & & & \\ \downarrow \qquad\quad \downarrow & & & & & & \\ L, \qquad\quad M_L & & & & & & \end{array}\right\rangle. \tag{5.45}$$

5.4 Dynamic symmetries

The construction of dynamic symmetries here is done in the same way as in Part I, by expanding the Hamiltonian in terms of Casimir invariants. For the combined $\pi+\nu$ groups, the Casimir invariants are still given by Table 2.7 with the generators G replaced by the corresponding sum $G_\pi+G_\nu$. An operator which has a special importance is the Majorana operator, $\hat{M}_{\pi\nu}$, defined in (4.43). This operator is related to the quadratic Casimir operator of $\mathrm{U}_{\pi+\nu}(6)$ by

$$\hat{M}_{\pi\nu}=\tfrac{1}{2}(\hat{N}(\hat{N}+5)-\mathscr{C}_2(\mathrm{U}_{\pi+\nu}6)). \tag{5.46}$$

5.4.1 *Energy eigenvalues for F-spin chains*

(i) Chain I_1

The Hamiltonian appropriate to this chain can be written as

$$\begin{aligned} H^{(\mathrm{I}_1)}=E_0&+a\mathscr{C}_1(\mathrm{U}_{\pi+\nu}6)+a'\mathscr{C}_2(\mathrm{U}_{\pi+\nu}6)+\varepsilon\mathscr{C}_1(\mathrm{U}_{\pi+\nu}5)\\ &+\alpha\mathscr{C}_2(\mathrm{U}_{\pi+\nu}5)+\beta\mathscr{C}_2(\mathrm{O}_{\pi+\nu}5)+\gamma\mathscr{C}_2(\mathrm{O}_{\pi+\nu}3), \end{aligned} \tag{5.47}$$

where E_0 is a quadratic function of N_π and N_ν. This Hamiltonian is diagonal in the basis (5.21). Its eigenvalues can be found by taking the expectation value of $H^{(\mathrm{I}_1)}$ in (5.21) and using Table 2.8. One has

$$\begin{aligned} &E^{(\mathrm{I}_1)}([N_\pi],[N_\nu],[N_1,N_2],(n_1,n_2),(v_1,v_2),\tilde{n}_{1\Delta},\tilde{n}_{2\Delta},L,M_L)\\ &\quad=E_0+a(N_1+N_2)+a'[N_1(N_1+5)+N_2(N_2+3)]+\varepsilon(n_1+n_2)\\ &\qquad+\alpha[n_1(n_1+4)+n_2(n_2+2)]+\beta 2[v_1(v_1+3)+v_2(v_2+1)]+\gamma 2L(L+1). \end{aligned} \tag{5.48}$$

The spectrum of states corresponding to (5.48) is shown in Fig. 5.1.

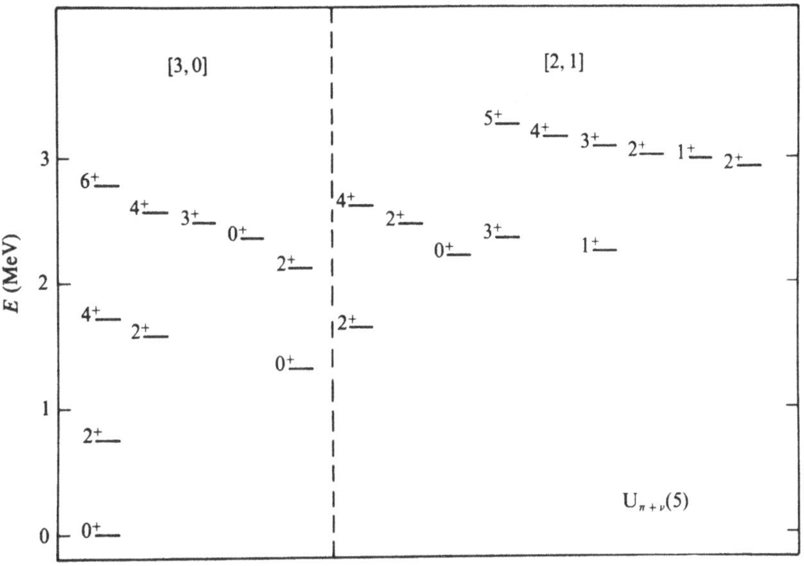

Fig. 5.1. A typical spectrum with $U_{\pi+\nu}(6) \supset U_{\pi+\nu}(5)$ symmetry and $N_\pi = 2$, $N_\nu = 1$.

(*ii*) *Chain* II_1

For this chain one may consider

$$H^{(\mathrm{II}_1)} = E_0 + a\mathscr{C}_1(\mathrm{U}_{\pi+\nu}6) + a'\mathscr{C}_2(\mathrm{U}_{\pi+\nu}6) + \delta\mathscr{C}_2(\mathrm{SU}_{\pi+\nu}3) + \gamma\mathscr{C}_2(\mathrm{O}_{\pi+\nu}3). \tag{5.49}$$

This Hamiltonian is diagonal in the basis (5.30) with eigenvalues

$$E^{(\mathrm{II}_1)}([N_\pi], [N_\nu], [N_1, N_2], \tilde{\chi}'_1, \tilde{\chi}'_2, \tilde{\chi}'_3, (\lambda, \mu), \tilde{\chi}, L, M_L) = E_0 + a(N_1 + N_2) + a'[N_1(N_1+5) + N_2(N_2+3)] + \delta\tfrac{2}{3}(\lambda^2 + \mu^2 + \lambda\mu + 3\lambda + 3\mu) + \gamma 2L(L+1). \tag{5.50}$$

The spectrum of states corresponding to (5.50) is shown in Fig. 5.2.

(*iii*) *Chain* III_1

For this chain we consider

$$H^{(\mathrm{III}_1)} = E_0 + a\mathscr{C}_1(\mathrm{U}_{\pi+\nu}6) + a'\mathscr{C}_2(\mathrm{U}_{\pi+\nu}6) + \beta\mathscr{C}_2(\mathrm{O}_{\pi+\nu}5) + \gamma\mathscr{C}_2(\mathrm{O}_{\pi+\nu}3) + \eta\mathscr{C}_2(\mathrm{O}_{\pi+\nu}6). \tag{5.51}$$

This Hamiltonian is diagonal in the basis (5.33) with eigenvalues

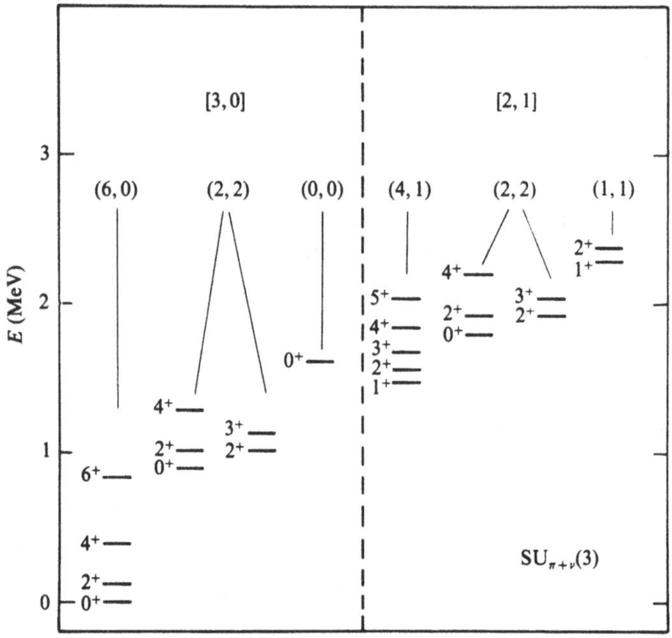

Fig. 5.2. A typical spectrum with $U_{\pi+\nu}(6) \supset SU_{\pi+\nu}(3)$ symmetry and $N_\pi = 2$, $N_\nu = 1$.

$$E^{(\mathrm{III}_1)}([N_\pi], [N_\nu], [N_1, N_2], (\sigma_1, \sigma_2), (\tau_1, \tau_2), \tilde{\nu}_{1\Delta}, \tilde{\nu}_{2\Delta}, L, M_L)$$
$$= E_0 + a(N_1 + N_2) + a'[N_1(N_1+5) + N_2(N_2+3)] + \beta 2[\tau_1(\tau_1+3) + \tau_2(\tau_2+1)]$$
$$+ \gamma 2L(L+1) + \eta 2[\sigma_1(\sigma_1+4) + \sigma_2(\sigma_2+2)]. \tag{5.52}$$

The spectrum of states corresponding to (5.52) is shown in Fig. 5.3. Since for all chains

$$N_1 + N_2 = N_\pi + N_\nu = N, \tag{5.53}$$

the term $a(N_1 + N_2)$ does not contribute to excitation energies. Furthermore, the contribution coming from $\mathscr{C}_2(U_{\pi+\nu}6)$ can be rewritten, if one wishes to do so, in terms of the Majorana operator $\hat{M}_{\pi\nu}$, (5.46). The eigenvalue of this operator, diagonal in all three chains, I_1, II_1 and III_1, can also be written in terms of N and of the F-spin value F as

$$\langle M_{\pi\nu} \rangle = (\tfrac{1}{2}N - F)(\tfrac{1}{2}N + F + 1). \tag{5.54}$$

The number of parameters needed to describe completely the three chains of the interacting boson model-2 are the same as the corresponding chains

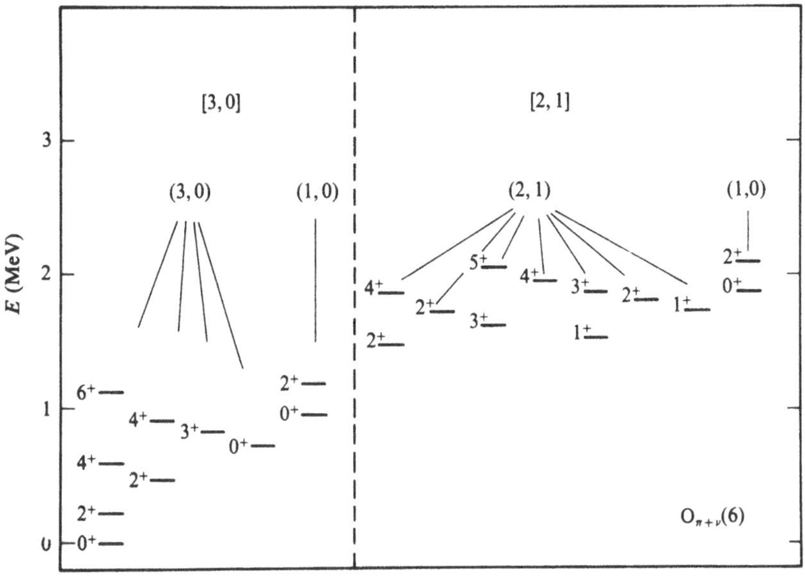

Fig. 5.3. A typical spectrum with $U_{\pi+\nu}(6) \supset O_{\pi+\nu}(6)$ symmetry and $N_\pi = 2$, $N_\nu = 1$.

of the interacting boson model-1 plus one. This extra parameter is either the coefficient a' in (5.47), (5.49), (5.51), or the strength of the Majorana interaction λ'. This parameter sets the location of states with $F < F_{\max} = N/2$ (asymmetric states).

5.4.2 *Energy eigenvalues for coupled chains*

(i) Chain I_2

For this chain, one may consider

$$H^{(I_2)} = E_0 + a_\pi \mathscr{C}_1(U_\pi 5) + a_\nu \mathscr{C}_1(U_\nu 5) + a'_\pi \mathscr{C}_2(U_\pi 5) + a'_\nu \mathscr{C}_2(U_\nu 5)$$
$$+ \varepsilon \mathscr{C}_1(U_{\pi+\nu}5) + \alpha \mathscr{C}_2(U_{\pi+\nu}5) + \beta \mathscr{C}_2(O_{\pi+\nu}5) + \gamma \mathscr{C}_2(O_{\pi+\nu}3), \tag{5.55}$$

with eigenvalues

$$E^{(I_2)}([N_\pi], [N_\nu], n_{d_\pi}, n_{d_\nu}, (n_1, n_2), (v_1, v_2), \tilde{n}_{1\Delta}, \tilde{n}_{2\Delta}, L, M_L)$$
$$= E_0 + a_\pi n_{d_\pi} + a_\nu n_{d_\nu} + a'_\pi [n_{d_\pi}(n_{d_\pi} + 4)] + a'_\nu [n_{d_\nu}(n_{d_\nu} + 4)]$$
$$+ \varepsilon(n_1 + n_2) + \alpha[n_1(n_1 + 4) + n_2(n_2 + 2)]$$
$$+ \beta 2[v_1(v_1 + 3) + v_2(v_2 + 1)] + \gamma 2L(L + 1). \tag{5.56}$$

The spectrum corresponding to (5.56) is shown in Fig. 5.4.

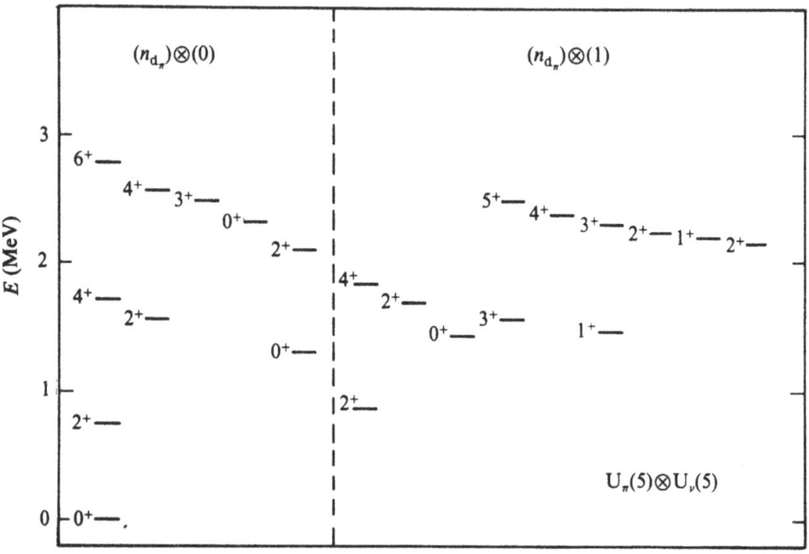

Fig. 5.4. A typical spectrum with $U_\pi(5)\otimes U_\nu(5)$ symmetry and $N_\pi=2$, $N_\nu=1$.

(*ii*) *Chains* II_2 *and* IV

For these chains one may consider

$$H^{(II_2)}=E_0+\delta_\pi\mathscr{C}_2(SU_\pi 3)+\delta_\nu\mathscr{C}_2(SU_\nu 3)+\delta\mathscr{C}_2(SU_{\pi+\nu}3)+\gamma\mathscr{C}_2(O_{\pi+\nu}3) \quad (5.57)$$

and

$$H^{(IV)}=E_0+\delta_\pi\mathscr{C}_2(SU_\pi 3)+\delta_\nu\mathscr{C}_2(\overline{SU_\nu 3})+\delta\mathscr{C}_2(SU_{\pi+\nu}3^*)+\gamma\mathscr{C}_2(O_{\pi+\nu}3). \quad (5.58)$$

Since the eigenvalues of the Casimir operators of both chains have similar expressions, both chains yield

$$\begin{aligned}E^{(II_2)-(IV)}&([N_\pi],[N_\nu],(\lambda_\pi,\mu_\pi),(\lambda_\nu,\mu_\nu),\tilde{\chi}'_1,\tilde{\chi}'_2,\tilde{\chi}'_3,(\lambda,\mu),\tilde{\chi},L,M_L)\\&=E_0+\delta_\pi\tfrac{2}{3}(\lambda_\pi^2+\mu_\pi^2+\lambda_\pi\mu_\pi+3\lambda_\pi+3\mu_\pi)\\&\quad+\delta_\nu\tfrac{2}{3}(\lambda_\nu^2+\mu_\nu^2+\lambda_\nu\mu_\nu+3\lambda_\nu+3\mu_\nu)\\&\quad+\delta\tfrac{2}{3}(\lambda^2+\mu^2+\lambda\mu+3\lambda+3\mu)+\gamma 2L(L+1).\end{aligned} \quad (5.59)$$

However, they give rise to different spectra since the appropriate SU(3) representations are different, as shown in Figs. 5.5 and 5.6.

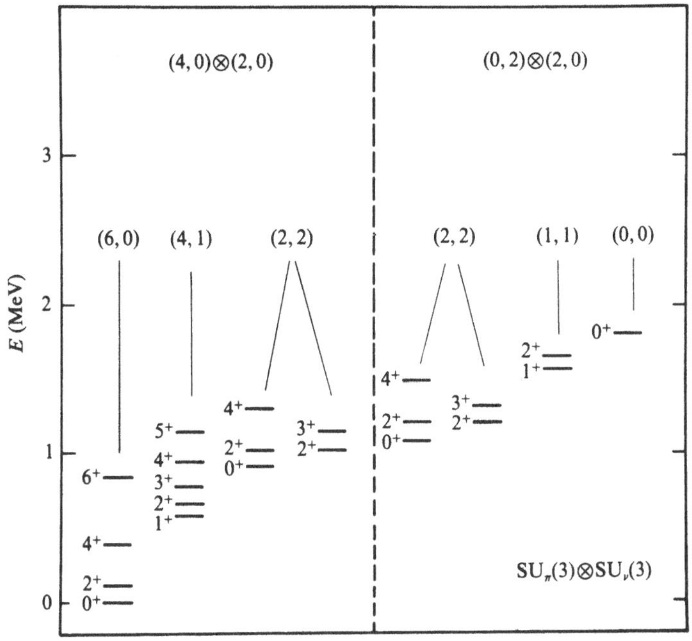

Fig. 5.5. A typical spectrum with $SU_\pi(3) \otimes SU_\nu(3)$ symmetry and $N_\pi = 2$, $N_\nu = 1$.

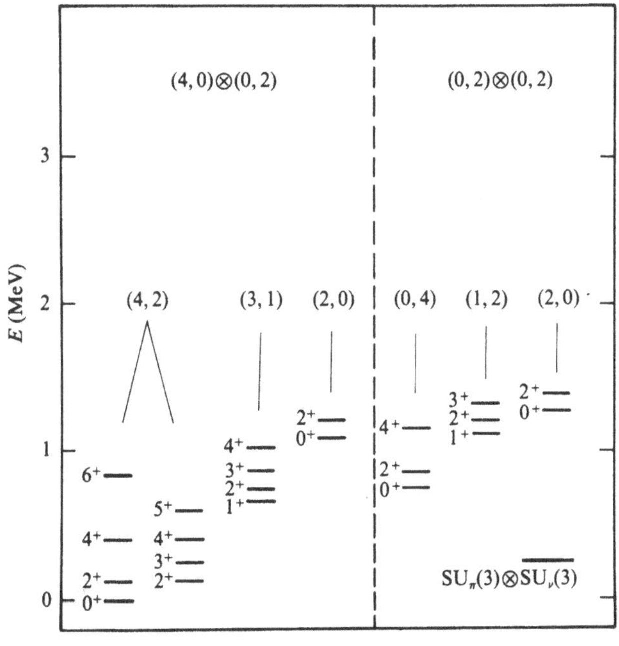

Fig. 5.6. A typical spectrum with $SU_\pi(3) \otimes \overline{SU_\nu(3)}$ symmetry and $N_\pi = 2$, $N_\nu = 1$.

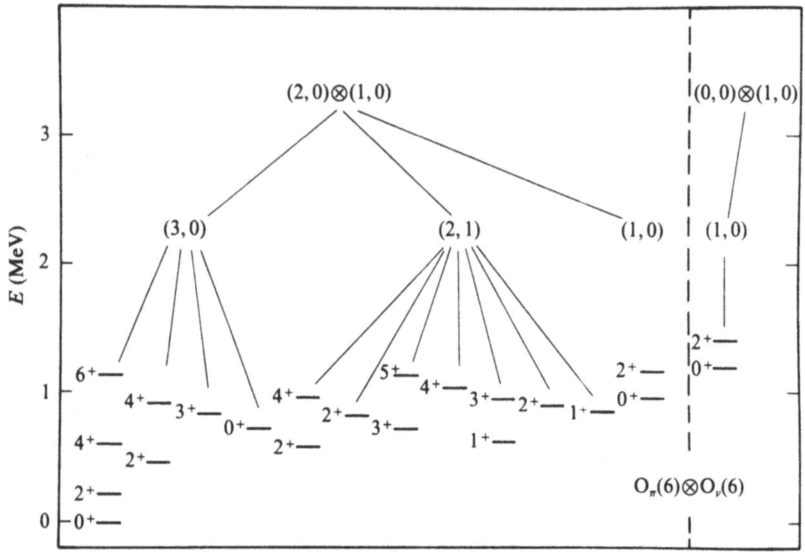

Fig. 5.7. A typical spectrum with $O_\pi(6) \otimes O_\nu(6)$ symmetry and $N_\pi = 2$, $N_\nu = 1$.

(*iii*) *Chain* III_2

Here one considers

$$H^{(\mathrm{III}_2)} = E_0 + \eta_\pi \mathscr{C}_2(\mathrm{O}_\pi 6) + \eta_\nu \mathscr{C}_2(\mathrm{O}_\nu 6) + \eta \mathscr{C}_2(\mathrm{O}_{\pi+\nu} 6)$$
$$+ \beta \mathscr{C}_2(\mathrm{O}_{\pi+\nu} 5) + \gamma \mathscr{C}_2(\mathrm{O}_{\pi+\nu} 3). \tag{5.60}$$

This operator has eigenvalues

$$E^{(\mathrm{III}_2)}([N_\pi], [N_\nu], \sigma_\pi, \sigma_\nu, (\sigma_1, \sigma_2), (\tau_1, \tau_2), \tilde{\nu}_{1\Delta}, \tilde{\nu}_{2\Delta}, L, M_L)$$
$$= E_0 + \eta_\pi 2[\sigma_\pi(\sigma_\pi + 4)] + \eta_\nu 2[\sigma_\nu(\sigma_\nu + 4)] + \eta 2[\sigma_1(\sigma_1 + 4) + \sigma_2(\sigma_2 + 2)]$$
$$+ \beta 2[\tau_1(\tau_1 + 3) + \tau_2(\tau_2 + 1)] + \gamma 2L(L + 1). \tag{5.61}$$

The corresponding spectrum is shown in Fig. 5.7.

It should be noted that the Majorana operator is not diagonal in the chains I_2, II_2, III_2 and IV.

5.5 Electromagnetic transitions and moments

Explicit expressions for electromagnetic transition rates and moments are more difficult to evaluate here than in the corresponding cases in the interacting boson model-1, and have been evaluated only recently (Scholten *et al.*, 1985a; van Isacker *et al.*, 1986). In quoting the

Table 5.4. *Notation of states for chain* I_1 ($\tilde{n}_{1\Delta}=\tilde{n}_\Delta$, $\tilde{n}_{2\Delta}=0$)

$\|L_i^+\rangle$	=	$\|[N_\pi]\otimes[N_\nu]$;	$[N_1,N_2]$,	(n_1,n_2),	(v_1,v_2),	$\tilde{n}_\Delta$,	$L\rangle$
$\|0_1^+\rangle$	=	$\|[N_\pi]\otimes[N_\nu]$;	$[N,0]$,	(0,0),	(0,0),	0,	$0\rangle$
$\|2_1^+\rangle$	=	$\|[N_\pi]\otimes[N_\nu]$;	$[N,0]$,	(1,0),	(1,0),	0,	$2\rangle$
$\|2_2^+\rangle$	=	$\|[N_\pi]\otimes[N_\nu]$;	$[N,0]$	(2,0),	(2,0),	0,	$2\rangle$
$\|2_3^+\rangle$	=	$\|[N_\pi]\otimes[N_\nu]$;	$[N,0]$	(3,0),	(1,0),	0,	$2\rangle$
$\|3_1^+\rangle$	=	$\|[N_\pi]\otimes[N_\nu]$;	$[N,0]$	(3,0),	(3,0),	0,	$3\rangle$
$\|1_1^+\rangle$	=	$\|[N_\pi]\otimes[N_\nu]$;	$[N-1,1]$,	(1,1),	(1,1),	0,	$1\rangle$
$\|2_4^+\rangle$	=	$\|[N_\pi]\otimes[N_\nu]$;	$[N-1,1]$,	(1,0),	(1,0),	0,	$2\rangle$

Table 5.5. *Notation of states for chain* II_1 ($\tilde{\chi}'_1=\tilde{\chi}'_2=\tilde{\chi}'_3=0$)

$\|L_i^+\rangle$	=	$\|[N_\pi]\otimes[N_\nu]$;	$[N_1,N_2]$,	(λ,μ),	$\tilde{\chi}$,	$L\rangle$
$\|0_1^+\rangle$	=	$\|[N_\pi]\otimes[N_\nu]$;	$[N,0]$,	$(2N,0)$,	0,	$0\rangle$
$\|0_\beta^+\rangle$	=	$\|[N_\pi]\otimes[N_\nu]$;	$[N,0]$,	$(2N-4,2)$,	0,	$0\rangle$
$\|2_1^+\rangle$	=	$\|[N_\pi]\otimes[N_\nu]$;	$[N,0]$,	$(2N,0)$,	0,	$2\rangle$
$\|2_\beta^+\rangle$	=	$\|[N_\pi]\otimes[N_\nu]$;	$[N,0]$,	$(2N-4,2)$,	0,	$2\rangle$
$\|2_\gamma^+\rangle$	=	$\|[N_\pi]\otimes[N_\nu]$;	$[N,0]$,	$(2N-4,2)$,	2,	$2\rangle$
$\|3_\gamma^+\rangle$	=	$\|[N_\pi]\otimes[N_\nu]$;	$[N,0]$,	$(2N-4,2)$,	2,	$3\rangle$
$\|1_1^+\rangle$	=	$\|[N_\pi]\otimes[N_\nu]$;	$[N-1,1]$,	$(2N-2,1)$,	1,	$1\rangle$
$\|2_4^+\rangle$	=	$\|[N_\pi]\otimes[N_\nu]$;	$[N-1,1]$,	$(2N-2,1)$,	1,	$2\rangle$
$\|3_2^+\rangle$	=	$\|[N_\pi]\otimes[N_\nu]$;	$[N-1,1]$,	$(2N-2,1)$,	1,	$3\rangle$
$\|3_3^+\rangle$	=	$\|[N_\pi]\otimes[N_\nu]$;	$[N-1,1]$,	$(2N-4,2)$,	2,	$3\rangle$

appropriate values, we shall use a compact notation, as follows:

(i) Chain I_1: states in this chain will be labeled in this section as in Table 5.4.

(ii) Chain II_1: the notation here will be as in Table 5.5.

(iii) Chain III_1: the notation here will be as in Table 5.6.

Because of their experimental accessibility, we begin our discussion from M1 transitions and moments.

5.5.1 *Magnetic dipole operators (M1)*

The magnetic dipole (M1) transition operator was written in (4.41) as

$$T^{(M1)}=\sqrt{\left(\frac{3}{4\pi}\right)}(g_\pi\hat{L}_\pi+g_\nu\hat{L}_\nu), \tag{5.62}$$

where g_π and g_ν are the proton and neutron boson g-factors. A major property of the interacting boson model-2 is the occurrence of 1^+ states. These states are missing in the interacting boson model-1 (see Chapter 2), since they belong to representations which are not fully symmetric. These states had not been discovered until recently. They have, however, now been found in several nuclei (Bohle *et al.*, 1984), in the region of excitation energy, $E_x \approx 3$ MeV. Matrix elements between the ground state, denoted here by 0_1^+, and the first excited 1^+ state, denoted here by 1_1^+, have been evaluated explicitly for the chains I_1, II_1, III_1 and IV by Scholten *et al.* (1985a).

The technique that has been used in this evaluation is as follows. For each chain one begins by decoupling protons from neutrons. For complex group chains such as (5.21), this is, in principle, a very difficult problem. However, a drastic simplification occurs through the use of Racah's factorization lemma (Wybourne, 1974), which states that the coupling coefficients can be factorized for each step. For example, the coupling in (5.21) can be written as the product of three coefficients, called isoscalar factors (IF) respectively associated with the reduction U(6) ⊃ U(5), U(5) ⊃ O(5) and O(5) ⊃ O(3) and an overall Clebsch–Gordan coefficient (CGC) associated with the group O(3), i.e.

$$
\begin{aligned}
&|[N_\pi],[N_\nu],[N_1,N_2],(n_1,n_2),(v_1,v_2),\tilde{n}_{1\Delta},\tilde{n}_{2\Delta},L,M_L\rangle \\
&\quad=\sum_{\{\pi\},\{\nu\}}\left\langle\begin{matrix}[N_\pi] & [N_\nu] \\ (n_{d_\pi}) & (n_{d_\nu})\end{matrix}\,\middle|\,\begin{matrix}[N_1,N_2] \\ (n_1,n_2)\end{matrix}\right\rangle\left\langle\begin{matrix}(n_{d_\pi}) & (n_{d_\nu}) \\ (v_\pi) & (v_\nu)\end{matrix}\,\middle|\,\begin{matrix}(n_1,n_2) \\ (v_1,v_2)\end{matrix}\right\rangle \\
&\qquad\times\left\langle\begin{matrix}(v_\pi) & (v_\nu) \\ \tilde{n}_{\Delta_\pi}L_\pi & \tilde{n}_{\Delta_\nu}L_\nu\end{matrix}\,\middle|\,\begin{matrix}(v_1,v_2) \\ \tilde{n}_{1\Delta},\tilde{n}_{2\Delta},L\end{matrix}\right\rangle(L_\pi M_{L_\pi}L_\nu M_{L_\nu}|LM_L) \\
&\qquad\times|[N_\pi],(n_{d_\pi}),(v_\pi),\tilde{n}_{\Delta_\pi},L_\pi,M_{L_\pi}\rangle|[N_\nu],(n_{d_\nu}),(v_\nu),\tilde{n}_{\Delta_\nu},L_\nu,M_{L_\nu}\rangle,
\end{aligned}
\tag{5.63}
$$

where $\{\pi\}$, $\{\nu\}$ denote all quantum numbers. Reduced matrix elements of operators acting on proton, or neutron, wave functions can be obtained from the reduction formulas (De Shalit and Talmi, 1963):

$$
\begin{aligned}
&\langle\tilde{\alpha}_\pi,L_\pi,\tilde{\alpha}_\nu,L_\nu,L|T_\pi^{(k)}|\tilde{\alpha}'_\pi,L'_\pi,\tilde{\alpha}'_\nu,L'_\nu,L'\rangle \\
&\quad=(-)^{L_\pi+L_\nu+L'+k}\sqrt{[(2L+1)(2L'+1)]}\langle\tilde{\alpha}_\pi L_\pi\|T_\pi^{(k)}\|\tilde{\alpha}'_\pi L'_\pi\rangle \\
&\qquad\times\begin{Bmatrix}L_\pi & L & L_\nu \\ L' & L'_\pi & k\end{Bmatrix}\delta_{\tilde{\alpha}_\nu,\tilde{\alpha}'_\nu},\delta_{L_\nu,L'_\nu},
\end{aligned}
\tag{5.64}
$$

Table 5.6. *Notation of states for chain III_1 ($\tilde{v}_{1\Delta}=\tilde{v}_\Delta, \tilde{v}_{2\Delta}=0$)*

$\|L_i^+\rangle$	=	$\|[N_\pi]\otimes[N_\nu];$	$[N_1,N_2],$	$(\sigma_1,\sigma_2),$	$(\tau_1,\tau_2),$	$\tilde{v}_\Delta,$	$L\rangle$
$\|0_1^+\rangle$	=	$\|[N_\pi]\otimes[N_\nu];$	$[N,0],$	$(N,0),$	$(0,0)$	0,	$0\rangle$
$\|0_2^+\rangle$	=	$\|[N_\pi]\otimes[N_\nu];$	$[N,0],$	$(N,0),$	$(3,0),$	1,	$0\rangle$
$\|0_3^+\rangle$	=	$\|[N_\pi]\otimes[N_\nu];$	$[N,0],$	$(N-2,0),$	$(0,0),$	0,	$0\rangle$
$\|2_1^+\rangle$	=	$\|[N_\pi]\otimes[N_\nu];$	$[N,0],$	$(N,0),$	$(1,0),$	0,	$2\rangle$
$\|2_2^+\rangle$	=	$\|[N_\pi]\otimes[N_\nu];$	$[N,0],$	$(2,0),$	$(2,0),$	0,	$2\rangle$
$\|2_3^+\rangle$	=	$\|[N_\pi]\otimes[N_\nu];$	$[N,0],$	$(N-2,0),$	$(1,0),$	0,	$2\rangle$
$\|2_4^+\rangle$	=	$\|[N_\pi]\otimes[N_\nu];$	$[N,0],$	$(N-2,0),$	$(2,0),$	0,	$2\rangle$
$\|3_1^+\rangle$	=	$\|[N_\pi]\otimes[N_\nu];$	$[N,0],$	$(N,0),$	$(3,0),$	0,	$3\rangle$
$\|1_1^+\rangle$	=	$\|[N_\pi]\otimes[N_\nu];$	$[N-1,1],$	$(N-1,1),$	$(1,1),$	0,	$1\rangle$
$\|2_5^+\rangle$	=	$\|[N_\pi]\otimes[N_\nu];$	$[N-1,1],$	$(N-1,1),$	$(1,0),$	0,	$2\rangle$
$\|3_2^+\rangle$	=	$\|[N^\pi]\otimes[\nabla_\nu];$	$[N-1,1],$	$(N-1,1),$	$(1,1),$	0,	$3\rangle$

Table 5.7. *The isoscalar factor* $\left\langle \begin{matrix}[N] & [1] \\ (n_\pi) & (n_\nu)\end{matrix} \middle| \begin{matrix}[N_1,N_2] \\ (n_1,n_2)\end{matrix}\right\rangle$

$(n_\pi),\ (n_\nu)$	$[N_1,N_2]$ (n_1,n_2)	$[N+1,0]$ $(n,0)$	$[N,1]$ $(n,0)$	$[N,1]$ $(n-1,1)$
$(n),\ (0)$		$\left(\frac{N-n+1}{N+1}\right)^{\frac{1}{2}}$	$-\left(\frac{n}{N+1}\right)^{\frac{1}{2}}$	—
$(n-1),\ (1)$		$\left(\frac{n}{N+1}\right)^{\frac{1}{2}}$	$\left(\frac{N-n+1}{N+1}\right)^{\frac{1}{2}}$	1

where

$$|\tilde{\alpha}_\pi, L_\pi, \tilde{\alpha}_\nu, L_\nu, L, M_L\rangle$$

$$= \sum_{M_{L_\pi}, M_{L_\nu}} (L_\pi M_{L_\pi} L_\nu M_{L_\nu}|LM_L)$$

$$\times |[N_\pi], (n_{d_\pi}), (v_\pi), \tilde{n}_{\Delta_\pi}, L_\pi, M_{L_\pi}\rangle |[N_\nu], (n_{d_\nu}), (v_\nu), \tilde{n}_{\Delta_\nu}, L_\nu, M_{L_\nu}\rangle, \quad (5.65)$$

and $\tilde{\alpha}_\pi$, $\tilde{\alpha}_\nu$ denote all quantum numbers except L_π and L_ν. The crucial ingredients in the calculations are thus the isoscalar factors (IF). These have been tabulated by Sun (Chen *et al.*, 1982; Sun *et al.*, 1982; Zhang *et al.*, 1982), and by van Isacker (1984) for some simple cases. Examples are given in Tables 5.7, 5.8 and 5.9. A similar expansion of the wave functions

Table 5.8. *The isoscalar factor* $\left\langle \begin{matrix} (n) & (1) \\ (v_\pi) & (1) \end{matrix} \middle| \begin{matrix} (n_1, n_2) \\ (v_1, v_2) \end{matrix} \right\rangle$

(v_π) \ (n_1,n_2) (v_1,v_2)	$(n+1,0)$ $(v,0)$	$(n,1)$ $(v,0)$	$(n,1)$ $(v-1,1)$
$(v+1)$	$\left(\frac{(n-v+1)(v+3)}{(n+1)(2v+3)}\right)^{\frac{1}{2}}$	$-\left(\frac{(n+v+4)v}{(n+1)(2v+3)}\right)^{\frac{1}{2}}$	—
$(v-1)$	$\left(\frac{(n+v+4)v}{(n+1)(2v+3)}\right)^{\frac{1}{2}}$	$\left(\frac{(n-v+1)(v+3)}{(n+1)(2v+3)}\right)^{\frac{1}{2}}$	1

Table 5.9. *The isoscalar factor* $\left\langle \begin{matrix} (v) & (1) \\ L_\pi & 2 \end{matrix} \middle| \begin{matrix} (v_1, v_2) \\ L \end{matrix} \right\rangle$ *for* $L=2v$

L_π \ (v_1,v_2) L	$(v+1,0)$ $2v$	$(v,1)$ $2v$
$2v$	$\left(\frac{2(2v+1)}{(v+1)(4v-1)}\right)^{\frac{1}{2}}$	$\left(\frac{(v-1)(4v+3)}{(v+1)(4v-1)}\right)^{\frac{1}{2}}$
$2v-2$	$\left(\frac{(v-1)(4v+3)}{(v+1)(4v-1)}\right)^{\frac{1}{2}}$	$-\left(\frac{2(2v+1)}{(v+1)(4v-1)}\right)^{\frac{1}{2}}$

can be done for all other chains. Using these expansions, Scholten *et al.* (1985a) have been able to evaluate matrix elements of the M1 operator.

(*i*) *Chain* I_1

Here

$$\langle 0_1^+ \| T^{(\mathrm{M1})} \| 1_1^+ \rangle = 0, \tag{5.66}$$

as it can also be obtained from elementary arguments, since

$$|0_1^+\rangle = \frac{1}{\mathcal{N}} s_\pi^{\dagger N_\pi} s_\nu^{\dagger N_\nu} |0\rangle,$$

$$|1_1^+\rangle = \frac{1}{\mathcal{N}'} s_\pi^{\dagger N_\pi - 1} s_\nu^{\dagger N_\nu - 1} [d_\pi^\dagger \times d_\nu^\dagger]^{(1)} |0\rangle. \tag{5.67}$$

Using the same elementary arguments, Hamilton *et al.* (1984) have also computed

$$\langle 2_1^+ \| T^{(M1)} \| 2_4^+ \rangle = \sqrt{\left(\frac{3}{4\pi}\right)} \sqrt{\left(\frac{30 N_\pi N_\nu}{N^2}\right)} (g_\pi - g_\nu). \tag{5.68}$$

From (5.66) one finds

$$B(\mathrm{M1}; 0_1^+ \to 1_1^+) = 0, \quad (\mathrm{I}_1). \tag{5.69}$$

(*ii*) *Chain II_1*
Here

$$\langle 0_1^+ \| T^{(M1)} \| 1_1^+ \rangle = \sqrt{\left(\frac{3}{4\pi}\right)} \sqrt{\left(\frac{8 N_\nu N_\pi}{2N-1}\right)} (g_\pi - g_\nu), \tag{5.70}$$

giving

$$B(\mathrm{M1}; 0_1^+ \to 1_1^+) = \left(\frac{3}{4\pi}\right)\left(\frac{8 N_\nu N_\pi}{2N-1}\right)(g_\pi - g_\nu)^2, \quad (\mathrm{II}_1). \tag{5.71}$$

(*iii*) *Chain III_1*
Here

$$\langle 0_1^+ \| T^{(M1)} \| 1_1^+ \rangle = \sqrt{\left(\frac{3}{4\pi}\right)} \sqrt{\left(\frac{3 N_\nu N_\pi}{N+1}\right)} (g_\pi - g_\nu), \tag{5.72}$$

giving

$$B(\mathrm{M1}; 0_1^+ \to 1_1^+) = \left(\frac{3}{4\pi}\right)\left(\frac{3 N_\nu N_\pi}{N+1}\right)(g_\pi - g_\nu)^2, \quad (\mathrm{III}_1). \tag{5.73}$$

(*iv*) *Chain IV*
Scholten *et al.* (1985a) have also been able to evaluate the matrix elements $0_1^+ \to 1_1^+$ for chain IV with the result

$$\langle 0_1^+ \| T^{(M1)} \| 1_1^+ \rangle = \sqrt{\left(\frac{3}{4\pi}\right)} \sqrt{\left(\frac{2 N_\nu N_\pi}{N}\right)} (g_\pi - g_\nu), \tag{5.74}$$

and thus

$$B(\mathrm{M1}; 0_1^+ \to 1_1^+) = \left(\frac{3}{4\pi}\right)\left(\frac{2 N_\nu N_\pi}{N}\right)(g_\pi - g_\nu)^2, \quad (\mathrm{IV}). \tag{5.75}$$

It is interesting to note that all matrix elements leading from symmetric to asymmetric states are proportional to the difference $(g_\pi - g_\nu)$. Conversely, all matrix elements between symmetric states are proportional to a combination of g_π and g_ν with a positive sign. For example, for chains

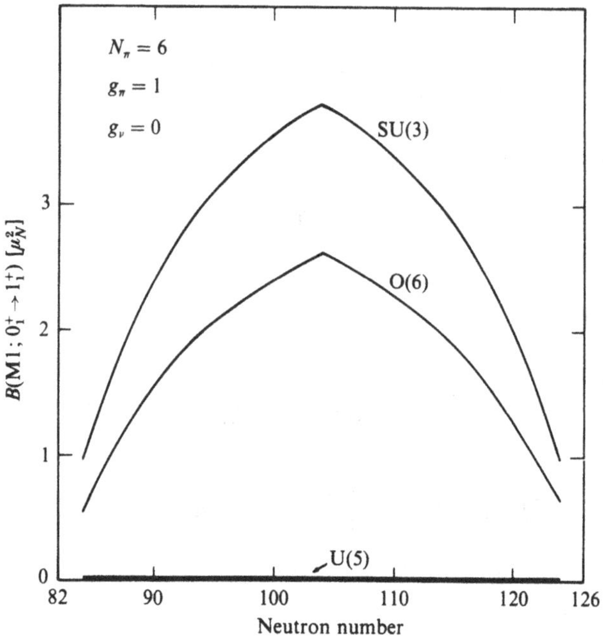

Fig. 5.8. Expected behavior of $B(\mathrm{M}1; 0_1^+ \to 1_1^+)$ as a function of N_ν for fixed $N_\pi = 6$.

I_1, II_1 and III_1 one obtains for the g-factors of the states 2_1^+,

$$g_{2_1^+} = g_\pi\left(\frac{N_\pi}{N}\right) + g_\nu\left(\frac{N_\nu}{N}\right). \tag{5.76}$$

The g-factors have been defined in Sect. 2.6.6.

Also, it is worth mentioning that microscopic calculations indicate that

$$\begin{aligned} g_\pi &\approx 1, \\ g_\nu &\approx 0. \end{aligned} \tag{5.77}$$

Using these values, one can calculate the expected behavior of $B(\mathrm{M}1; 0_1^+ \to 1_1^+)$ and of $g_{2_1^+}$, as shown in Figs. 5.8 and 5.9.

Equations (5.71) and (5.76) should be contrasted with those appropriate to the case in which the nucleus behaves as a rigid body. For this case, one obtains (Bohr and Mottelson, 1975)

$$g_{2_1^+} = Z/A, \tag{5.78}$$

where Z is the total number of protons and A the total mass number. The value Z/A is also shown in Fig. 5.9. Similarly, a rigid motion produces

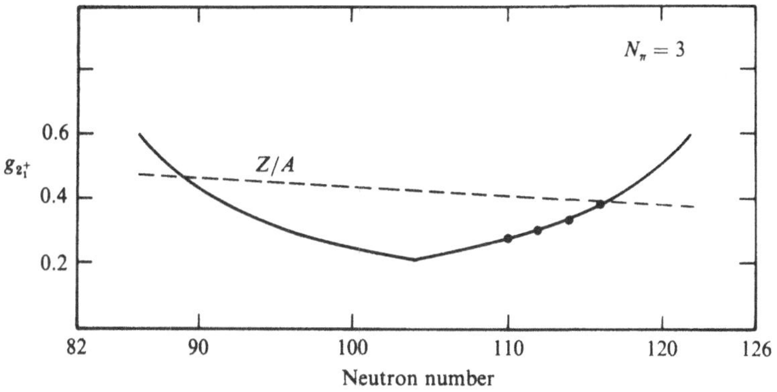

Fig. 5.9. Expected behavior of $g_{2_1^+}$ as a function of N_ν for fixed $N_\pi=3$. This behavior is compared with the experimental values in $_{76}$Os, $N_\pi=3$. The values predicted by (5.78) are also shown for comparison.

Table 5.10. B(M1) *values and magnetic moments for chain* I_1

L_i	L_f	$B(\mathrm{M1}; L_i \to L_f)$
0_1^+	1_1^+	0
1_1^+	0_2^+	$\frac{3}{4\pi}(g_\nu - g_\pi)^2 \frac{4}{N(N-1)} N_\nu N_\pi$
1_1^+	2_2^+	$\frac{3}{4\pi}(g_\nu - g_\pi)^2 \frac{7}{N(N-1)} N_\nu N_\pi$
2_1^+	2_2^+	$\frac{3}{4\pi}(g_\nu - g_\pi)^2 \frac{6}{N^2} N_\nu N_\pi$
	L_i	μ
	2_1^+	$(g_\nu N_\nu + g_\pi N_\pi)\frac{2}{N}$
	2_2^+	$(g_\nu N_\nu + g_\pi N_\pi)\frac{2}{N}$
	1_1^+	$\frac{1}{2}(g_\nu + g_\pi)$

$B(\mathrm{M1}; 0_1^+ \to 1_1^+)$ values much larger than those in (5.71) (Lo Iudice and Palumbo, 1978). They can be obtained from (5.71) by replacing $(8N_\nu N_\pi)/(2N-1)$ with NZ/A, where N and Z are the total neutron and proton number and A is the total mass number. The experimental strengths of M1 transitions and the measured magnetic moments, all appear to indicate

Table 5.11. *B*(M1) *values and magnetic moments for chain* II_1

L_i	L_f	$B(\mathrm{M1}; L_i \to L_f)$
0_1^+	1_1^+	$\frac{3}{4\pi}(g_\nu - g_\pi)^2 \frac{8}{2N-1} N_\nu N_\pi$
1_1^+	2_1^+	$\frac{3}{4\pi}(g_\nu - g_\pi)^2 \frac{2(2N+3)}{3N(2N-1)} N_\nu N_\pi$
1_1^+	0_β^+	$\frac{3}{4\pi}(g_\nu - g_\pi)^2 \frac{4(2N+1)}{3N(2N-3)(2N-1)} N_\nu N_\pi$
1_1^+	2_β^+	$\frac{3}{4\pi}(g_\nu - g_\pi)^2 \frac{(4N^2-8N+7)^2}{3(N-1)N(2N-3)(2N-1)(4N^2-8N+1)} N_\nu N_\pi$
1_1^+	2_γ^+	$\frac{3}{4\pi}(g_\nu - g_\pi)^2 \frac{8(N-2)(2N+1)}{N(2N-3)(4N^2-8N+1)} N_\nu N_\pi$

L_i	μ
2_1^+	$(g_\nu N_\nu + g_\pi N_\pi)\frac{2}{N}$
2_β^+	$(g_\nu N_\nu + g_\pi N_\pi)\frac{2}{N}$
2_γ^+	$(g_\nu N_\nu + g_\pi N_\pi)\frac{2}{N}$
1_1^+	$\frac{1}{2}(g_\nu + g_\pi)$

that valence particles (i.e. those outside the major closed shells) rather than the entire nucleus, mostly take part in the collective motion. The available *B*(M1) values and magnetic moment are summarized in Tables 5.10, 5.11 and 5.12.

5.5.2 *Electric quadrupole operators (E2)*

The electric quadrupole operator was written in (4.39) as

$$T^{(\mathrm{E2})} = e_\pi \hat{Q}_\pi^\chi + e_\nu \hat{Q}_\nu^\chi, \tag{5.79}$$

where e_π and e_ν are the boson effective charges. Matrix elements of this operator have been evaluated in several cases.

(i) Chain I_1

Hamilton *et al.* (1984) have evaluated, using elementary methods, the

Table 5.12. B(M1) *values and magnetic moments for chain* III_1

L_i	L_f	$B(\mathrm{M1}; L_i \to L_f)$
0_1^+	1_1^+	$\frac{3}{4\pi}(g_\nu - g_\pi)^2 \frac{3}{N+1} N_\nu N_\pi$
1_1^+	0_3^+	$\frac{3}{4\pi}(g_\nu - g_\pi)^2 \frac{(N+2)(N+3)}{(N-1)N^2(N+1)} N_\nu N_\pi$
1_1^+	2_2^+	$\frac{3}{4\pi}(g_\nu - g_\pi)^2 \frac{(N+4)(N+5)}{2(N-1)N(N+1)} N_\nu N_\pi$
1_1^+	2_4^+	$\frac{3}{4\pi}(g_\nu - g_\pi)^2 \frac{(N-3)(N-2)}{2(N-1)N^2(N+1)} N_\nu N_\pi$
L_i		μ
2_1^+		$(g_\nu N_\nu + g_\pi N_\pi)\frac{2}{N}$
2_2^+		$(g_\nu N_\nu + g_\pi N_\pi)\frac{2}{N}$
1_1^+		$\frac{1}{2}(g_\nu + g_\pi)$

matrix elements

$$\langle 0_1^+ \| T^{(\mathrm{E2})} \| 2_1^+ \rangle = \sqrt{\left(\frac{5}{N}\right)}(e_\pi N_\pi + e_\nu N_\nu),$$

$$\langle 0_1^+ \| T^{(\mathrm{E2})} \| 2_4^+ \rangle = \sqrt{\left(\frac{5N_\nu N_\pi}{N}\right)}(e_\nu - e_\pi),$$

$$\langle 2_1^+ \| T^{(\mathrm{E2})} \| 2_4^+ \rangle = \sqrt{\left(\frac{5N_\pi N_\nu}{N^2}\right)}(e_\nu \chi_\nu - e_\pi \chi_\pi). \tag{5.80}$$

Again here, transitions between symmetric and asymmetric states are proportional to a combination of effective charges with negative sign, while transitions between two symmetric states are proportional to a combination with positive sign. From (5.80) one obtains

$$\begin{aligned} B(\mathrm{E2}; 0_1^+ \to 2_1^+) &= \frac{5}{N}(e_\pi N_\pi + e_\nu N_\nu)^2 \\ B(\mathrm{E2}; 0_1^+ \to 2_4^+) &= \frac{5N_\pi N_\nu}{N}(e_\nu - e_\pi)^2. \end{aligned} \tag{5.81}$$

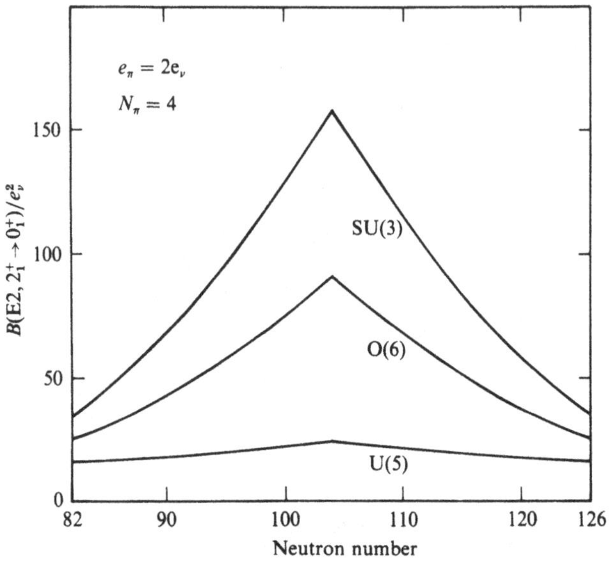

Fig. 5.10. The behavior of $B(\mathrm{E2}; 2_1^+ \to 0_1^+)$ for chains I_1, II_1 and III_1 as function of N_ν and fixed $N_\pi = 4$, $e_\pi = 2e_\nu$.

For $N_\pi = N_\nu$,

$$\frac{B(\mathrm{E2}; 0_1^+ \to 2_4^+)}{B(\mathrm{E2}; 0_1^+ \to 2_1^+)} = \left(\frac{e_\nu - e_\pi}{e_\nu + e_\pi}\right)^2. \tag{5.82}$$

A measurement of the ratio (5.82) offers the possibility to determine e_ν and e_π. Microscopic calculations indicate that $e_\pi \approx e_\nu$ in deformed nuclei (close to the limit II_1 or IV) and $e_\pi \approx 2e_\nu$ in spherical nuclei (close to the limit I_1). The behavior of $B(\mathrm{E2};\, 2_1^+ \to 0_1^+)$ is shown in Fig. 5.10.

(ii) Chain II_1

Several matrix elements in this chain have been evaluated by van Isacker *et al.* (1986) using the operator (5.79) with $\chi_\pi = \chi_\nu = -\sqrt{7}/2$. Particularly important here are the matrix elements

$$\langle 0_1^+ \| T(\mathrm{E2}) \| 2_1^+ \rangle = (e_\pi N_\pi + e_\nu N_\nu)\sqrt{\left(\frac{2N+3}{N}\right)},$$

$$\langle 0_1^+ \| T^{(\mathrm{E2})} \| 2_4^+ \rangle = (e_\nu - e_\pi)\sqrt{\left(\frac{3(N-1)}{N(2N-1)} N_\nu N_\pi\right)}. \tag{5.83}$$

The behavior of $B(\mathrm{E2}; 2_1^+ \to 0_1^+)$ is shown in Fig. 5.10.

Table 5.13. B(E2) *values and quadrupole moments for chain* I_1

L_i	L_f	$B(\mathrm{E2}; L_i \to L_f)$
0_1^+	2_1^+	$(e_\nu N_\nu + e_\pi N_\pi)^2 \dfrac{5}{N}$
0_1^+	2_4^+	$(e_\nu - e_\pi)^2 \dfrac{5}{N} N_\nu N_\pi$
1_1^+	2_1^+	$(e_\nu - e_\pi)^2 \dfrac{1}{N} N_\nu N_\pi$
1_1^+	2_2^+	0
1_1^+	2_4^+	$(e_\nu N_\nu + e_\pi N_\pi)^2 \dfrac{1}{N}$
1_1^+	3_1^+	0
L_i		Q
2_1^+		$\sqrt{\left(\dfrac{32\pi}{35}\right)}(e_\nu \chi_\nu N_\nu + e_\pi \chi_\pi N_\pi)\dfrac{1}{N}$

(*iii*) *Chain* III_1

Also for this chain, van Isacker *et al.* (1986) have evaluated several matrix elements. Once more, of particular importance are

$$\langle 0_1^+ \| T^{(\mathrm{E2})} \| 2_1^+ \rangle = (e_\pi N_\pi + e_\nu N_\nu)\sqrt{\left(\frac{N+4}{N}\right)},$$

$$\langle 0_1^+ \| T^{(\mathrm{E2})} \| 2_5^+ \rangle = (e_\nu - e_\pi)\sqrt{\left(\frac{2(N+2)}{N(N+1)} N_\nu N_\pi\right)}. \qquad (5.84)$$

The behavior of $B(\mathrm{E2}; 2_1^+ \to 0_1^+)$ is shown in Fig. 5.10.

The available B(E2) values and quadrupole moments are summarized in Tables 5.13, 5.14 and 5.15.

5.5.3 *Magnetic octupole operators* (*M3*)

The magnetic octupole operator is defined by (4.42)

$$T^{(\mathrm{M3})} = \sqrt{\left(\frac{7}{4\pi}\right)}(m_\pi \hat{\mathrm{U}}_\pi + m_\nu \hat{\mathrm{U}}_\nu). \qquad (5.85)$$

Matrix elements of this operator have been evaluated by van Isacker *et al.* (1986) in some cases and are summarized in Table 5.16.

Table 5.14. *B*(E2) *values and quadrupole moments for chain* II_1

L_i	L_f	$B(\mathrm{E2}; L_i \rightarrow L_f)$
0_1^+	2_1^+	$(e_\nu N_\nu + e_\pi N_\pi)^2 \left(\frac{2N+3}{N}\right)$
0_1^+	2_4^+	$(e_\nu - e_\pi)^2 \frac{3(N-1)}{N(2N-1)} N_\nu N_\pi$
1_1^+	2_1^+	$(e_\nu - e_\pi)^2 \frac{3(2N+3)}{4N(2N-1)} N_\nu N_\pi$
1_1^+	2_β^+	$(e_\nu - e_\pi)^2 \frac{3(4N^2-8N-1)^2}{8(N-1)N(2N-3)(2N-1)(4N^2-8N+1)} N_\nu N_\pi$
1_1^+	2_γ^+	$(e_\nu - e_\pi)^2 \frac{(N-2)(2N+1)}{N(2N-3)(4N^2-8N+1)} N_\nu N_\pi$
1_1^+	3_γ^+	$(e_\nu - e_\pi)^2 \frac{(N-2)}{(N-1)N(2N-3)} N_\nu N_\pi$

L_i	Q
2_1^+	$-\sqrt{\left(\frac{2\pi}{5}\right)}(e_\nu N_\nu + e_\pi N_\pi)\frac{2}{7}\frac{(4N+3)}{N}$
1_1^+	$\sqrt{\left(\frac{\pi}{250}\right)}[e_\nu(4N_\nu+3)+e_\pi(4N_\pi+3)]$

Table 5.15. *B*(E2) *values and quadrupole moments of chain* III_1

L_i	L_f	$B(\mathrm{E2}; L_i \rightarrow L_f)$
0_1^+	2_1^+	$(e_\nu N_\nu + e_\pi N_\pi)^2 \frac{N+4}{N}$
0_1^+	2_5^+	$(e_\nu - e_\pi)^2 \frac{2(N+2)}{N(N+1)} N_\nu N_\pi$
1_1^+	2_1^+	$(e_\nu - e_\pi)^2 \frac{N+4}{2N(N+1)} N_\nu N_\pi$
1_1^+	2_3^+	$(e_\nu - e_\pi)^2 \frac{(N-2)(N+3)}{(2N-1)N^2(N+1)} N_\nu N_\pi$

L_i	Q
2_1^+	$\sqrt{\left(\frac{2\pi}{35}\right)}(e_\nu \chi_\nu N_\nu + e_\pi \chi_\pi N_\pi)\frac{4(N^2+4N+9)}{7(N+1)}$

Table 5.16. B(M3) *values for some states in chains* I_1, II_1 *and* III_1

L_i	L_f	$B(\mathrm{M3}; L_i \rightarrow L_f)$
(i) Chain I_1		
0_1^+	3_1^+	0
(ii) Chain II_1		
0_1^+	3_γ^+	$\frac{35}{8\pi}(m_\nu N_\nu + m_\pi N_\pi)^2 \frac{8(N-2)(N-1)}{3N(2N-3)(2N-1)}$
0_1^+	3_2^+	$\frac{35}{8\pi}(m_\nu - m_\pi)^2 \frac{4(2N+3)}{15(N-1)(2N-1)} N_\nu N_\pi$
0_1^+	3_3^+	$\frac{35}{8\pi}(m_\nu - m_\pi)^2 \frac{4(N-2)^2}{3(N-1)N(2N-3)} N_\nu N_\pi$
(iii) Chain III_1		
0_1^+	3_2^+	$\frac{35}{8\pi}(m_\nu - m_\pi)^2 \frac{7}{10(N+1)} N_\nu N_\pi$

5.6 Transitional classes

An interesting new feature arises in the interacting boson model-2 as compared with the interacting boson model-1. The spectrum of the symmetry IV is qualitatively different from those obtained in the interacting boson model-1. As it will be shown in Chapter 6, it corresponds to nuclei with a mass distribution which, on the average, is not axially symmetric (triaxial shapes). Thus, there are here four possible different situations. This can be schematically indicated by constructing a tetrahedron, at the vertices of which one places the four symmetries, I, II, III and IV, as suggested by Dieperink (1983) and shown in Fig. 5.11. The transitional classes correspond now to the sides of the tetrahedron. Only a few studies of these classes have been performed so far. We only mention the transitional class II_2–IV. This can be done by using the Hamiltonian

$$H^{(\mathrm{II}_2)\text{-}(\mathrm{IV})} = c_1(\hat{L}\cdot\hat{L}) + c_2(\hat{Q}^\chi\cdot\hat{Q}^\chi) \tag{5.86}$$

in the consistent-Q formalism, where

$$\hat{L} = \hat{L}_\pi + \hat{L}_\nu,$$
$$\hat{Q}^\chi = \hat{Q}^\chi_\pi + \hat{Q}^\chi_\nu. \tag{5.87}$$

For fixed $\chi_\pi = -\sqrt{7}/2$, by varying χ_ν between $-\sqrt{7}/2$ and $+\sqrt{7}/2$ one obtains the transitional class II_2–IV. The control parameter is thus here χ_ν.

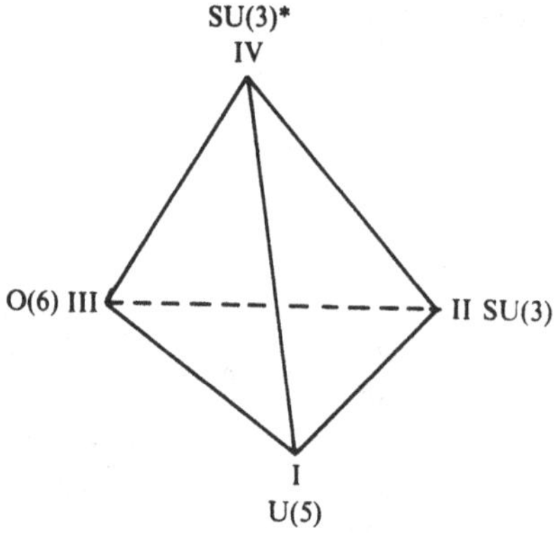

Fig. 5.11. Dieperink's tetrahedron.

Another transitional class that can be analyzed in this way is that between III_2 and IV. This can be studied using (5.86) with $\chi_\nu = -\chi_\pi$ varying between 0 and $-\sqrt{7}/2$ (Dieperink, 1983). It appears that some nuclei, such as ^{104}Ru, are in this transitional class.

5.7 Numerical studies

Most studies of nuclear properties within the framework of the interacting boson model-2 have been done numerically. These studies have made use of Talmi's Hamiltonian in the form (4.44)

$$H = E_0 + \varepsilon(\hat{n}_{d_\pi} + \hat{n}_{d_\nu}) + \kappa \hat{Q}^\chi_\pi \cdot \hat{Q}^\chi_\nu + \lambda' \hat{M}_{\pi\nu}, \tag{5.88}$$

with

$$E_0 = E_c + A_\pi N_\pi + A_\nu N_\nu + \tfrac{1}{2} B_\pi N_\pi (N_\pi - 1) + \tfrac{1}{2} B_\nu N_\nu (N_\nu - 1) + C N_\pi N_\nu. \tag{5.89}$$

The Hamiltonian (5.88) contains *four* parameters ε, κ, χ_π, χ_ν determining the structure of the low-lying excitation, *one* parameter λ' determining the location of the asymmetric states, in addition to the terms, A_π, A_ν, B_π, B_ν, C which contribute only to binding energies. Since the location of the asymmetric states was not known until recently, in most calculations the parameter λ' is set to a large value in order to remove those states from the low-lying part of the spectrum.

In addition to energies, in all studies, electromagnetic E2 properties are calculated. The operator used is (4.45)

$$T^{(E2)} = e_B(\hat{Q}^\chi_\pi + \hat{Q}^\chi_\nu), \tag{5.90}$$

with $e_\pi = e_\nu = e_B$. Again, the reason why this choice was made was the absence of information on F-spin excitations, which are sensitive to $(e_\pi - e_\nu)$. In light of recent experiments, it appears that, in some nuclei, $e_\pi \neq e_\nu$.

In some studies, electromagnetic M1 properties are calculated, with

$$T^{(\mathrm{M1})} = \sqrt{\left(\frac{3}{4\pi}\right)}(g_\pi \hat{L}_\pi + g_\nu \hat{L}_\nu). \tag{5.91}$$

While diagonal matrix elements of this operator are insensitive to the choice of parameters, off-diagonal matrix elements are very sensitive to the value of λ' since admixtures of asymmetric states into the symmetric states strongly influence these matrix elements.

Some calculations also address the problem of proton radii, using

$$r_\pi^2 = r_{c\pi}^2 + \gamma_{\pi\pi}\hat{N}_\pi + \gamma_{\pi\nu}\hat{N}_\nu + \beta_{\pi\pi}\hat{n}_{\mathrm{d}_\pi} + \beta_{\pi\nu}\hat{n}_{\mathrm{d}_\nu}, \tag{5.92}$$

and, in particular, the problem of isomer and isotope shifts, (4.54) and (4.55). In these calculations the quantity $\beta_{\pi\nu}$ is usually set to zero, since there does not appear at present any direct way of evaluating it.

Finally, two-neutron transfer reactions are often calculated, using, for $L=0$ transfer, (4.64),

$$P^{(0)}_{+,\nu,0} = p_{\nu,0} s_\nu^\dagger \sqrt{(\Omega_\nu - N_\nu + \hat{n}_{\mathrm{d}_\nu})},$$
$$P^{(0)}_{-,\nu,0} = p_{\nu,0} \sqrt{(\Omega_\nu - N_\nu + \hat{n}_{\mathrm{d}_\nu})}\tilde{s}_\nu. \tag{5.93}$$

To these standard forms of the operators, variations are often applied. For example, in many calculations, an interaction between identical bosons is added to the Hamiltonian (5.88). The interaction used is of the type

$$V_{\rho\rho} = \sum_{L=0,2,4} \tfrac{1}{2}(2L+1)^{\frac{1}{2}} c_L^{(\rho)} [[d_\rho^\dagger \times d_\rho^\dagger]^{(L)} \times [\tilde{d}_\rho \times \tilde{d}_\rho]^{(L)}]_0^{(0)}; \quad \rho = \pi, \nu. \tag{5.94}$$

The structure of this interaction is dictated by microscopic considerations and arises from the fact that the interaction between identical nucleons is mostly seniority-conserving (Talmi, 1971). In other calculations, the condition $e_\pi = e_\nu$ is relaxed, etc. In Table 5.17 we list all calculations available up to now, and present in the following figures some illustrative results. In Figs. 5.12, 5.13 and 5.14 a comparison between the calculated and experimental energy spectra in $_{54}$Xe, $_{56}$Ba and $_{58}$Ce is shown. These spectra show a gradual change from a situation typical of the symmetry I towards the beginning of the major shell (neutron number $\approx$54–58), to a situation typical of symmetry II towards the middle of the major shell (neutron number$\approx$62–66), and finally to a situation typical of symmetry

Table 5.17. *List of available numerical calculations of nuclear properties using the interacting boson model-2*

Proton number		Neutron number	Mass number	Reference
32	Ge[a]	36–44	68–76	(Duval *et al.*, 1983)
34	Se	38–46	72–80	(Kaup *et al.*, 1983)
36	Kr	38–46	74–82	(Kaup and Gelberg, 1979)
38	Sr	36	84	(Dewald *et al.*, 1982)
42	Mo[a]	54–62	96–104	(Sambataro and Molnar, 1982)
44	Ru	54–78	98–122	(van Isacker and Puddu, 1980)
46	Pd	54–78	100–124	(van Isacker and Puddu, 1980)
48	Cd[a]	54–78	102–126	(Sambataro, 1982)
52	Te	54–78	106–130	(Sambataro, 1982)
54	Xe	54–78	108–132	(Puddu *et al.*, 1980)
56	Ba	54–78	110–134	(Puddu *et al.*, 1980)
58	Ce	54–78	112–136	(Puddu *et al.*, 1980)
60	Nd	86–92	146–152	(Scholten, 1980)
62	Sm	86–92	148–154	(Scholten, 1980)
64	Gd	86–92	150–156	(Scholten, 1980)
74	W	94–118	168–192	(Duval and Barrett, 1981)
76	Os	108–118	184–194	(Bijker *et al.*, 1980)
78	Pt	90–118	168–196	(Bijker *et al.*, 1980)
80	Hg[a]	104–114	184–194	(Barfield *et al.*, 1983)

[a] Including configuration mixing.

III towards the end of the major shell (neutron number ≈74–78). These properties are dictated by a competition between the pairing term ε in (5.88) and the quadrupole term κ. It is interesting to note that the matrix elements of the operator $\hat{Q}_\pi^\chi \cdot \hat{Q}_\nu^\chi$ are, to a rough approximation, proportional to the product $N_\pi N_\nu$. This product appears to determine the observed properties, as remarked recently by Casten (1985a,b). Another important property of the interacting boson model-2 is that its parameters ε, κ, χ_π, χ_ν appear to vary in a smooth way. This allows one to make extrapolations and thus to compute properties of nuclei not yet known, as shown in Figs. 5.12, 5.13 and 5.14.

Examples of calculations of electromagnetic E2 properties are shown in Figs. 5.15 and 5.16. These results have been obtained by using a constant boson effective charge $e_\pi = e_\nu = e_B = 0.12$ eb. Nonetheless the calculated values show a clear structure. This structure is a straightforward consequence of the introduction of proton and neutron bosons. Particularly interesting is the behavior shown in Fig. 5.16. The ratio presented in this figure, denoted by R''' in (2.253), is quite indicative of the

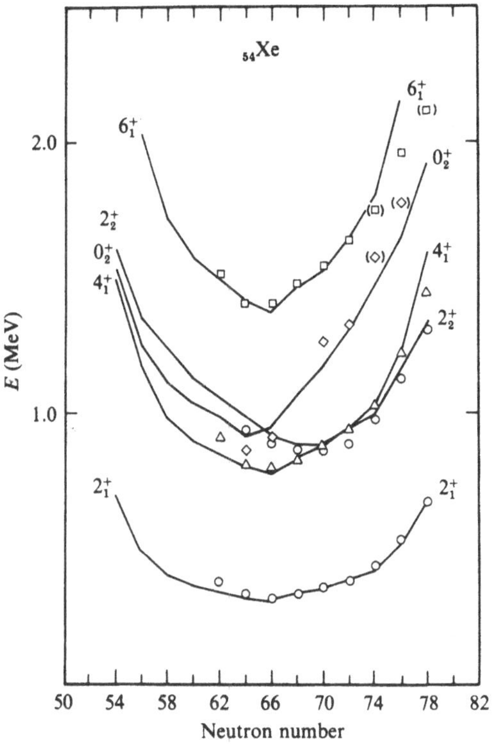

Fig. 5.12. Comparison between calculated (lines) and experimental (circles, squares and triangles) energy spectra of ${}_{54}$Xe (Puddu *et al.*, 1980).

structure of the nucleus. One can see from this figure that the ratio remains small for ${}_{54}$Xe, in accordance with the interpretation of these nuclei as belonging to the transitional class C as discussed in Sect. 2.10. However, for ${}_{56}$Ba, and especially for ${}_{58}$Ce, the ratio R''' begins to approach the SU(3) value 0.7 in the middle of the major shell (neutron number 66), indicating that these nuclei begin to acquire an axially-deformed shape.

Another property which has been calculated extensively is magnetic moments. In Fig. 5.17 the results for nuclei in the vicinity of proton number 50 are shown. The main noticeable effect here is the deviation of $g_{2_1^+}$ from the value Z/A expected on the basis of rigid rotations. The fact that the g-factors are essentially proportional to the proton boson number

$$g_{2_1^+} \approx g_\pi \frac{N_\pi}{N_\pi + N_\nu}, \quad (g_\nu \approx 0), \tag{5.95}$$

as one obtains numerically, and that it is exactly true for the dynamic

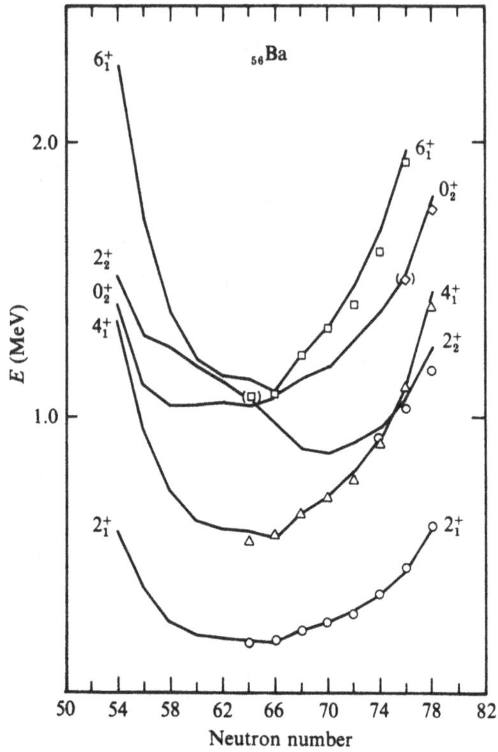

Fig. 5.13. Comparison between calculated (lines) and experimental (circles, squares and triangles) energy spectra of $_{56}$Ba (Puddu *et al.*, 1980).

symmetries I_1, II_1 and III_1, (5.71), has been used by Wolf *et al.* (1983) to extract information on N_π in region where partial shell closures occur.

Finally, extensive calculations of isomer and isotope shifts have been performed. An example is shown in Fig. 5.18. Here again one sees a considerable structure in the isotope shifts, as evidenced from the fact that isotope shifts even become negative in $_{58}$Ce (i.e. the nucleus shrinks by adding neutrons). The calculations listed in Table 5.18 refer, with few exceptions, to nuclei in transitional regions, or close to the U(5) or O(6) symmetries. Until recently, calculations for strongly-deformed nuclei were not available due to the large dimensions of the boson spaces to diagonalize. However, recently a new computer program has been developed by Novoselsky (1984), which allows one to diagonalize larger matrices. This program has been used to calculate the spectra of ^{168}Er and ^{178}Hf within the framework of the interacting boson model-2 (Novoselski, 1985).

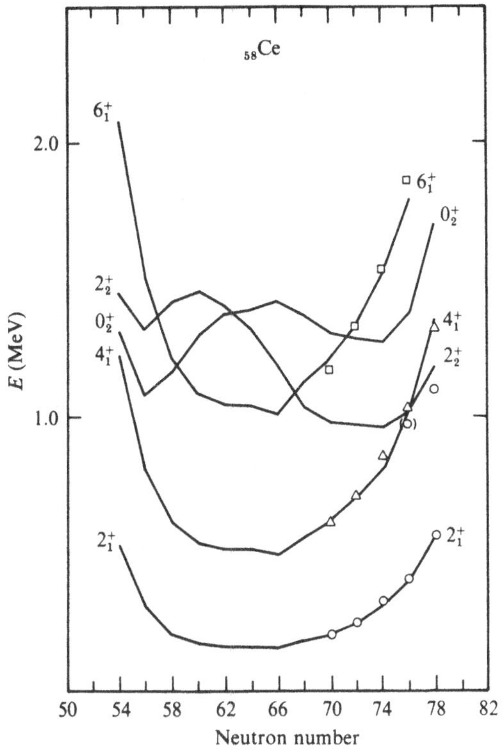

Fig. 5.14. Comparison between calculated (lines) and experimental (circles, squares and triangles) energy spectra of $_{58}$Ce (Puddu *et al.*, 1980).

The discovery of asymmetric states in ^{156}Gd has also provided information on the strength of the Majorana interaction, λ'. Calculations of ^{156}Gd and ^{158}Gd have been performed by Bohle *et al.* (1984) using the value $\lambda' = 0.2$ MeV.

5.8 Transition densities

For calculations of transition densities one must use the expressions (4.65).

5.8.1 *Scalar densities*

In this case it is convenient to consider separately proton and neutron densities, since proton densities are those directly measured in electron-scattering experiments. The proton scalar transition operator can be

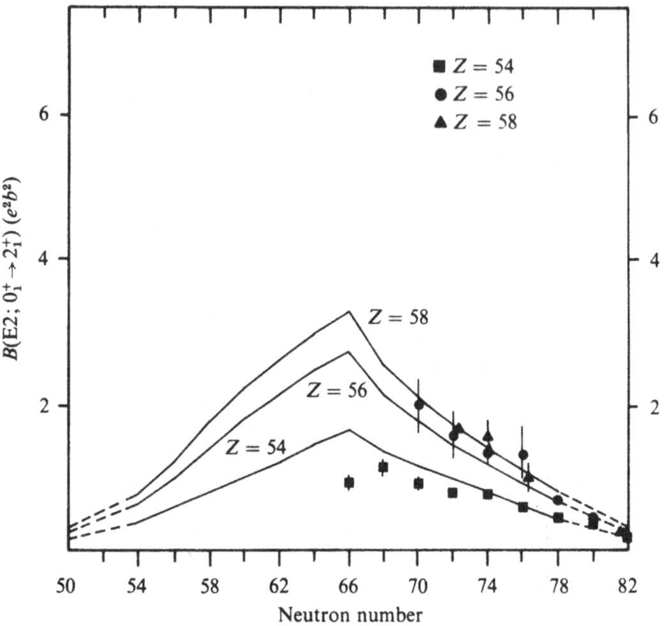

Fig. 5.15. Comparison between experimental $B(\mathrm{E2}; 0_1^+ \to 2_1^+)$ values in $_{54}$Xe, $_{56}$Ba and $_{58}$Ce and those calculated in the interacting boson model-2 (Puddu *et al.*, 1980).

written as

$$T_\pi^{(0)}(r) = \gamma_\pi(r) + \alpha_{\pi\pi}(r)N_\pi + \alpha_{\pi\nu}(r)N_\nu + \beta_{\pi\pi}(r)\hat{n}_{d_\pi} + \beta_{\pi\nu}(r)\hat{n}_{d_\nu} \tag{5.96}$$

and the neutron operator as

$$T_\nu^{(0)}(r) = \gamma_\nu(r) + \alpha_{\nu\pi}(r)N_\pi + \alpha_{\nu\nu}(r)N_\nu + \beta_{\nu\pi}(r)\hat{n}_{d_\pi} + \beta_{\nu\nu}(r)\hat{n}_{d_\nu}. \tag{5.97}$$

Equations (5.96) and (5.97) can be obtained from (4.65) using the technique discussed in Sect. 1.4.5. Proton and neutron transition densities can be obtained by taking matrix elements of the operators (5.96) and (5.97). Particularly important here are the transition densities for states with $L^P = 0^+$. If we denote these states by $|N_\pi, N_\nu, 0_i^+\rangle$, the diagonal densities are given by

$$\begin{aligned}
\rho^{(0)}_{\pi,0_i^+ \to 0_i^+}(r) &= \gamma_\pi(r) + \alpha_{\pi\pi}(r)N_\pi + \alpha_{\pi\nu}(r)N_\nu \\
&\quad + \beta_{\pi\pi}(r)B^{(0)}_{\pi,\,ii} + \beta_{\pi\nu}(r)B^{(0)}_{\nu,\,ii}, \\
\rho^{(0)}_{\nu,0_i^+ \to 0_i^+}(r) &= \gamma_\nu(r) + \alpha_{\nu\pi}(r)N_\pi + \alpha_{\nu\nu}(r)N_\nu \\
&\quad + \beta_{\nu\pi}(r)B^{(0)}_{\pi,\,ii} + \beta_{\nu\nu}(r)B^{(0)}_{\nu,\,ii},
\end{aligned} \tag{5.98}$$

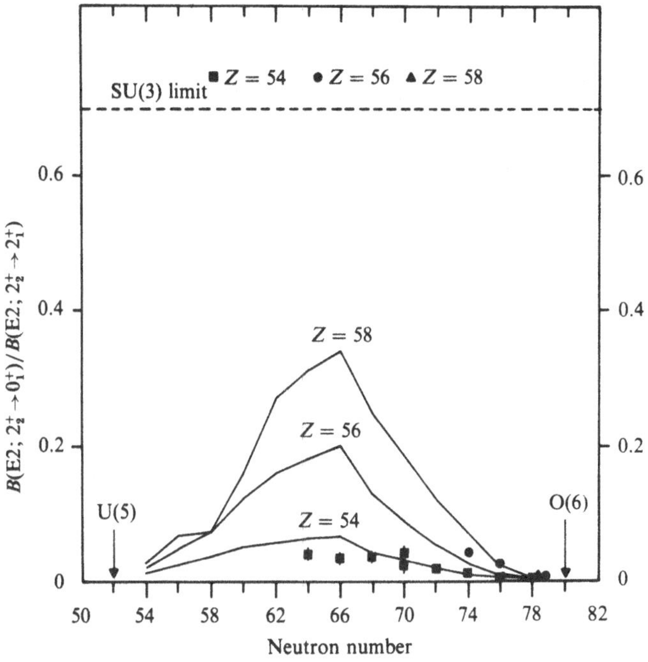

Fig. 5.16. Calculated and experimental branching ratios $B(\mathrm{E2}; 2_2^+ \rightarrow 0_1^+)/B(\mathrm{E2}; 2_2^+ \rightarrow 2_1^+)$ in ${}_{54}$Xe, ${}_{56}$Ba and ${}_{58}$Ce (Puddu *et al.*, 1980).

where

$$B^{(0)}_{\rho,\,ii} = \langle N_\pi, N_\nu, 0_i^+ | \hat{n}_{d_\rho} | N_\pi, N_\nu, 0_i^+ \rangle. \tag{5.99}$$

Transition densities from the ground state to the excited 0^+ states are given by

$$\begin{aligned} \rho^{(0)}_{\pi,0_1^+\rightarrow 0_i^+}(r) &= \beta_{\pi\pi}(r) B^{(0)}_{\pi,\,1i} + \beta_{\pi\nu}(r) B^{(0)}_{\nu,\,1i}, \\ \rho^{(0)}_{\nu,0_1^+\rightarrow 0_i^+}(r) &= \beta_{\nu\pi}(r) B^{(0)}_{\pi,\,1i} + \beta_{\nu\nu}(r) B^{(0)}_{\nu,\,1i}. \end{aligned} \tag{5.100}$$

The coefficients $B^{(0)}_{\pi,\,ij}$, $B^{(0)}_{\nu,\,ij}$ contain the nuclear structure information and can be calculated using the program NPBOS mentioned above (Otsuka, 1977).

The densities (5.98) and (5.100) can be used to calculate electron-scattering cross-sections. They can also be used to calculate static properties. These properties will involve only certain integrals over the densities. For example, proton mean-square radii are given by

$$\langle r_\pi^2 \rangle_{0_1^+} = \frac{4\pi}{Z} \int_0^\infty r^4 \rho^{(0)}_{\pi,0_1^+\rightarrow 0_i^+}(r)\, \mathrm{d}r, \tag{5.101}$$

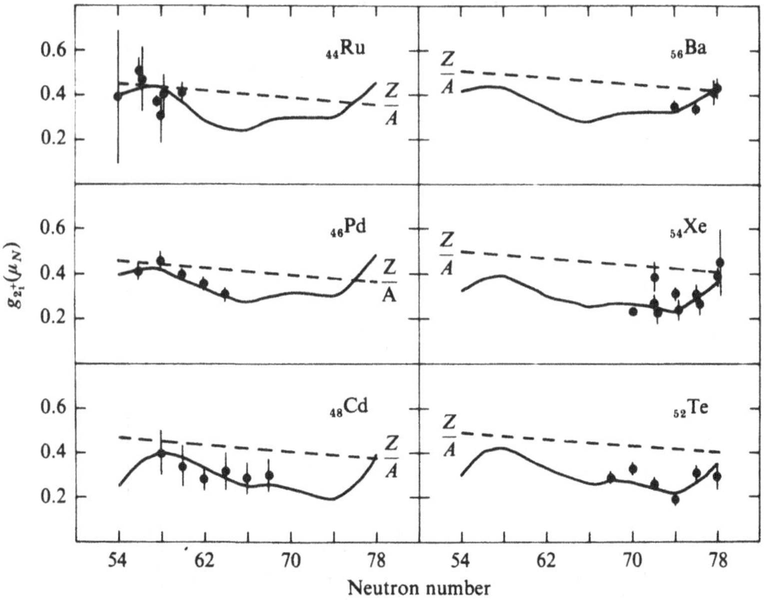

Fig. 5.17. The g-factors of the 2_1^+ states (in μ_N) as a function of neutron number for various isotopes around proton number 50. The solid lines represent the results of the calculations (Sambataro and Dieperink, 1981).

if the diagonal matrix elements are normalized to

$$\int_0^\infty 4\pi r^2 \rho_{\pi,\, 0_1^+ \to 0_1^+}(r)\, \mathrm{d}r = Z. \tag{5.102}$$

A similar expression holds for $\langle r_\nu^2 \rangle$. Accordingly, proton mean-square radii can be written as

$$\langle r_\pi^2 \rangle_{0_1^+} = r_{c\pi}^2 + \alpha_{\pi\pi} N_\pi + \alpha_{\pi\nu} N_\nu + \beta_{\pi\pi} B_{\pi,\,11}^{(0)} + \beta_{\pi\nu} B_{\nu,\,11}^{(0)}, \tag{5.103}$$

where $\alpha_{\pi\pi}$, $\alpha_{\pi\nu}$, $\beta_{\pi\pi}$, $\beta_{\pi\nu}$ are integrals over $\alpha_{\pi\pi}(r)$, $\alpha_{\pi\nu}(r)$, $\beta_{\pi\pi}(r)$, $\beta_{\pi\nu}(r)$. With appropriate change of notation, this expression is identical to that used in Sect. 4.5.9. From this and similar expressions, one can compute isomer and isotope shifts.

Measurements of cross-sections in elastic and inelastic proton scattering provide direct information on the functions $\gamma_\pi(r)$, $\alpha_{\pi\pi}(r)$, $\alpha_{\pi\nu}(r)$, $\beta_{\pi\pi}(r)$ and $\beta_{\pi\nu}(r)$. In electromagnetic decays, one instead measures the monopole transition moment

$$m_{1i}(\mathrm{E0}) = \frac{4\pi \int_0^\infty r^4 \rho_{\pi,0_1^+ \to 0_i^+}^{(0)}(r)\, \mathrm{d}r}{R^2}, \tag{5.104}$$

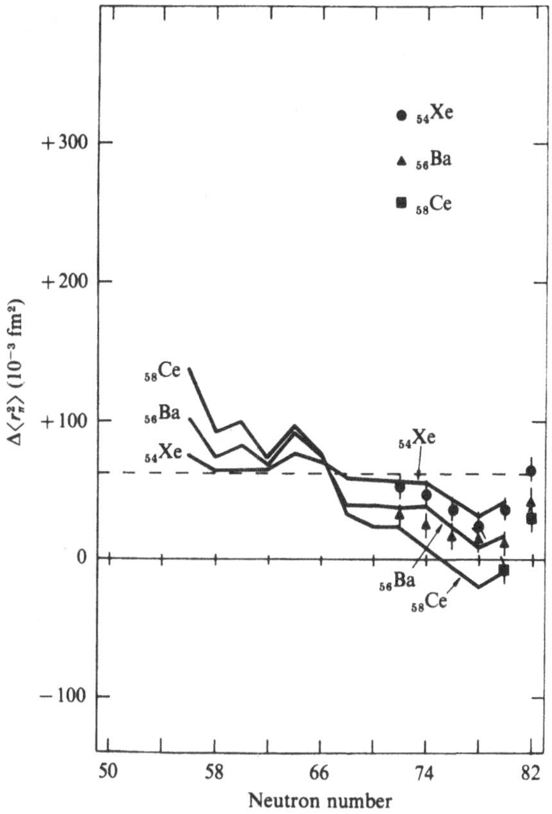

Fig. 5.18. Comparison between calculated and experimental isotope shifts in $_{54}$Xe, $_{56}$Ba and $_{58}$Ce (Iachello, 1981).

where R^2 is some convenient normalization, usually taken as $R^2 = R_0^2 A^{2/3}$ ($R_0 = 1.2$ fm). In electromagnetic transitions and electron-scattering experiments only the proton scalar densities are relevant, since the neutrons *cannot* acquire a monopole effective charge. Neutron scalar densities and radii can only be measured in experiments with hadronic probes.

5.8.2 *Quadrupole densities*

Quadrupole densities can be discussed in a similar way. The appropriate operator can be written as

$$T^{(E2)}(r) = e_\pi^{(2)}\{\alpha_\pi^{(2)}(r)[d_\pi^\dagger \times \tilde{s}_\pi + s_\pi^\dagger \times \tilde{d}_\pi]^{(2)} + \beta_\pi^{(2)}(r)[d_\pi^\dagger \times \tilde{d}_\pi]^{(2)}\}$$
$$+ e_\nu^{(2)}\{\alpha_\nu^{(2)}(r)[d_\nu^\dagger \times \tilde{s}_\nu + s_\nu^\dagger \times \tilde{d}_\nu]^{(2)} + \beta_\nu^{(2)}(r)[d_\nu^\dagger \times \tilde{d}_\nu]^{(2)}\}, \qquad (5.105)$$

where we have introduced the quadrupole effective charges $e_\pi^{(2)}, e_\nu^{(2)}$. Introducing the notation $|N_\pi, N_\nu, 2_i^+\rangle$ for 2^+ states, one obtains

$$\rho^{(2)}_{0_1^+\to 2_i^+}(r) = e_\pi^{(2)}\{\alpha_\pi^{(2)}(r)A^{(2)}_{\pi,\,1i} + \beta_\pi^{(2)}(r)B^{(2)}_{\pi,\,1i}\} + e_\nu^{(2)}\{\alpha_\nu^{(2)}(r)A^{(2)}_{\nu,\,1i} + \beta_\nu^{(2)}(r)B^{(2)}_{\nu,\,1i}\}, \tag{5.106}$$

where $A^{(2)}_{\pi,\,1i}$, $B^{(2)}_{\pi,\,1i}$, $A^{(2)}_{\nu,\,1i}$, $B^{(2)}_{\nu,\,1i}$ are matrix elements of the appropriate operators. From (5.106) one can compute $B(\mathrm{E2})$ values

$$B(\mathrm{E2}; 0_1^+ \to 2_i^+) = \left[\int_0^\infty r^4 \rho^{(2)}_{0_1^+\to 2_i^+}(r)\,\mathrm{d}r\right]^2. \tag{5.107}$$

In the case of E2 transitions, both protons and neutrons contribute since the neutrons can acquire a quadrupole effective charge.

5.8.3 *Hexadecapole densities*

The transition operator here is

$$T^{(\mathrm{E4})}(r) = e_\pi^{(4)}\beta_\pi^{(4)}(r)[d_\pi^\dagger \times \tilde{d}_\pi]^{(4)} + e_\nu^{(4)}\beta_\nu^{(4)}(r)[d_\nu^\dagger \times \tilde{d}_\nu]^{(4)}, \tag{5.108}$$

where $e_\pi^{(4)}$ and $e_\nu^{(4)}$ are hexadecapole effective charges. Introducing the notation $|N_\pi, N_\nu, 4_i^+\rangle$ for 4^+ states, one obtains

$$\rho_{0_1^+\to 4_i^+}(r) = e_\pi^{(4)}\beta_\pi^{(4)}(r)B^{(4)}_{\pi,\,1i} + e_\nu^{(4)}\beta_\nu^{(4)}(r)B^{(4)}_{\nu,\,1i}, \tag{5.109}$$

where $B^{(4)}_{\pi,\,1i}$ and $B^{(4)}_{\nu,\,1i}$ are appropriate matrix elements. From (5.109), one can derive $B(\mathrm{E4})$ values

$$B(\mathrm{E4}; 0_1^+ \to 4_i^+) = \left[\int_0^\infty r^6 \rho^{(4)}_{0_1^+\to 4_i^+}(r)\,\mathrm{d}r\right]^2. \tag{5.110}$$

5.8.4 *Magnetic dipole densities*

The appropriate operator is

$$T^{(\mathrm{M1})}(r) = g_\pi^{(1)}\beta_\pi^{(1)}(r)[d_\pi^\dagger \times \tilde{d}_\pi]^{(1)} + g_\nu^{(1)}\beta_\nu^{(1)}(r)[d_\nu^\dagger \times \tilde{d}_\nu]^{(1)}, \tag{5.111}$$

where $g_\pi^{(1)}$ and $g_\nu^{(1)}$ are effective boson g-factors. From (5.111) one can calculate transition densities. Denoting by $|N_\pi, N_\nu, 1_i^+\rangle$ the 1^+ states, one obtains

$$\rho^{(1)}_{0_1^+\to 1_i^+}(r) = g_\pi^{(1)}\beta_\pi^{(1)}(r)B^{(1)}_{\pi,\,1i} + g_\nu^{(1)}\beta_\nu^{(1)}(r)B^{(1)}_{\nu,\,1i}, \tag{5.112}$$

where $B^{(1)}_{\pi,\,1i}$ and $B^{(1)}_{\nu,\,1i}$ are the appropriate matrix elements. This expression has been used by Bohle *et al.* (1984) in the analysis of their experiment. The

$B(\mathrm{M}1)$ values are given by

$$B(\mathrm{M}1; 0_1^+ \to 1_i^+) = \left[\int_0^\infty r^2 \rho^{(1)}_{0_1^+ \to 1_i^+}(r)\,\mathrm{d}r\right]^2. \tag{5.113}$$

One can also use (5.111) to compute magnetic moments. One then obtains expressions identical to those discussed in Sect. 5.5.1.

5.8.5 *Magnetic octupole densities*

Finally, we briefly discuss magnetic octupole densities. The appropriate operator here is

$$T^{(\mathrm{M}3)}(r) = g_\pi^{(3)} \beta_\pi^{(3)}(r)[d_\pi^\dagger \times \tilde{d}_\pi]^{(3)} + g_\nu^{(3)} \beta_\nu^{(3)}(r)[d_\nu^\dagger \times \tilde{d}_\nu]^{(3)}, \tag{5.114}$$

where we have introduced octupole effective g-factors, $g_\pi^{(3)}$ and $g_\nu^{(3)}$. Transition densities to 3^+ states, $|N_\pi, N_\nu, 3_i^+\rangle$, are given by

$$\rho^{(3)}_{0_1^+ \to 3_i^+} = g_\pi^{(3)} \beta_\pi^{(3)}(r) B^{(3)}_{\pi,\,1i} + g_\nu^{(3)} \beta_\nu^{(3)}(r) B^{(3)}_{\nu,\,1i}, \tag{5.115}$$

where $B^{(3)}_{\pi,\,1i}$, $B^{(3)}_{\nu,\,1i}$ are appropriate matrix elements. $B(\mathrm{M}3)$ values can be obtained from (5.115) as usual

$$B(\mathrm{M}3; 0_1^+ \to 3_i^+) = \left[\int_0^\infty r^4 \rho^{(3)}_{0_1^+ \to 3_i^+}(r)\,\mathrm{d}r\right]^2. \tag{5.116}$$

5.8.6 *Experimental determinations of transition densities*

Several experimental studies of transition densities in medium-mass and heavy nuclei have been performed recently (Bazantay *et al.*, 1985; Goutte, 1984; de Jager, 1984; van der Laan *et al.*, 1985; Borghols *et al.*, 1985). From these studies, it appears that the boson transition densities vary very little within a major shell and thus that they can be analyzed within the framework discussed in this section. In studies performed up to now the proton and neutron densities have been assumed to be identical. An example of experimentally-extracted boson quadrupole densities

$$\alpha^{(2)}(r) = \alpha_\pi^{(2)}(r) = \alpha_\nu^{(2)}(r)$$

and

$$\beta^{(2)}(r) = \beta_\pi^{(2)}(r) = \beta_\nu^{(2)}(r)$$

is shown in Fig. 5.19. It would be interesting to devise experiments to determine the proton and neutron densities separately.

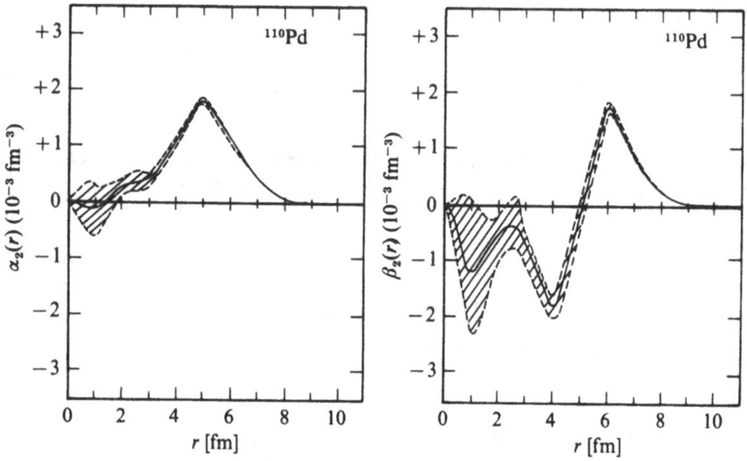

Fig. 5.19. Quadrupole boson densities in ^{110}Pd deduced from the transition charge densities of the two lowest 2^+ states (van der Laan *et al.*, 1985; de Jager, 1984).

5.9 Relation between the interacting boson models-1 and -2

Properties of even–even nuclei have been extensively analyzed using the interacting boson model-1. This model has a simple structure which allows one to make straight-forward comparisons with experiment. The microscopic interpretation of the bosons as nucleon pairs has led to the introduction of the interacting boson model-2. This model is more realistic but also more complex. For this reason, one may wish to relate the two models in such a way that calculations in one model can be translated into calculations in the other. Since the interacting boson model-2 contains more states than the model-1, a relation between the two models can be found by projecting the interacting boson model-2 operators on the set of states of the interacting boson model-1, i.e. the states with maximum F-spin, $F = F_{\text{max}} = N/2$. Starting from a general IBM-2 Hamiltonian, one can, for example, construct a projected Hamiltonian in the space of F_{max} which can be identified with an IBM-1 Hamiltonian (Harter *et al.*, 1985; Novoselski and Talmi, 1985).

Starting from the interacting boson model-2 Hamiltonian

$$H = \varepsilon(\hat{n}_{d_\pi} + \hat{n}_{d_\nu}) + \kappa \hat{Q}_\pi^\chi \cdot \hat{Q}_\nu^\chi + \lambda' \hat{M}_{\pi\nu}, \tag{5.117}$$

one obtains the projected Hamiltonian

$$H^P = C + \varepsilon' \hat{n}_d + \kappa' \hat{Q}^\chi \cdot \hat{Q}^\chi + \kappa' \chi'^2 \hat{Q}' \cdot \hat{Q}', \tag{5.118}$$

where

$$
\begin{aligned}
C &= -\frac{5N_\pi N_\nu}{N-1}\kappa, \\
\varepsilon' &= \varepsilon + (4-\chi_\pi\chi_\nu)\frac{N_\pi N_\nu}{N(N-1)}\kappa, \\
\kappa' &= \frac{N_\pi N_\nu}{N(N-1)}\kappa, \\
\chi &= \tfrac{1}{2}(\chi_\pi + \chi_\nu), \\
\chi' &= \tfrac{1}{2}(\chi_\pi - \chi_\nu).
\end{aligned}
\tag{5.119}
$$

The projected Hamiltonian does not depend on the strength of the Majorana operator, λ'.

Another question that one can investigate is to what extent states calculated within the framework of the interacting boson model-2 have a definite value of F-spin. This obviously depends on the values of the parameters in (5.119). Novoselski and Talmi (1985) find that for realistic values of the parameters used to describe the spectrum of ^{178}Hf, the 0^+ ground state of this nucleus contains 82.1 % of $F=F_{\max}$ and 16.4 % of $F=F_{\max}-1$. However, an interesting result is that the mixture of states with $F<F_{\max}$ appears to be constant for all states in the ground-state band. Harter *et al.* (1985) also find that a large number of nuclei can be arranged into F-spin multiplets. This, in principle, requires a Hamiltonian which is F-spin invariant, while (5.119) is not. However, the results of Novoselski and Talmi appear to indicate that this result may also be due to the fact that admixtures of states with $F<F_{\max}$ are constant in an F-spin multiplet.

6
Geometry

6.1 Introduction

Geometric properties of the interacting boson model-2 are a straightforward generalization of those of the interacting boson model-1. Since there are here two types of particles, protons and neutrons, one obtains two types of geometric variables $\alpha_{\pi,\mu}$ and $\alpha_{\nu,\mu}$. This implies that, in terms of shape variables one has now a two-fluid model, a proton fluid and a neutron fluid (Greiner, 1965, 1966).

Several properties of the interacting proton–neutron system can be investigated using coherent state techniques. However, a systematic exploitation of these techniques has begun only recently and they have been mostly concerned with the study of F-vector excitations in deformed nuclei.

6.2 Coherent states

We follow here the discussion of Chapter 3 and introduce coset spaces for protons and neutrons. Since each of them is described by the group U(6), each coset space is defined by five complex variables, which we denote by $\alpha_{\pi,\mu}$ and $\alpha_{\nu,\mu}$. The geometric space of the interacting boson model-2 is thus the product of two cosets,

$$\frac{\mathrm{U}_\pi(6)}{\mathrm{U}_\pi(5)\otimes\mathrm{U}_\pi(1)}\otimes\frac{\mathrm{U}_\nu(6)}{\mathrm{U}_\nu(5)\otimes\mathrm{U}_\nu(1)}. \tag{6.1}$$

Projective coherent states can be defined over the variables $\alpha_{\pi,\mu}$ and $\alpha_{\nu,\mu}$ ($\mu=0,\ \pm1,\ \pm2$) by

$$|N_\pi, N_\nu; \alpha_{\pi,\mu}, \alpha_{\nu,\mu}\rangle=\left(s_\pi^\dagger+\sum_\mu \alpha_{\pi,\mu}d_{\pi,\mu}^\dagger\right)^{N_\pi}\left(s_\nu^\dagger+\sum_\mu \alpha_{\nu,\mu}d_{\nu,\mu}^\dagger\right)^{N_\nu}|0\rangle, \tag{6.2}$$

where $|0\rangle$ is a boson vacuum both for protons and neutrons.

In studying static properties, the ten complex variables $\alpha_{\pi,\mu}$ and $\alpha_{\nu,\mu}$ can be chosen to be real. Furthermore, instead of using $\alpha_{\pi,\mu}$ and $\alpha_{\nu,\mu}$, one can introduce intrinsic variables for protons and neutrons by the transformations

$$\alpha_{\pi,\mu}=\sum_{\mu'} a_{\pi,\mu'}\mathscr{D}^{(2)}_{\mu\mu'}(\Omega_\pi),$$
$$\alpha_{\nu,\mu}=\sum_{\mu'} a_{\nu,\mu'}\mathscr{D}^{(2)}_{\mu\mu'}(\Omega_\nu), \tag{6.3}$$

where $\Omega_\pi\equiv(\theta_{1\pi},\ \theta_{2\pi},\ \theta_{3\pi})$ and $\Omega_\nu\equiv(\theta_{1\nu},\ \theta_{2\nu},\ \theta_{3\nu})$ and

$$a_{\rho,0}=\beta_\rho\cos\gamma_\rho,$$
$$a_{\rho,\pm 2}=\frac{1}{\sqrt{2}}\beta_\rho\sin\gamma_\rho, \tag{6.4}$$
$$a_{\rho,\pm 1}=0, \quad \rho=\pi,\nu.$$

One is thus reduced to four intrinsic variables , $\beta_\pi, \gamma_\pi, \beta_\nu, \gamma_\nu$ and six Euler angles. It is convenient to perform a further transformation introducing the Euler angles $\Omega\equiv(\theta_1,\theta_2,\theta_3)$ describing the orientation of the mass distribution and those $\Phi\equiv(\phi_1,\phi_2,\phi_3)$ describing the relative orientation of the neutron intrinsic system with respect to the proton intrinsic system (Dieperink, 1983). This is schematically shown in Fig. 6.1.

Using the coherent state (6.2) it is possible to evaluate a large number of properties of the interacting boson model-2. For example, one may evaluate the energy functional

$$E(N_\pi,N_\nu;\beta_\pi,\gamma_\pi,\beta_\nu,\gamma_\nu,\Phi)=\frac{\langle N_\pi,N_\nu;\beta_\pi,\gamma_\pi,\beta_\nu,\gamma_\nu,\Phi|H|N_\pi,N_\nu;\beta_\pi,\gamma_\pi,\beta_\nu,\gamma_\nu,\Phi\rangle}{\langle N_\pi,N_\nu;\beta_\pi,\gamma_\pi,\beta_\nu,\gamma_\nu,\Phi|N_\pi,N_\nu;\beta_\pi,\gamma_\pi,\beta_\nu,\gamma_\nu,\Phi\rangle}, \tag{6.5}$$

where

$$H=H_\pi+H_\nu+V_{\pi\nu}. \tag{6.6}$$

Minimizing E with respect to β_π, γ_π, β_ν, γ_ν, $\Phi=(\phi_1,\ \phi_2,\ \phi_3)$, gives the equilibrium 'shape' corresponding to the boson Hamiltonian, H (Hartree–Bose procedure). This equilibrium shape can be represented in a plot similar to that of Fig. 6.2.

If one wished to do so, one could relate the variables $\beta_\pi, \gamma_\pi, \beta_\nu, \gamma_\nu$ to Bohr variables describing the distortion of the protons $\hat{\delta}_\pi, \hat{\gamma}_\pi$, and of the neutrons

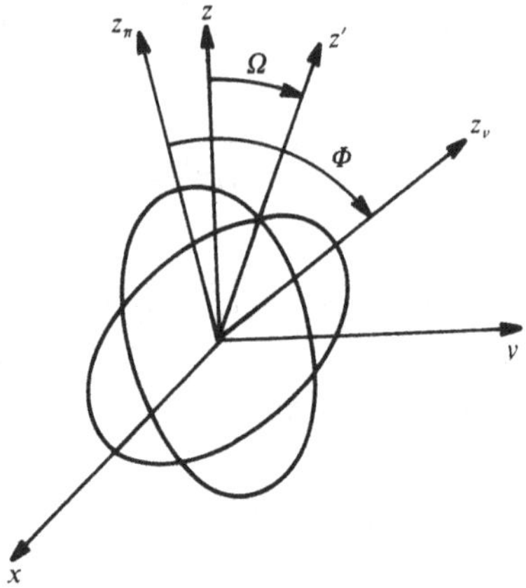

Fig. 6.1. Schematic representation of the Euler angles Ω and Φ, needed to describe the proton–neutron system.

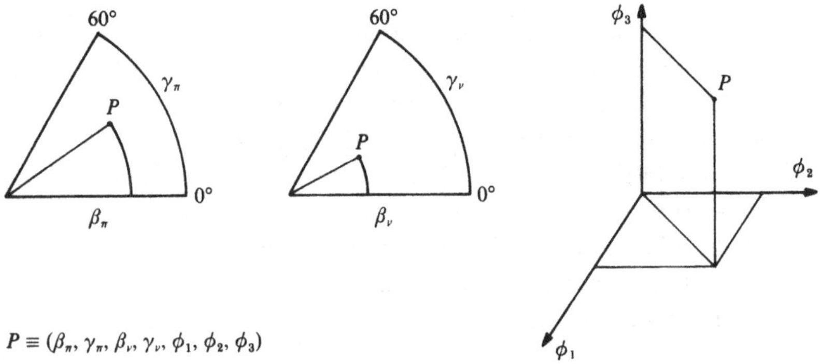

Fig. 6.2. The β_π, γ_π, β_ν, γ_ν, ϕ space for the interacting boson model-2.

$\hat{\delta}_\nu, \hat{\gamma}_\nu$, or the deformation parameters of protons and neutrons,

$$
\begin{aligned}
R_\pi(\theta_\pi, \phi_\pi) &= R_{\pi 0}\left(1 + \sum_\mu \hat{\alpha}_{\pi,2\mu} Y^*_{2\mu}(\theta_\pi, \phi_\pi)\right), \\
R_\nu(\theta_\nu, \phi_\nu) &= R_{\nu 0}\left(1 + \sum_\mu \hat{\alpha}_{\nu,2\mu} Y^*_{2\mu}(\theta_\nu, \phi_\nu)\right), \\
\hat{\beta}^2_{\pi 2} &= \sum_\mu |\hat{\alpha}_{\pi,2\mu}|^2, \\
\hat{\beta}^2_{\nu 2} &= \sum_\mu |\hat{\alpha}_{\nu,2\mu}|^2.
\end{aligned}
\tag{6.7}
$$

These can be related to the variables $\beta_\pi, \gamma_\pi, \beta_\nu, \gamma_\nu$ by

$$\hat{\gamma}_\pi = \gamma_\pi; \quad \hat{\gamma}_\nu = \gamma_\nu;$$
$$\hat{\beta}_\pi \approx c\beta_\pi; \quad \hat{\beta}_\nu \approx c\beta_\nu; \tag{6.8}$$

where c is an average scale parameter which again depends on the definition of $\hat{\beta}$. The scale parameter c can be obtained by a relation similar to (3.27). For rare-earth nuclei,

$$\hat{\delta}_\pi \approx 0.15\beta_\pi,$$
$$\hat{\delta}_\nu \approx 0.15\beta_\nu. \tag{6.9}$$

Dieperink has analyzed, using (6.5), the intrinsic shapes corresponding to the limits II_2 and IV.

(i) Chain II_2

The equilibrium values occur, in the limit of large N_π and N_ν, for $\beta_\pi = \beta_\nu = \sqrt{2}$, and for any N, at $\gamma_\pi = \gamma_\nu = 0°$. The two intrinsic shapes are axially symmetric. There is only one relative angle between the two symmetry axes that suffices to specify the relative orientation. This can be taken, for example, as ϕ_2. The equilibrium value is $\phi_2 = 0°$. This limit thus corresponds to an axially-symmetric deformed rotor, with the same axis for protons and neutrons. This situation is similar to that described by the interacting boson model-1, in which no distinction is made between protons and neutrons. However, in contrast to that case, there are now possible oscillations of the proton versus the neutron deformation. These will be discussed in Sect. 6.3.

(ii) Chain IV

The equilibrium values occur, in the limit of large N_π and N_ν, for $\beta_\pi = \beta_\nu = \sqrt{2}$ and for any N at $\gamma_\pi = 0°, \gamma_\nu = 60°$. Again, there is only one relative angle specifying the relative orientation. The equilibrium value is $\phi_2 = 90°$. The geometric interpretation is that of a prolate proton and an oblate neutron axial rotor (or vice-versa) coupled in such a way as to maximize their overlap. The resulting mass distribution can be parametrized by an asymmetry parameter, $\bar{\gamma}$, given by (Bohr and Mottelson, 1975)

$$\tan \bar{\gamma} = \sqrt{2}\, \bar{Q}_{+2}/\bar{Q}_0, \tag{6.10}$$

where $\bar{Q}$ represents the expectation value of the mass quadrupole operator (i.e. sum of protons and neutrons). By taking a prolate proton distribution with respect to the z-axis, and an oblate neutron distribution with respect to the x-axis, one finds (Dieperink, 1983)

$$\bar{Q}_0 = Q_{\pi,0} - \tfrac{1}{2} Q_{\nu,0},$$
$$\bar{Q}_{+2} = -\tfrac{1}{2}\sqrt{\tfrac{3}{2}}\, Q_{\nu,0}. \tag{6.11}$$

For $N_\pi = N_\nu$, one finds

$$Q_{\pi,0} = -Q_{\nu,0}, \tag{6.12}$$

and therefore, from (6.11) and (6.10),

$$\bar{\gamma} = 30^\circ. \tag{6.13}$$

Thus the mass distribution has a triaxial deformation.

6.3 Boson condensates

The coherent states (6.2) and their generalizations can be used to derive a variety of results. An interesting application has been done by Dieperink, who has evaluated the $B(\mathrm{M}1)$ value from the ground state of a deformed nucleus to the first excited 1^+ state with $F = F_{\max} - 1$. As will be shown in the following section, this state corresponds to oscillations of the proton versus the neutron deformation. Dieperink has considered a ground state of the type

$$|g\rangle = (N_\pi!\, N_\nu!)^{-\frac{1}{2}} (b^\dagger_{\pi,g})^{N_\pi} (b^\dagger_{\nu,g})^{N_\nu} |0\rangle \tag{6.14}$$

where

$$b^\dagger_{\rho,g} = 1/\sqrt{3}\, (s^\dagger_\rho + \sqrt{2}\, d^\dagger_{\rho,0}), \quad \rho = \pi, \nu, \tag{6.15}$$

corresponding to chain II_2, $\mathrm{SU}_\pi(3) \otimes \mathrm{SU}_\nu(3)$, and thus describing two axially symmetric rotors. The 1^+ state, which we shall call here t-vibration (for twisting), can be described by a boson condensate of the type

$$|t\rangle = \frac{1}{\sqrt{2}} \left(\sqrt{\left(\frac{N_\pi}{N_\pi + N_\nu}\right)}\, d^\dagger_{\pi,+1} b_{\pi,g} - \sqrt{\left(\frac{N_\nu}{N_\pi + N_\nu}\right)}\, d^\dagger_{\nu,+1} b_{\nu,g} \right) |g\rangle. \tag{6.16}$$

Using the operator $T^{(\mathrm{M}1)}$ of (4.41), one then finds

$$\langle t | T^{(\mathrm{M}1)}_{+1} + T^{(\mathrm{M}1)}_{-1} | g \rangle = 2\sqrt{\left(\frac{3}{4\pi}\right)} \sqrt{\left(\frac{N_\pi N_\nu}{N_\pi + N_\nu}\right)} (g_\pi - g_\nu), \tag{6.17}$$

from which one can obtain (Dieperink, 1983)

$$B(\mathrm{M}1; 0^+_1 \to 1^+_1) = \frac{3}{4\pi} \left(\frac{4 N_\pi N_\nu}{N_\pi + N_\nu} \right) (g_\pi - g_\nu)^2. \tag{6.18}$$

Comparing this with the exact result (5.66), one finds that they agree to order $1/N$.

6.4 Dynamic properties

The techniques described in Sect. 3.10 can be generalized to the interacting boson model-2, thus providing a way to solve problems in this case. It is convenient here to use group coherent states defined by

$$|N_\pi, N_\nu; \zeta_{\pi,\mu}, \zeta_{\nu,\mu}\rangle = (N_\pi!\, N_\nu!)^{-\frac{1}{2}} \left[\sqrt{\left(1-\sum_\mu \zeta^*_{\pi,\mu}\zeta_{\pi,\mu}\right)}\, s^\dagger_\pi + \zeta_{\pi,\mu} d^\dagger_{\pi,\mu}\right] \times \left[\sqrt{\left(1-\sum_\mu \zeta^*_{\nu,\mu}\zeta_{\nu,\mu}\right)}\, s^\dagger_\nu + \zeta_{\nu,\mu} d^\dagger_{\nu,\mu}\right]|0\rangle, \tag{6.19}$$

in terms of ten complex variables $\zeta_{\pi,\mu}, \zeta_{\nu,\mu}$. One then introduces coordinates and momenta through

$$\zeta_{\pi,\mu} = q_{\pi,\mu} + \mathrm{i}\, p_{\pi,\mu}, \qquad \zeta_{\nu,\mu} = q_{\nu,\mu} + \mathrm{i}\, p_{\nu,\mu}. \tag{6.20}$$

Evaluating the expectation value of H in the intrinsic state (6.19), gives the classical Hamiltonian H_{cl}, as in (3.69). The rotational invariance of the problem can be exploited by transforming to proton and neutron intrinsic frames, through

$$q_{\rho,\mu} = \sum_{\mu'} u_{\rho,\mu'} \mathscr{D}^{(2)}_{\mu\mu'}(\Omega_\rho), \qquad p_{\rho,\mu} = \sum_{\mu'} v_{\rho,\mu'} \mathscr{D}^{*(2)}_{\mu\mu'}(\Omega_\rho), \quad \rho = \pi, \nu. \tag{6.21}$$

One can then introduce proton and neutron intrinsic variables through

$$u_{\rho,0} = \frac{1}{\sqrt{2}} \bar{\beta}_\rho \cos\gamma_\rho, \qquad u_{\rho,\pm 2} = \tfrac{1}{2}\bar{\beta}_\rho \sin\gamma_\rho, \qquad u_{\rho,\pm 1} = 0, \quad \rho = \pi, \nu, \tag{6.22}$$

where

$$\bar{\beta}_\rho = \frac{\sqrt{2}\beta_\rho}{\sqrt{(1+\beta^2_\rho)}}, \quad \rho = \pi, \nu, \tag{6.23}$$

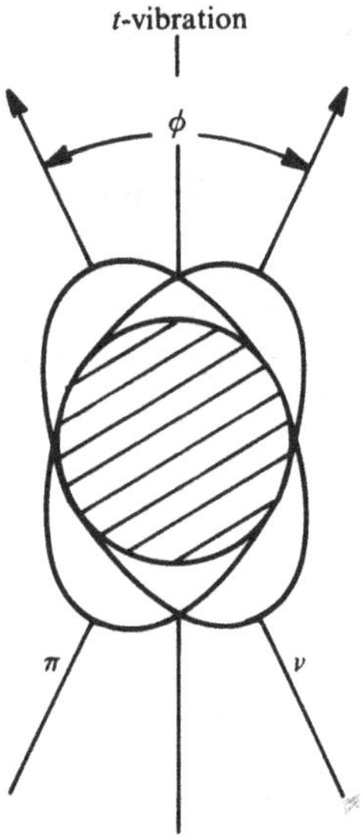

Fig. 6.3. The geometry of the t-vibration in axially deformed nuclei.

as in (3.72). In addition to the four intrinsic variables, one has six Euler angles, which can be chosen as in Sect. 6.2 as defining the orientation of the mass distribution $\Omega \equiv (\theta_1, \theta_2, \theta_3)$ and the relative orientation of the proton versus the neutron distribution, $\Phi \equiv (\phi_1, \phi_2, \phi_3)$. Associated with all these variables there are the momenta $p_{\beta\pi}, p_{\beta\nu}, p_{\gamma\pi}, p_{\gamma\nu}, p_{\theta 1}, p_{\theta 2}, p_{\theta 3}, p_{\phi 1}, p_{\phi 2}, p_{\phi 3}$, which can be obtained as in (3.73).

Bijker (1985) has evaluated the excitation energy of the t-vibration, using this technique by deriving H_{cl} with the following assumptions: (i) both the proton and neutron distributions have prolate axial symmetry ($\gamma_{e,\pi} = \gamma_{e,\nu} = 0°$); (ii) the shape of the neutron and proton distributions does not change ($p_{\beta\pi} = p_{\beta\nu} = p_{\gamma\pi} = p_{\gamma\nu} = 0$); and (iii) the oscillation of the neutron versus the proton distribution is in the second Euler angle $\phi_2 = \phi$. This implies that one can write for the proton and neutron angles, $\theta_{\nu 2} = -\phi/2$, $\theta_{\pi 2} = +\phi/2$, $\theta_{\nu 1} = \theta_{\nu 3} = p_{\theta_{\nu 1}} = p_{\theta_{\nu 3}} = 0$, $\theta_{\pi 1} = \theta_{\pi 3} = p_{\theta_{\pi 1}} = p_{\theta_{\pi 3}} = 0$. The geometry of this problem is shown in Fig. 6.3.

Under the assumptions (i)–(iii), the classical Hamiltonian H_{cl} depends only on four variables.

$$H_{\mathrm{cl}} = H_{\mathrm{cl}}(\tilde{\beta}_\pi, \tilde{\beta}_\nu, \phi, p_\phi), \tag{6.24}$$

where

$$\phi = \theta_{\pi 2} - \theta_{\nu 2}, \tag{6.25}$$

and

$$p_\phi = p_{\theta\pi 2} - p_{\theta\nu 2} \tag{6.26}$$

is the momentum associated with the ϕ-variable. The energy of the t-vibration, which, in this approximation, is the lowest excitation of a harmonic oscillator in ϕ, can be obtained by expanding $H_{\rm cl}$ around the minimum ($\bar\beta_\pi = \bar\beta_{e,\pi}$; $\bar\beta_\nu = \bar\beta_{e,\nu}$; $\phi = 0°$) of the classical Hamiltonian,

$$H_{\rm cl} = H_{\rm cl}(\text{equilibrium}) + \frac{1}{2C} p_\phi^2 + \frac{1}{2} D_{\phi^2}, \tag{6.27}$$

where

$$C = \left.\frac{\partial^2 H_{\rm cl}}{\partial p_\phi^2}\right|_{\phi = 0°,\ \bar\beta_\rho = \bar\beta_{e,\rho}}. \tag{6.28}$$

The excitation energy of the t-vibration is then

$$E_t = (D/C)^{\frac{1}{2}}. \tag{6.29}$$

Bijker considered the Hamiltonian

$$\begin{aligned} H = {} & \varepsilon_\pi \hat n_{d\pi} + \varepsilon_\nu \hat n_{d\nu} + \kappa \hat Q_\pi^\chi \cdot \hat Q_\nu^\chi + \lambda' \hat M_{\pi\nu} \\ & + \sum_{L=1,3} \kappa_L [d_\pi^\dagger \times \tilde d_\pi]^{(L)} \cdot [d_\nu^\dagger \times \tilde d_\nu]^{(L)}, \end{aligned} \tag{6.30}$$

which is slightly more general than (4.44). He found

$$\begin{aligned} D = 3N_\pi N_\nu \bar\beta_{e,\pi} \bar\beta_{e,\nu} \Big\{ & \kappa \Big[\frac{1}{\sqrt7} \chi_\nu \bar\beta_{e,\nu} \sqrt{(1 - \tfrac12 \bar\beta_{e,\pi}^2)} + \frac{1}{\sqrt7} \chi_\pi \bar\beta_{e,\pi} \sqrt{(1 - \tfrac12 \bar\beta_{e,\nu}^2)} \\ & - 2\sqrt{[(1 + \tfrac12 \bar\beta_{e,\pi}^2)(1 - \tfrac12 \bar\beta_{e,\nu}^2)]} - \tfrac{1}{14} \chi_\pi \chi_\nu \bar\beta_{e,\nu} \bar\beta_{e,\pi} \Big] \\ & + \lambda' [\sqrt{[(1 - \tfrac12 \bar\beta_{e,\pi}^2)(1 - \tfrac12 \bar\beta_{e,\nu}^2)]} + \tfrac12 \bar\beta_{e,\pi} \bar\beta_{e,\nu}] \Big\}, \end{aligned} \tag{6.31}$$

and

$$\begin{aligned} \frac{1}{C} = {} & \frac{\varepsilon_\pi}{N_\pi} \frac{1}{3\bar\beta_{e,\pi}^2} + \frac{\varepsilon_\nu}{N_\nu} \frac{1}{3\bar\beta_{e,\nu}} - \frac16 (2\kappa - \lambda') \Bigg[\frac{N_\nu}{N_\pi} \frac{\bar\beta_{e,\nu}}{\bar\beta_{e,\pi}} \sqrt{\left(\frac{1 - \frac12 \bar\beta_{e,\nu}^2}{1 - \frac12 \bar\beta_{e,\pi}^2} \right)} \\ & + \frac{N_\pi}{N_\nu} \frac{\bar\beta_{e,\pi}}{\bar\beta_{e,\nu}} \sqrt{\left(\frac{1 - \frac12 \bar\beta_{e,\pi}^2}{1 - \frac12 \bar\beta_{e,\nu}^2} \right)} \Bigg] \\ & + \kappa \chi_\nu \Bigg[\frac{N_\nu}{N_\pi} \frac{\bar\beta_{e,\nu}^2}{6\sqrt7\, \bar\beta_{e,\pi} \sqrt{(1 - \frac12 \beta_{e,\pi}^2)}} - \frac{N_\pi}{N_\nu} \frac{\bar\beta_{e,\pi} \sqrt{(1 - \frac12 \bar\beta_{e,\pi}^2)}}{3\sqrt7\, \bar\beta_{e,\nu}^2} \Bigg] \end{aligned}$$

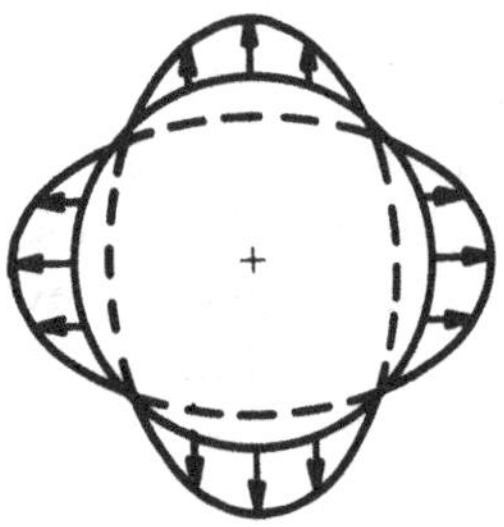

Fig. 6.4. The geometry of F-vector excitations in spherical nuclei.

$$
\begin{aligned}
&+\kappa\chi_\pi\left[\frac{N_\pi}{N_\nu}\frac{\tilde{\beta}^2_{e,\pi}}{6\sqrt{7}\,\tilde{\beta}_{e,\nu}\sqrt{(1-\frac{1}{2}\tilde{\beta}^2_{e,\nu})}}-\frac{N_\nu}{N_\pi}\frac{\tilde{\beta}_{e,\nu}\sqrt{(1-\frac{1}{2}\tilde{\beta}^2_{e,\nu})}}{3\sqrt{7}\,\tilde{\beta}^2_{e,\pi}}\right]\\
&+\frac{1}{42}\kappa\chi_\pi\chi_\nu\left[\frac{N_\nu}{N_\pi}\frac{\tilde{\beta}^2_{e,\nu}}{\tilde{\beta}^2_{e,\pi}}+\frac{N_\pi}{N_\nu}\frac{\tilde{\beta}^2_{e,\pi}}{\tilde{\beta}^2_{e,\nu}}\right]\\
&+\frac{1}{3}\lambda'\left[\frac{N_\nu}{N_\pi}\frac{(1-\frac{1}{2}\tilde{\beta}^2_{e,\nu})}{\tilde{\beta}^2_{e,\pi}}+\frac{N_\pi}{N_\nu}\frac{(1-\frac{1}{2}\tilde{\beta}^2_{e,\pi})}{\tilde{\beta}^2_{e,\nu}}+2\frac{\sqrt{[(1-\frac{1}{2}\tilde{\beta}^2_{e,\pi})(1-\frac{1}{2}\tilde{\beta}^2_{e,\nu})]}}{\tilde{\beta}^2_{e,\pi}\tilde{\beta}^2_{e,\nu}}\right]\\
&-\tfrac{1}{15}(3\kappa_1+2\kappa_3-5\lambda'). \qquad (6.32)
\end{aligned}
$$

It is interesting to note that Eqs. (6.29)–(6.31) and (6.32) give the exact results in the limit of the dynamic symmetries II_1 and III_1,

$$
\begin{aligned}
&\mathrm{II}_1 \quad E_t=N(6\kappa+\lambda'),\\
&\mathrm{III}_1 \quad E_t=N(2\kappa+\lambda').
\end{aligned} \qquad (6.33)
$$

These formulas provide also a good estimate of the excitation energy even in the absence of a dynamic symmetry. For example, using realistic parameters for ^{156}Gd, one obtains $E_t=3.03$ MeV by means of Eqs. (6.29), (6.31) and (6.32), to be compared with the value 3.09 MeV obtained from the exact numerical diagonalization of the Hamiltonian.

Another possible application of the technique described here is to the study of 2^+ asymmetric states in nuclei close to the dynamic symmetry I_1, spherical nuclei. This study is best done by first introducing coordinates and momenta as above and then analyzing the vibrations by a transformation to center of mass and relative $\tilde{\beta}$-variables,

$$
\begin{aligned}
\tilde{\beta}_{\mathrm{CM}}&=\frac{N_\pi}{N}\tilde{\beta}_\pi+\frac{N_\nu}{N}\tilde{\beta}_\nu,\\
\tilde{\beta}_{\mathrm{REL}}&=\tilde{\beta}_\pi-\tilde{\beta}_\nu.
\end{aligned} \qquad (6.34)
$$

The oscillations in the relative variable have a different frequency than those in the center-of-mass variable. The two 2^+ states (symmetric and asymmetric) then correspond to the geometric situation depicted in Fig. 6.4 (Iachello, 1984, 1985).

Part III

The interacting boson model-k

7

The interacting boson models-3 and -4

7.1 Introduction

The techniques discussed in Parts I and II can be extended to include other degrees of freedom, if necessary. These extensions will be generically called interacting boson models-k. In them, particularly important is the extension to include isospin degrees of freedom, since it allows one to treat light nuclei where protons and neutrons occupy the same single-particle orbits and thus where isospin plays a major role. There are two versions of this model, introduced by Elliott and White (1980) and Elliott and Evans (1981) which we shall call interacting boson models-3 and -4. Although, in principle, one should repeat for these models the entire discussion of Parts I and II, we shall confine ourselves to a brief account only, since most properties of the interacting boson models-3 and -4 can be deduced from those of the models-1 and -2.

7.2 The interacting boson model-3

This model is devised to describe the situation in light nuclei, where protons and neutrons occupy the same orbits, Fig. 7.1. Elliott and White (1980) have suggested that in order to treat the isospin degree of freedom properly, one must introduce in these nuclei a third boson, called δ, formed by a proton–neutron (pn) pair, in addition to those of the interacting boson model-2, called π and ν, formed by proton–proton (pp) and neutron–neutron (nn) pairs. The new bosons, with angular momenta and parities $J^P=0^+$ and $J^P=2^+$, will be denoted by s_δ, d_δ. There are thus here $6\times 3=18$ building blocks. The corresponding creation and annihilation operators can be written as

$$\begin{cases} s_\pi^\dagger, d_{\pi,\mu}^\dagger, \quad s_\nu^\dagger, d_{\nu,\mu}^\dagger, \quad s_\delta^\dagger, d_{\delta,\mu}^\dagger, \quad (\mu=0,\pm 1,\pm 2), \\ s_\pi, d_{\pi,\mu}, \quad s_\nu, d_{\nu,\mu}, \quad s_\delta, d_{\delta,\mu}, \quad (\mu=0,\pm 1,\pm 2). \end{cases} \tag{7.1}$$

Operators of different kinds commute, while each set π, ν, δ satisfies Bose commutation relations (1.2).

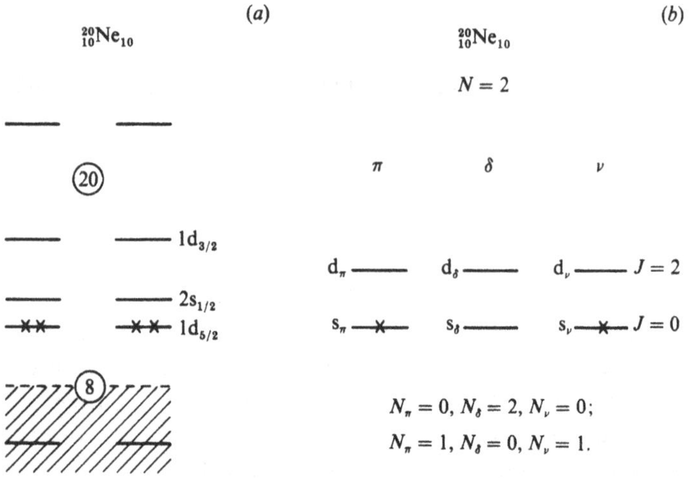

Fig. 7.1. (*a*) Schematic representation of the shell-model problem for $^{20}_{10}Ne_{10}$; (*b*) The boson problem-3 which replaces the shell-model problem for $^{20}_{10}Ne_{10}$. There are in this case two possible boson configurations (bottom right). Only one is schematically represented in the top part of the figure.

A more compact notation is either

$$b^\dagger_{\rho,l,m};\quad b_{\rho,l,m};\quad (\rho=\pi,\delta,\nu;\ l=0,2;\ -l\leqslant m\leqslant +l), \tag{7.2}$$

or

$$b^\dagger_{\rho,\alpha};\quad b_{\rho,\alpha};\quad (\rho=\pi,\delta,\nu;\ \alpha=1,\ldots,6). \tag{7.3}$$

Properties of these operators are a straightforward generalization of those of Part II, Sect. 4.3, and thus we shall not repeat them here.

7.2.1 T-spin

Instead of explicitly writing the label π, ν, δ, one can use here the concept of isotopic spin T, since in order to take into account isospin properties of nuclei the proton–neutron, δ, pair must be in an isospin $T=1$ state. Thus the three bosons, π, δ, ν form an isospin triplet,

$$\begin{aligned}|\pi\rangle &= |1,+1\rangle,\\ |\delta\rangle &= |1,0\rangle,\\ |\nu\rangle &= |1,-1\rangle.\end{aligned} \tag{7.4}$$

The use of the isospin formalism here is much more convenient than the explicit use of the label π, ν, δ.

7.2.2 *Basis states*

The basis states can be written as product of a 'space' part and of an 'isospin' part,

$$\Psi = \Phi \cdot X. \tag{7.5}$$

The states Φ are representations of a group U(6) formed by s and d bosons. Since the three basic bosons form an isospin triplet, the 'isospin' parts for many-boson systems are representations of a group U(3), which we denote by $U_T(3)$. This group contains as a subgroup the usual isospin group $SU_T(2)$. A natural decomposition of $U_T(3)$ is thus

$$U_T(3) \supset SU_T(2) \supset O_T(2). \tag{7.6}$$

The representations of $U_T(3)$ are three-rowed and characterized by the Young tableau

$$[f_1, f_2, f_3] = \begin{array}{l} \overbrace{\square\,\square \cdots \square}^{f_1} \\ \overbrace{\square\,\square}^{f_2} \\ \overbrace{\square}^{f_3} \end{array}. \tag{7.7}$$

Instead of the three quantum numbers f_1, f_2, f_3 one can use the quantum numbers

$$\begin{aligned} N &= f_1 + f_2 + f_3, \\ \lambda &= f_1 - f_2, \\ \mu &= f_2 - f_3, \end{aligned} \tag{7.8}$$

Elliott quantum numbers. The decomposition of $U_T(3)$ into $SU_T(2)$ is then done in the standard way and is reported in Table 7.1. Thus

$$|X\rangle = |N, (\lambda, \mu), T, M_T\rangle. \tag{7.9}$$

Returning to (7.5), we note that since the total wave function Ψ must be totally symmetric, and the isospin part X is a three-rowed tableau, so must be the space part Φ. In other words, the two tableaux Φ and X must be identical for boson systems. This statement is different from the corresponding one for fermion systems (totally antisymmetric wave functions) for which the two tableaux must have conjugate symmetry. As a

Table 7.1. *Partial classification scheme for isovector bosons (Elliott and White, 1980)*

N	(λ,μ)	T
1	(1,0)	1
2	(2,0)	2, 0
	(0, 1)	1
3	(3,0)	3, 1
	(1, 1)	2, 1
	(0,0)	0
4	(4,0)	4, 2, 0
	(2, 1)	3, 2, 1
	(0,2)	2, 0
	(1,0)	1

result, the representation of U(6) must be of the form

$$[N_1,N_2,N_3]=\begin{array}{l}\overbrace{\square\,\square\cdots\square}^{N_1}\\ \overbrace{\square\,\square}^{N_2}\\ \overset{N_3}{\square}\end{array}. \tag{7.10}$$

The decomposition of U(6) into subgroups (2.25) is then done using the techniques discussed in Parts I and II.

7.2.3 *Physical operators*

The structure of all physical operators is as in Parts I and II. However, in the same way in which in Part II the boson–boson interaction could depend on proton or neutron degrees of freedom, here it can be explicitly isospin-dependent. This would reflect the strong isospin dependence of the nucleon–nucleon interaction. This interaction is very different in the states of two nucleons with $T=0$ and $T=1$. The structure of the Hamiltonian is thus the same as in (1.21) but with coefficients that now depend on T. For example,

$$v_2=v_{20}+v_{21}T(T+1), \tag{7.11}$$

or, in general,

$$v_2=v_2(T). \tag{7.12}$$

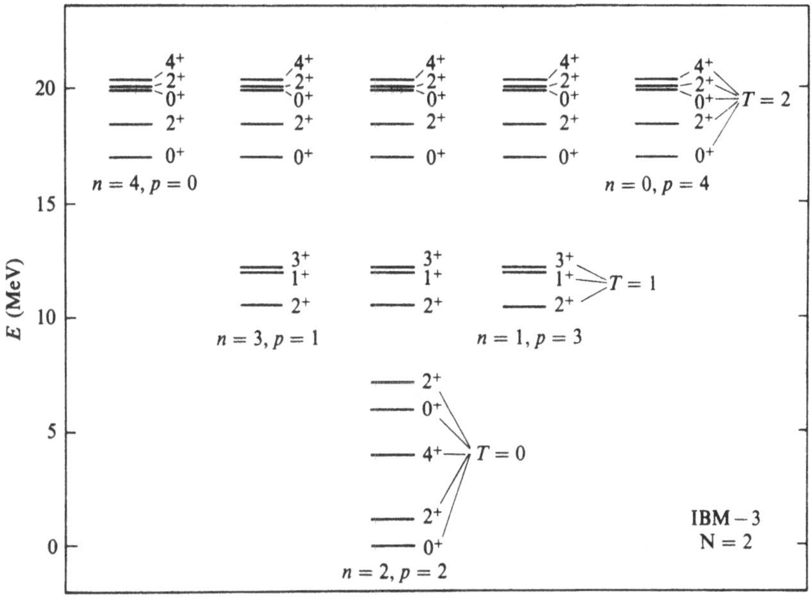

Fig. 7.2. Schematic representation of the spectrum of the interacting boson model-3, for $N=2$. The five isobars, with neutron number n and proton number p, are shown. An experimental example of this type of spectrum is provided by the isobars with mass number $A=20$, ^{20}O, ^{20}F, ^{20}Ne, ^{20}Na, ^{20}Mg.

Elliott and Evans (1981) have analyzed binding energies in light nuclei using this idea.

7.2.4 Energy spectra

It is possible to repeat for this model the same analysis of dynamic symmetries as in Parts I and II. We shall not do this here, but only present a schematic representation of the spectrum of two bosons, $N=2$, Fig. 7.2.

7.3 The interacting boson model-4

The introduction of isotopic spin is necessary in light nuclei in order to take into account isospin properties. One may go even further and introduce, in addition to $T=1$ bosons, $T=0$ bosons also. For $T=1$ bosons the choice of 0^+ and 2^+ angular momenta is clear, being dictated by simple microscopic considerations. For $T=0$ bosons this choice is not so clear. Elliott and Evans (1981) have assumed that both $T=1$ and $T=0$ bosons carry a 'space' part with angular momentum s and d, but that $T=0$ bosons have an 'intrinsic' spin $S=1$, while $T=1$ bosons have $S=0$. This model has thus

$6 \times 6 = 36$ bosonic building blocks with angular momenta, spins and isospins which we now write in the spectroscopic notation, ${}^{2S+1}l_j$,

$$\begin{aligned} T=1:&\quad {}^1s_0, {}^1d_2, \\ T=0:&\quad {}^3s_1, {}^3d_1, {}^3d_2, {}^3d_3. \end{aligned} \tag{7.13}$$

7.3.1 *Basis states*

The basis states here can still be written as in (7.5) but where now X denotes a 'spin–isospin' part. Since the basic bosons form a spin–isospin sextet, $T=1, S=0$; $T=0, S=1$, the states X are representations of a group U(6), which we shall denote by $U_{ST}(6)$. This U(6) group is important because it contains as a subgroup $U_T(3)$ discussed in the previous section, but, most importantly, it contains the Wigner supermultiplet group SU(4) (Wigner, 1937), which is known to have physical significance.

In order to classify the states X, a possible subgroup chain is thus

$$U_{ST}(6) \supset SU(4) \supset SU_S(2) \otimes SU_T(2) \supset O_S(2) \otimes O_T(2). \tag{7.14}$$

The representations of $U_{ST}(6)$ are six-rowed. Elliott and Evans (1981) have considered their decomposition into (7.14), part of which is shown in Table 7.2. For $U_{ST}(6)$ the notation is the usual Young tableaux

$$[N_1, N_2, N_3, N_4, N_5, N_6] \equiv \begin{array}{llll} \square & \square & \square\cdots & \square \\ \square & \square & \square & \\ \square & \square & & \\ \square & \square & & \\ \square & & & \\ \square. & & & \end{array} \tag{7.15}$$

For SU(4), one can use either the Young tableaux $[n_1, n_2, n_3, n_4]$ or the equivalent notation

$$\begin{aligned} \lambda &= n_1 - n_2, \\ \mu &= n_2 - n_3, \\ \nu &= n_3 - n_4. \end{aligned} \tag{7.16}$$

It is worth noting that the eigenvalues of the quadratic Casimir operator of SU(4) are given by

$$\langle \mathscr{C}_2(\mathrm{SU4}) \rangle \equiv g(\lambda, \mu, \nu) = 3\lambda(\lambda+4) + 3\nu(\nu+4) + 4\mu(\mu+4) + 4\mu(\lambda+\nu) + 2\lambda\nu. \tag{7.17}$$

Table 7.2. *Partial classification scheme for sextet bosons (Elliott and Evans, 1981)*

N	$U_{ST}(6)$	$(\lambda\mu\nu)$	$g(\lambda\mu\nu)$	(TS)
1	[1]	(010)	20	(01)(10)
2	[2]	(000)	0	(00)
		(020)	48	(00)(02)(11)(20)
	[11]	(101)	32	(01)(10)(11)
3	[3]	(010)	20	(01)(10)
		(030)	84	(01)(03)(10)(12)(21)(30)
	[21]	(010)	20	(01)(10)
		(111)	60	$(01)(02)(10)(11)^2(12)(20)(21)$
	[111]	(200)	36	(00)(11)
		(002)	36	(00)(11)
4	[4]	(000)	0	(00)
		(020)	48	(00)(02)(11)(20)
		(040)	128	(00)(02)(04)(11)(13)(20)(22)(31)(40)
	[31]	(101)	32	(01)(10)(11)
		(020)	48	(00)(02)(11)(20)
		(121)	96	$(01)(02)(03)(10)(11)^2(12)^2(13)(20)(21)^2$ (22)(30)(31)
	[22]	(000)	0	(00)
		(020)	48	(00)(02)(11)(20)
		(202)	80	$(00)(02)(11)^2(12)(20)(21)(22)$
	[211]	(101)	32	(01)(10)(11)
		(210)	64	(01)(10)(11)(12)(21)
		(012)	64	(01)(10)(11)(12)(21)
	[1111]	(101)	32	(01)(10)(11)

These values are also given in Table 7.2. Thus, in conclusion, the complete classification scheme for the 'spin–isospin' part X is

$$|X\rangle = |[N_1, N_2, N_3, N_4, N_5, N_6], (\lambda, \mu, \nu), S, T, M_S, M_T\rangle. \quad (7.18)$$

Returning to (7.5), we find now that the 'space' part Φ is described by a six-rowed representation of U(6), which is identical to the six-rowed representation of $U_{ST}(6)$, given by (7.15). This representation can be decomposed into those of the subgroups (2.25) as in Parts I and II.

7.3.2 *Physical operators*

Here, as in Sect. 7.2.3, the structure of all operators is as in Parts I and II. However, the coefficients in the Hamiltonian may now depend on (λ, μ, ν), S and T. For example,

$$v_2 = v_{20} + v_{21}T(T+1) + v_{22}S(S+1). \quad (7.19)$$

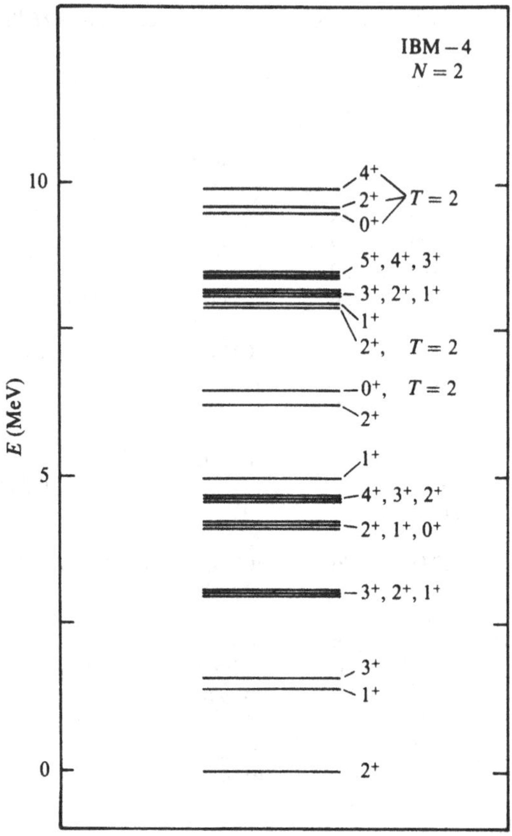

Fig. 7.3. Schematic representation of the spectrum of an odd–odd nucleus, $N=2$, in the interacting boson model-4. An experimental example of this type of spectrum is provided by ${}^{20}_{9}F_{11}$.

7.3.3 *Energy spectra*

An interesting application both of the interacting boson model-3 and -4 is to the spectra of odd–odd nuclei. These spectra can now be included, since one has proton–neutron (pn) pairs. In Fig. 7.3 we show the expected spectrum for the odd–odd nucleus ${}^{20}F$. The interacting boson models-3 and -4 have, at present, been applied only to a few nuclei in the mass region $A \sim 20$.

8

The interacting boson models-G, -F, -CM

8.1 Introduction

Another set of degrees of freedom neglected in the interacting boson models -1 and -2 is that of pairs with values of the angular momenta different from 0 and 2 (s and d bosons). We discuss briefly in this chapter those versions of the model which include other pairs. We note that from microscopic considerations, it appears to be important to include $J^P = 4^+$ pairs in strongly deformed nuclei. The effects of 4^+ pairs (G-pairs) can often be introduced as a renormalization (Otsuka and Ginocchio, 1985) of the parameters of the interacting boson models-1 and -2; for certain properties explicit inclusion of $J^P = 4^+$ bosons (g bosons) may be necessary. The corresponding models will be denoted by interacting boson models-1G and -2G.

From another side, negative-parity collective states appear in the spectra of nuclei at an excitation energy ~ 1 MeV. Some of these states can be described by introducing $J^P = 3^-$ pairs (F-pairs). The effects of these pairs can be taken into account by means of $J^P = 3^-$ bosons (f bosons). The corresponding models will be denoted by interacting boson models-1F and -2F.

Finally, the presence of subshell gaps in the shell-model levels may cause the occurrence of excited modes with angular momenta $J^P = 0^+$ and 2^+ which we denote by S′ and D′ and describe by s′ and d′ bosons. Thus, the complete situation may be as complex as that indicated in Fig. 8.1. We shall now briefly discuss some of these extensions.

8.2 The interacting boson model-1G

Here, one introduces, in addition to s and d bosons, also g bosons,

$$g^\dagger_\mu, \quad g_\mu; \quad (\mu = 0, \pm 1, \pm 2, \pm 3, \pm 4). \tag{8.1}$$

The total space now is $1 + 5 + 9 = 15$ dimensional. There are two types of studies that can be made. In the first, one starts from the algebraic structure

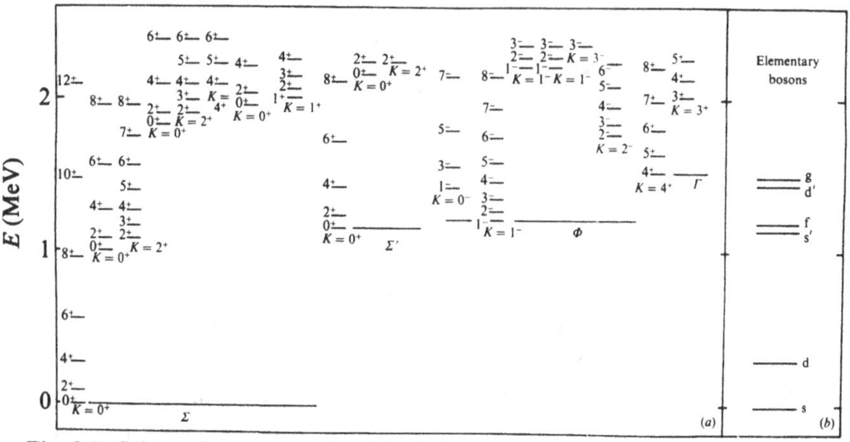

Fig. 8.1. Schematic representation (*a*) of the spectrum of a medium mass nucleus ($A \sim 156$) when the excitation modes shown in (*b*) are included.

of U(15) and attempts to construct its subgroup chains as in Part I. In the second, one instead studies the problem numerically.

The algebraic structure of the U(15) boson model has been studied by Kota (1984), by Wu and Zhou (1984) and others (Ratna Raju, 1981, 1982; Lin, 1982, 1983; Wu, 1982). Particularly interesting here is the chain

$$\mathrm{U}(15) \supset \mathrm{SU}(3) \supset \mathrm{O}(3) \supset \mathrm{O}(2). \tag{8.2}$$

The eight generators of SU(3) in (8.2) can be constructed from s, d and g creation and annihilation operators. Wu and Zhou (1984) use the combination

$$\begin{aligned} X^{(1)}_\mu &= \sqrt{\tfrac{1}{7}}\,[d^\dagger \times \tilde{d}]^{(1)}_\mu + \sqrt{\tfrac{6}{7}}\,[g^\dagger \times \tilde{g}]^{(1)}_\mu, \\ X^{(2)}_\mu &= \sqrt{\tfrac{1}{70}}\,\{4\sqrt{\tfrac{7}{15}}\,[d^\dagger \times \tilde{s} + s^\dagger \times \tilde{d}]^{(2)}_\mu - 11\sqrt{\tfrac{2}{21}}\,[d^\dagger \times \tilde{d}]^{(2)}_\mu \\ &\quad + 36\sqrt{\tfrac{1}{105}}\,[d^\dagger \times \tilde{g} + g^\dagger \times \tilde{d}]^{(2)}_\mu - 2\sqrt{\tfrac{33}{7}}[g^\dagger \times \tilde{g}]^{(2)}_\mu\}. \end{aligned} \tag{8.3}$$

These operators are proportional to the angular momentum $\hat{L}$ and quadrupole operator $\hat{Q}$. States in the chain (8.2) can still be characterized by

$$|[N], \tilde{\chi}', (\lambda, \mu), \tilde{\chi}, L, M_L\rangle, \tag{8.4}$$

where $\tilde{\chi}'$ denotes additional quantum numbers needed to classify the states uniquely. There are, however, two important differences between (8.4) and (2.48):

(i) the representations (λ, μ) start with $(4N, 0)$ and not with $(2N, 0)$;
(ii) among these representations, there is one with $(\lambda, \mu) = (4N-6, 3)$

with $K=1$ and 3 and thus a state with $J^P=1^+$. This state is absent in (2.48).

Wu and Zhou have applied this scheme to ^{168}Er. However, the SU(3) symmetry is badly broken here because of the large single-particle energy of the g boson, Fig. 8.1. Thus, if one wants to improve on it, one must include the effects of a large single-boson energy, ε_g, either by perturbation theory or by direct straightforward diagonalization. Akiyama (1985) and Yoshinaga *et al.* (1986) have recently presented results of the former type of calculations, while van Isacker *et al.* (1981) and Heyde *et al.* (1982, 1983) have presented results of the latter type.

The calculation of van Isacker *et al.* includes states of the type $(sd)^N$ and $(sd)^{N-1}g$, i.e. states in which only one g boson is present. They use the Hamiltonian

$$H=H_{sd}+\varepsilon_g(g^\dagger\cdot\tilde{g})-\kappa_g(\hat{Q}_g\cdot\hat{Q})$$
$$+\xi[[d^\dagger\times d^\dagger]^{(4)}\times[\tilde{g}\times\tilde{s}]^{(4)}+[g^\dagger\times s^\dagger]^{(4)}\times[\tilde{d}\times\tilde{d}]^{(4)}]^{(0)}_0, \qquad (8.5)$$

where H_{sd} is the Hamiltonian describing the s–d bosons, $\hat{Q}$ is the operator (1.50) and

$$\hat{Q}_g=[g^\dagger\times\tilde{g}]^{(2)}. \qquad (8.6)$$

These authors have also investigated the effects of g bosons in the other limits of the interacting boson model-1.

8.3 The interacting boson model-2G

With the addition of more degrees of freedom, the numerical difficulty of the problem increases, and straightforward diagonalization becomes difficult. One of the most promising ways to solve this situation is the use of mean field approximation techniques. Such techniques have been applied to the study of the interacting boson model with proton and neutron g bosons (2G), in regions of strongly deformed nuclei. These techniques are a generalization of those used in Chapters 3 and 6.

For the ground state band, one starts from a trial function

$$|g\rangle=(N_\pi!\,N_\nu!)^{-\frac{1}{2}}(\Gamma^\dagger_{\pi,0})^{N_\pi}(\Gamma^\dagger_{\nu,0})^{N_\nu}|0\rangle, \qquad (8.7)$$

where

$$\Gamma^\dagger_{\rho,0}=\sum_l\eta_{\rho,l0}b^\dagger_{\rho,l,0}, \quad (\rho=\pi,\nu) \qquad (8.8)$$

and axial symmetry has been assumed. The quantitites $\eta_{\rho,l0}$ are treated as variational parameters, which are obtained in Hartree–Bose approximation by minimizing the expectation value of the Hamiltonian H. Pittel *et al.* (1984) take a Hamiltonian of the form

$$H^{2G}=\varepsilon_{\pi,d}\hat{n}_{d_\pi}+\varepsilon_{\nu,d}\hat{n}_{d_\nu}+\varepsilon_{\pi,g}\hat{n}_{g_\pi}+\varepsilon_{\nu,g}\hat{n}_{g_\nu}+F_{\pi\pi}\hat{Q}_\pi\cdot\hat{Q}_\pi$$
$$+F_{\nu\nu}\hat{Q}_\nu\cdot\hat{Q}_\nu+F_{\pi\nu}\hat{Q}_\pi\cdot\hat{Q}_\nu+G_{\pi\nu}\hat{V}_\pi\cdot\hat{V}_\nu, \tag{8.9}$$

where $\hat{Q}$ and $\hat{V}$ are the quadrupole and hexadecapole operators

$$\hat{Q}_{\rho,\mu}=\sum_{ll'}x_{\rho,ll'}[b^\dagger_{\rho,l}\times\tilde{b}_{\rho,l'}]^{(2)}_\mu,$$
$$\hat{V}_{\rho,\mu}=\sum_{ll'}y_{\rho,ll'}[b^\dagger_{\rho,l}\times\tilde{b}_{\rho,l'}]^{(4)}_\mu. \tag{8.10}$$

The Hartree–Bose equations for the variational parameters $\eta_{\rho,l0}$ are

$$\sum_{l'}h_{\rho,ll'}\eta_{\rho,l'0}=E_{\rho,0}\eta_{\rho,l0}, \tag{8.11}$$

where, with the Hamiltonian (8.9),

$$h_{\rho,ll'}=\tilde{\varepsilon}_{\rho,l}\delta_{l,l'}+2F_{\rho\rho}(N_\rho-1)Q_{\rho,00}x_{\rho,ll'}(l0l'0|20)$$
$$+F_{\pi\nu}N_{\rho'}Q_{\rho',00}x_{\rho',ll'}(l0l'0|20)$$
$$+G_{\pi\nu}N_{\rho'}R_{\rho',00}y_{\rho',ll'}(l0l'0|40)$$

$$E_{\rho,0}=\sum_{ll'}h_{\rho,ll'}\eta_{\rho,l0}\eta_{\rho,l'0}, \tag{8.12}$$

$$Q_{\rho,00}=\sum_{ll'}x_{\rho,ll'}(l0l'0|20)\eta_{\rho,l0}\eta_{\rho,l'0},$$

$$R_{\rho,00}=\sum_{ll'}y_{\rho,ll'}(l0l'0|40)\eta_{\rho,l0}\eta_{\rho,l'0},$$

$$\tilde{\varepsilon}_{\rho,l}=\varepsilon_{\rho,l}+\frac{5F_{\rho\rho}}{2l+1}\sum_{l'}x^2_{\rho,ll'}.$$

In these equations, $\rho'=\nu$ if $\rho=\pi$ and $\rho'=\pi$ if $\rho=\nu$. Equations (8.11)–(8.12) represent coupled eigenvalue equations for neutrons and protons, which can be solved using standard iterative techniques.

The Hartree–Bose procedure provides an approximation to the ground-state band. Low-lying excited bands can be generated as in Sect. 3.8 by raising bosons from the condensate. However, here there are several possibilities and thus a matrix must be constructed and diagonalized. (Tamm–Dancoff approximation). The basis states are of the type

$$|\rho, l, K\rangle = [(N_\rho - 1)!\, N_{\rho'}!]^{-\frac{1}{2}} b^\dagger_{\rho,lK} (\Gamma^\dagger_{\rho,0})^{N_\rho - 1} (\Gamma^\dagger_{\rho',0})^{N_{\rho'}} |0\rangle, \quad (8.13)$$

for $K \neq 0$. For $K = 0$, they are of the type

$$|\rho, l, 0\rangle = [(N_\rho - 1)!\, N_{\rho'}!]^{-\frac{1}{2}} \Gamma'_{\rho,0} (\Gamma^\dagger_{\rho,0})^{N_\rho - 1} (\Gamma^\dagger_{\rho',0})^{N_{\rho'}} |0\rangle, \quad (8.14)$$

where $\Gamma'_{\rho,0}$ is as in (8.8), but with all possible orthogonal combinations.

For each K, the matrix elements of the Tamm–Dancoff matrix are (Pittel *et al.*, 1984)

$$\begin{aligned} t_{\rho l,\rho l'} &= \tilde{\varepsilon}_{\rho,l} \delta_{ll'} + 2(-)^K F_{\rho\rho}(N_\rho - 1) Q_{\rho,00} x_{\rho,ll'} (lKl' - K|20) \\ &\quad + (-)^K F_{\rho\rho'} N_{\rho'} Q_{\rho',00} x_{\rho,ll'} (lKl' - K|20) \\ &\quad + (-)^K G_{\rho\rho'} N_{\rho'} R_{\rho',00} y_{\rho,ll'} (lKl' - K|40) \\ &\quad + \sum_{jj'} x_{\rho,lj} y_{\rho,l'j'} \eta_{\rho,j0} \eta_{\rho,j'0} (lKj0|2K)(l'Kj'0|2K), \end{aligned}$$

$$\begin{aligned} t_{\rho,l,\rho'l'} &= F_{\rho\rho'} \sqrt{(N_\rho N_{\rho'})} \sum_{jj'} x_{\rho,lj} x_{\rho',l'j'} \eta_{\rho,j0} \eta_{\rho',j'0} \\ &\quad \times (lKj0|2K)(l'Kj'0|2K) \\ &\quad + G_{\rho\rho'} \sqrt{(N_\rho N_{\rho'})} \sum_{jj'} y_{\rho,lj} y_{\rho',l'j'} \eta_{\rho,j0} \eta_{\rho',j'0} \\ &\quad \times (lKj0|4K)(l'Kj'0|4K). \end{aligned} \quad (8.15)$$

Assuming specific values of the coefficients appearing in the Hamiltonian, Pittel *et al.* (1984) have calculated the spectrum of excitations shown in Fig. 8.2. In this spectrum, there is also included a spurious 1^+ band corresponding to a rotation of the condensate. This band must be removed before comparing with experiment.

On top of each state there is built a rotational band. The moment of inertia of these bands can be obtained using the technique described in Sect. 3.9. Pittel *et al.* (1984) use the Inglis formula

$$\mathscr{I}_x = 2 \sum_\phi \frac{\langle \phi | \hat{L}_x | g \rangle^2}{E_\phi - E_g}, \quad (8.16)$$

where ϕ is any state reached by the operator $\hat{L}_x$, to evaluate it.

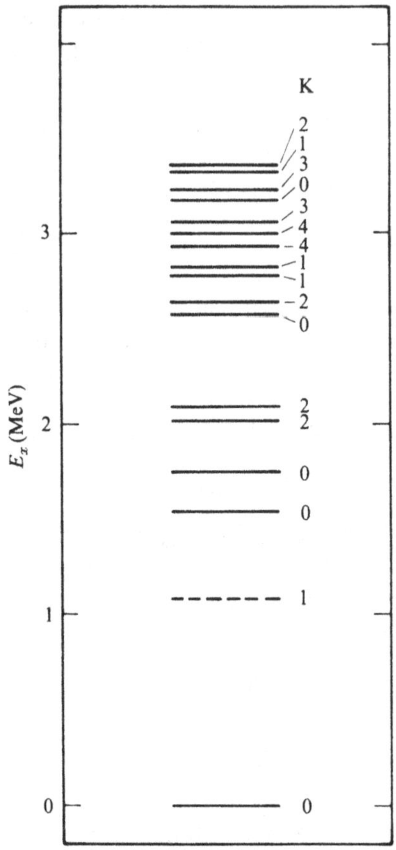

Fig. 8.2. Excitation spectrum in the Hartree–Bose plus Tamm–Dancoff limit of the interacting boson model-2G for $N_\pi = N_\nu = 7$. The dotted line represents the spurious 1^+ band discussed in the text.

8.4 The interacting boson model-1S′D′

van Isacker *et al.* (1981) have considered the possibility of introducing excited bosons with angular momenta 0 and 2 (s′ and d′ bosons), Fig. 8.1. The Hamiltonian H including these bosons is diagonalized numerically. These authors have interpreted the second excited 0^+ band in ^{156}Gd as arising from s′ and d′ bosons. The same band has been interpreted recently by Scholten *et al.* (1985b) as arising from states of mixed proton–neutron symmetry, an interpretation consistent also with the results of Fig. 8.2. It is thus not clear, at present, whether or not s′ and d′ bosons are needed to describe excited states in nuclei.

8.5 The interacting boson model-1F

We have discussed so far only bosons with positive parity. In many nuclei, there occur also collective states with negative parity. These can be interpreted as arising from $J^P=3^-$ pairs (F-pairs). One thus may wish to introduce bosons with angular momentum and parity $J^P=3^-$ (f or octupole bosons). If these bosons originate, as it is in many heavy even–even nuclei, from the valence shell, they can be treated in the same framework as discussed above for G-pairs. A model in which a system of s, d and f bosons are introduced has been considered (Arima and Iachello, 1978). This set of elementary bosons and their single-particle energies is shown in Fig. 8.3.

In this model, the fundamental building blocks are $1+5+7=13$

$$\begin{cases} s^\dagger, d^\dagger_\mu(\mu=0, \pm 1, \pm 2), f^\dagger_\mu(\mu=0, \pm 1, \pm 2, \pm 3), \\ s, d_\mu(\mu=0, \pm 1, \pm 2), f_\mu(\mu=0, \pm 1, \pm 2, \pm 3). \end{cases} \tag{8.17}$$

This case has been investigated in some detail. The most general Hamiltonian, written in terms of s, d, f creation and annihilation operators, is

$$\begin{aligned} H^{\mathrm{1F}} = {} & E_0+\varepsilon_s(s^\dagger\cdot\tilde{s})+\varepsilon_d(d^\dagger\cdot\tilde{d})+\sum_{L=0,2,4}\tfrac{1}{2}(2L+1)^{\frac{1}{2}}c_L[[d^\dagger\times d^\dagger]^{(L)}\times[\tilde{d}\times\tilde{d}]^{(L)}]^{(0)}_0 \\ & +\frac{1}{\sqrt{2}}v_2[[d^\dagger\times d^\dagger]^{(2)}\times[\tilde{d}\times\tilde{s}]^{(2)}+[s^\dagger\times d^\dagger]^{(2)}\times[\tilde{d}\times\tilde{d}]^{(2)}]^{(0)}_0 \\ & +\tfrac{1}{2}v_0[[d^\dagger\times d^\dagger]^{(0)}\times[\tilde{s}\times\tilde{s}]^{(0)}+[s^\dagger\times s^\dagger]^{(0)}\times[\tilde{d}\times\tilde{d}]^{(0)}]^{(0)}_0 \\ & +u_2[[d^\dagger\times s^\dagger]^{(2)}\times[\tilde{d}\times\tilde{s}]^{(2)}]^{(0)}_0+\tfrac{1}{2}u_0[[s^\dagger\times s^\dagger]^{(0)}\times[\tilde{s}\times\tilde{s}]^{(0)}]^{(0)}_0 \\ & +\varepsilon_f(f^\dagger\cdot\tilde{f})+\sum_{L=0,2,4,6}\tfrac{1}{2}(2L+1)^{\frac{1}{2}}c_{Lf}[[f^\dagger\times f^\dagger]^{(L)}\times[\tilde{f}\times\tilde{f}]^{(L)}]^{(0)}_0 \\ & +\frac{1}{\sqrt{2}}v_{2f}[[f^\dagger\times f^\dagger]^{(2)}\times[\tilde{d}\times\tilde{s}]^{(2)}+[d^\dagger\times s^\dagger]^{(2)}\times[\tilde{f}\times\tilde{f}]^{(2)}]^{(0)}_0 \\ & +\sum_{L=0,2,4}\tfrac{1}{2}(2L+1)^{\frac{1}{2}}c_{\mathrm{df}L}[f^\dagger\times f^\dagger]^{(L)}\times[\tilde{d}\times\tilde{d}]^{(L)}+[d^\dagger\times d^\dagger]^{(L)}\times[\tilde{f}\times\tilde{f}]^{(L)}]^{(0)}_0 \\ & +\tfrac{1}{2}v_{0f}[[f^\dagger\times f^\dagger]^{(0)}\times[\tilde{s}\times\tilde{s}]^{(0)}+[s^\dagger\times s^\dagger]^{(0)}\times[\tilde{f}\times\tilde{f}]^{(0)}]^{(0)}_0 \\ & +u_{2f}[[f^\dagger\times s^\dagger]^{(3)}\times[\tilde{f}\times\tilde{s}]^{(3)}]^{(0)}_0 \\ & +\sum_{L=1,2,3,4,5}x_L[[f^\dagger\times d^\dagger]^{(L)}\times[\tilde{f}\times\tilde{d}]^{(L)}]^{(0)}_0+w_2[[f^\dagger\times\tilde{f}]^{(2)} \\ & \qquad\times[d^\dagger\times\tilde{s}+s^\dagger\times\tilde{d}]^{(2)}]^{(0)}_0. \end{aligned} \tag{8.18}$$

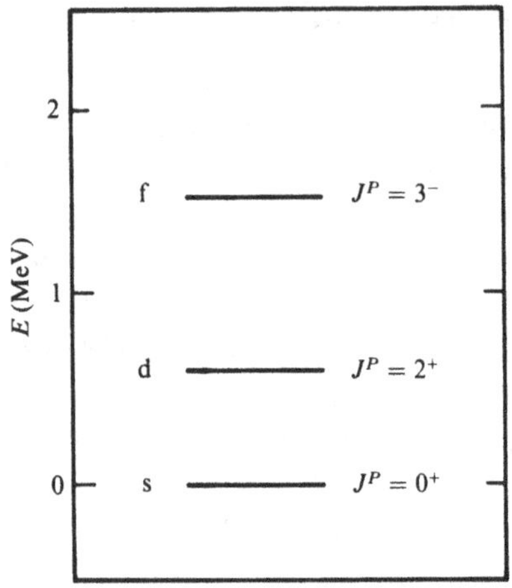

Fig. 8.3. Elementary bosons in the interacting boson model-1F.

The basis states can be written in the form

$$\mathscr{B}_{\mathrm{f}}\colon (s^\dagger)^{n_s}(d^\dagger)^{n_d}(f^\dagger)^{n_f}|0\rangle, \tag{8.19}$$

where, as in (1.17), further quantum numbers are needed to label the states. In these states, it is assumed that the total boson number is conserved

$$N = n_s + n_d + n_f. \tag{8.20}$$

A full treatment of (8.19) requires a construction of the states of the type $(d^\dagger)^{n_d}|0\rangle$, which has been done in Part I, and, in addition, states of the type $(f^\dagger)^{n_f}|0\rangle$. This construction has been done explicitly by Rohozinski (1978), Góźdź and Szpikowski (1978) and Góźdź, Szpikowski and Zajac (1980). States here can be characterized by the group chain

$$\left|\begin{matrix} \mathrm{U}(7) \supset & \mathrm{O}(7) & \supset \mathrm{O}(3) \supset & \mathrm{O}(2) \\ n_f, & v_f, \nu_1\nu_2\nu_3, & L, & M_L \end{matrix}\right\rangle \tag{8.21}$$

where ν_1, ν_2 and ν_3 are quantum numbers needed to classify the states uniquely. In Table 8.1 we give a partial classification scheme of these states.

Since the f bosons have negative parity, the 13-dimensional space splits into the direct product

$$\mathrm{U}(13) \supset \mathrm{U}(6) \otimes \mathrm{U}(7). \tag{8.22}$$

One may then inquire about possible dynamic symmetries for this coupled system. Unfortunately, there are no common subalgebras, except the rotation algebra O(3). Thus the only possible coupling can occur at the level

Table 8.1. *Partial classification scheme for states of the f^{n_f} configuration (Rohozinski, 1978)*

U(7)	O(3)
n_f	L
0	0
1	3
2	6, 4, 2, 0
3	$9, 7, 6, 5, 4, 3^2, 1$
4	$12, 10, 9, 8^2, 7, 6^3, 5, 4^3, 3, 2^2, 0^2$
5	$15, 13, 12, 11^2, 10^2, 9^3, 8^2, 7^4, 6^3, 5^4, 4^2, 3^4, 2, 1^2$
6	$18, 16, 15, 14^2, 13^2, 12^4, 11^2, 10^5, 9^4, 8^6, 7^4, 6^7, 5^3, 4^6, 3^3, 2^4, 0^3$

of the angular momenta and one can only consider the lattices of algebras

$$\begin{array}{ccccccc} & & & \mathrm{U}(5)\supset\mathrm{O}(5) & & & \\ & & \nearrow & & \searrow & & \\ & \mathrm{U}(6)\rightarrow\mathrm{SU}(3) & & & \rightarrow\mathrm{O}_{sd}(3) & & \\ \nearrow & & \searrow & \nearrow & & \searrow & \\ \mathrm{U}(13) & & & \mathrm{O}(6)\supset\mathrm{O}(5) & & & \mathrm{O}(3)\supset\mathrm{O}(2). \\ \searrow & & & & & \nearrow & \\ & \mathrm{U}(7)\supset\mathrm{O}(7) & & \supset & & \mathrm{O}_f(3) & \end{array} \quad (8.23)$$

Basis states for the first branch of the lattice can be taken as

$$|[n_s+n_d], n_d, v, \tilde{n}_\Delta, L_{sd};\ [n_f], v_f, \nu_1, \nu_2, \nu_3, L_f;\ L, M_L\rangle. \quad (8.24)$$

Most calculations up to now have considered the case $n_f \leqslant 1$, corresponding physically to situations in which $\varepsilon_f \gg \varepsilon_s, \varepsilon_d$, octupole vibrations. In this case, one can omit all the quantum numbers in the f-part of the wave function and take $L_f = 3$. In the space $n_f \leqslant 1$, the calculation splits into two disjoint pieces, $n_f = 0$ and $n_f = 1$, since these pieces have opposite parity and the Hamiltonian H of (8.18) conserves parity. Within this space, one can use a simple version of H, i.e.

$$\begin{aligned} H'^{1F} &= H_{sd} + \varepsilon_f(f^\dagger\cdot\tilde{f}) + u_{2f}[[f^\dagger\times s^\dagger]^{(3)}\times[\tilde{f}\times\tilde{s}]^{(3)}]^{(0)}_0 \\ &\quad + \sum_{L=1,2,3,4,5} x_L[[f^\dagger\times d^\dagger]^{(L)}\times[\tilde{f}\times\tilde{d}]^{(L)}]^{(0)}_0 \\ &\quad + w_2[[f^\dagger\times\tilde{f}]^{(2)}\times[d^\dagger\times\tilde{s}+s^\dagger\times\tilde{d}]^{(2)}]^{(0)}_0, \end{aligned} \quad (8.25)$$

where H_{sd} denotes the Hamiltonian (1.21). For purposes of discussion, it is convenient to rewrite (8.25) in the form

$$H'^{1F} = H_{sd} + \varepsilon_f(f^\dagger \cdot \tilde{f}) + u_{2f}[[f^\dagger \times \tilde{f}]^{(0)} \times [s^\dagger \times \tilde{s}]^{(0)}]_0^{(0)}$$
$$+ \sum_{L'=0,1,2,3,4} x'_{L'}[[f^\dagger \times \tilde{f}]^{(L')} \times [d^\dagger \times \tilde{d}]^{(L')}]_0^{(0)} \tag{8.26}$$
$$+ w_2[[f^\dagger \times \tilde{f}]^{(2)} \times [d^\dagger \times \tilde{s} + s^\dagger \times \tilde{d}]^{(2)}]_0^{(0)}.$$

The coefficients x'_L are related to x_L by a change of coupling transformation

$$x'_{L'} = \sum_L (2L+1)^{\frac{1}{2}}(2L'+1)^{\frac{1}{2}}(-)^{3+L+L'} \begin{Bmatrix} 3 & 3 & L' \\ 2 & 2 & L \end{Bmatrix} x_L. \tag{8.27}$$

The Hamiltonian (8.26) contains a single one-body term, ε_f, and seven two-body terms, u_{2f}, $x'_{L'}$ ($L'=0$, 1, 2, 3, 4) and w_2 in addition to those of H_{sd}. However, one of these parameters can be eliminated, since

$$[[f^\dagger \times \tilde{f}]^{(0)} \times [s^\dagger \times \tilde{s}]^{(0)}]_0^{(0)} = \frac{1}{\sqrt{7}} \hat{n}_f \cdot \hat{n}_s. \tag{8.28}$$

Using the conservation of total boson number, this yields,

$$H'^{1F} = H_{sd} + \varepsilon'_f(f^\dagger \cdot \tilde{f}) + \sum_{L'=0,1,2,3,4} x''_{L'}[[f^\dagger \times \tilde{f}]^{(L')} \times [d^\dagger \times \tilde{d}]^{(L')}]_0^{(0)}$$
$$+ w_2[[f^\dagger \times \tilde{f}]^{(2)} \times [d^\dagger \times \tilde{s} + s^\dagger \times \tilde{d}]^{(2)}]_0^{(0)}, \tag{8.29}$$

where

$$\varepsilon'_f = \varepsilon_f + \frac{1}{\sqrt{7}} u_{2f} N,$$
$$x''_0 = x'_0 - \sqrt{5}\, u_{2f}, \tag{8.30}$$
$$x''_{L'} = x'_{L'}, \quad L' \neq 0.$$

There are thus only six independent two-body terms.

8.5.1 *Energy spectra*

Although no solution of (8.29) can be found in closed form, except the trivial one of no coupling, an intermediate analysis is possible. This is when H_{sd} has one of the three dynamic symmetries, (2.25), discussed in Part I.

(*i*) *Chain I*

Consider the case in which

$$H^{(I)1F} = H_{sd}^{(I)} + \varepsilon'_f + \sum_{L'} x''_{L'}[[f^\dagger \times \tilde{f}]^{(L')} \times [d^\dagger \times \tilde{d}]^{(L')}]_0^{(0)}. \tag{8.31}$$

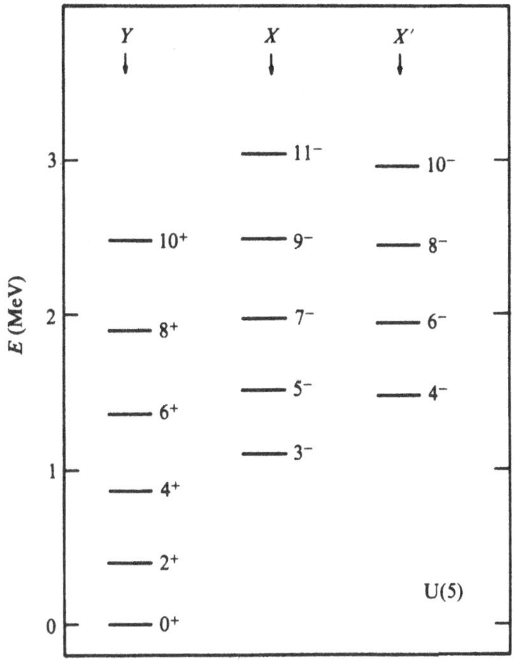

Fig. 8.4. Typical octupole spectrum in the U(5) limit of the sd boson part.

The eigenvalue of this Hamiltonian, at least for some states, can be found analytically (Arima and Iachello, 1976). If we denote by $E(n_d, L_d = 2n_d)$ the energies of the states $|[n_s + n_d], n_d, v, \tilde{n}_\Delta, L_d, M_{L_d}\rangle$ and use a similar notation for the coupled states, we find

$$E(n_d, L_d = 2n_d; n_f = 1, L_f = 3; L = 2n_d + 3) = E(n_d, L_d = 2n_d) + \varepsilon_f' + n_d x_5, \quad (8.32)$$

$$E(n_d, L_d = 2n_d; n_f = 1, L_f = 3; L = 2n_d + 2)$$
$$= E(n_d, L_d = 2n_d) + \varepsilon_f' + n_d x_5 + \tfrac{1}{5}(2n_d + 3)(x_4 - x_5), \quad (8.33)$$

where the coefficients x_4 and x_5 are obtained from the x_L'' by inverting the transformation (8.27). The structure of the corresponding spectra is shown in Fig. 8.4.

(ii) Chain II

In this case, one considers the Hamiltonian,

$$H^{(II)1F} = H_{sd}^{(II)} + \varepsilon_f'(f^\dagger \cdot \tilde{f}) + z'(\hat{L}_f \cdot \hat{L}) + z(\hat{Q}_f \cdot \hat{Q}), \quad (8.34)$$

where

$$\hat{L}_f = (28)^{\frac{1}{2}}[f^\dagger \times \tilde{f}]^{(1)},$$
$$\hat{Q}_f = [f^\dagger \times \tilde{f}]^{(2)}, \tag{8.35}$$

and $\hat{L}$ and $\hat{Q}$ are given by (1.50).

The eigenvalue problem for (8.34) cannot be solved analytically. However, it turns out that, for N and z large, the wave functions of an f boson coupled to the ground-state band of the $(N-1)$ s and d bosons can be written as

$$|K_f, L, M_L\rangle = \sum_{R=0,2,\ldots,(2N-2)} \left(\frac{2R+1}{2L+1}\right)^{\frac{1}{2}} \left(\frac{2}{1+\delta_{K_f 0}}\right)^{\frac{1}{2}} (3K_f R0|LK_f)$$
$$\times |[N-1], (\lambda = 2N-2, \mu = 0), K=0, R; f; L, M_L\rangle, \tag{8.36}$$

where

$$K_f = 0, 1, 2, 3 \tag{8.37}$$

and

$$L = K_f, K_f+1, K_f+2, \ldots \tag{8.38}$$

except for $K_f = 0$ for which

$$L = 1, 3, 5, \ldots. \tag{8.39}$$

One can then evaluate the expectation value of (8.34) in (8.36), with the result

$$E(K_f, L, M_L) = \mathscr{E} + (\tfrac{3}{4}\kappa - \kappa')L(L+1) + \alpha_f K_f^2 + \beta_f, \tag{8.40}$$

where β_f depends on ε_f', z, z' and α_f on z and z' (Arima and Iachello, 1978). The results of a numerical diagonalization of (8.34) as a function of z are shown in Fig. 8.5. One sees that, for large positive and negative z, the pattern (8.40) appears.

(iii) Chain III

In this case, one may consider the Hamiltonian,

$$H^{(\mathrm{III})1\mathrm{F}} = H^{(\mathrm{III})}_{\mathrm{sd}} + \varepsilon_f'(f^\dagger \cdot \tilde{f}) + z'(\hat{L}_f \cdot \hat{L}) + z''(\hat{Q}_f \cdot \hat{Q}^{\chi=0}), \tag{8.41}$$

where

$$\hat{Q}^{\chi=0} = [d^\dagger \times \tilde{s} + s^\dagger \times \tilde{d}]^{(2)}. \tag{8.42}$$

Properties of energies in this chain have not been investigated in detail.

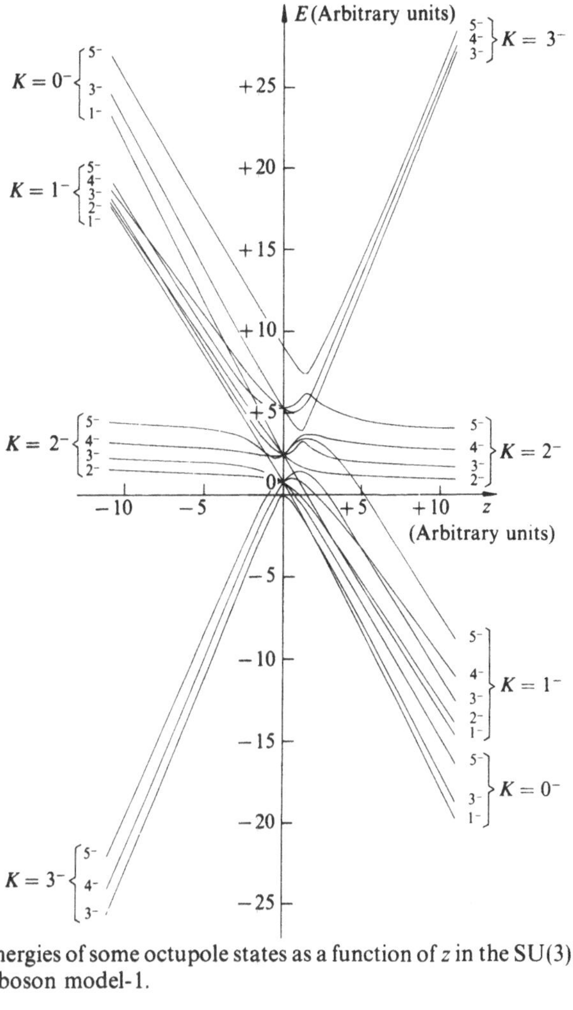

Fig. 8.5. Energies of some octupole states as a function of z in the SU(3) limit of the interacting boson model-1.

8.5.2 Electromagnetic transition rates

An interesting property of this model is that there appear a variety of new electromagnetic transition operators. These are,

$$
\begin{aligned}
T_\mu^{(E1)} &= \gamma_1[d^\dagger \times \tilde{f} + f^\dagger \times \tilde{d}]_\mu^{(1)}, \\
T_\mu^{(M2)} &= \gamma_2[d^\dagger \times \tilde{f} - f^\dagger \times \tilde{d}]_\mu^{(2)}, \\
T_\mu^{(E3)} &= \gamma_3[d^\dagger \times \tilde{f} + f^\dagger \times \tilde{d}]_\mu^{(3)} + \delta_3[s^\dagger \times \tilde{f} + f^\dagger \times \tilde{s}]_\mu^{(3)}, \\
T_\mu^{(M4)} &= \gamma_4[d^\dagger \times \tilde{f} - f^\dagger \times \tilde{d}]_\mu^{(4)}, \\
T_\mu^{(E5)} &= \gamma_5[d^\dagger \times \tilde{f} + f^\dagger \times \tilde{d}]_\mu^{(5)}.
\end{aligned} \tag{8.43}
$$

Furthermore, the operators in (1.24) now become

$$
\begin{aligned}
T_\mu^{(\mathrm{E0})} &= \gamma_0 + \alpha_0[s^\dagger \times \tilde{s}]_0^{(0)} + \beta_0[d^\dagger \times \tilde{d}]_0^{(0)} + \delta_0[f^\dagger \times \tilde{f}]_0^{(0)}, \\
T_\mu^{(\mathrm{M1})} &= \beta_1[d^\dagger \times \tilde{d}]_\mu^{(1)} + \delta_1[f^\dagger \times \tilde{f}]_\mu^{(1)}, \\
T_\mu^{(\mathrm{E2})} &= \alpha_2[d^\dagger \times \tilde{s} + s^\dagger \times \tilde{d}]_\mu^{(2)} + \beta_2[d^\dagger \times \tilde{d}]_\mu^{(2)} + \delta_2[f^\dagger \times \tilde{f}]_\mu^{(2)}, \\
T_\mu^{(\mathrm{M3})} &= \beta_3[d^\dagger \times \tilde{d}]_\mu^{(3)} + \eta_3[f^\dagger \times \tilde{f}]_\mu^{(3)}, \\
T_\mu^{(\mathrm{E4})} &= \beta_4[d^\dagger \times \tilde{d}]_\mu^{(4)} + \delta_4[f^\dagger \times \tilde{f}]_\mu^{(4)},
\end{aligned}
\tag{8.44}
$$

and

$$
\begin{aligned}
T_\mu^{(\mathrm{M5})} &= \delta_5[f^\dagger \times \tilde{f}]_\mu^{(5)}, \\
T_\mu^{(\mathrm{E6})} &= \delta_6[f^\dagger \times \tilde{f}]_\mu^{(6)}.
\end{aligned}
\tag{8.45}
$$

The matrix elements of some of these operators have been evaluated explicitly in a few cases.

(i) Chain I

The matrix elements of the E1 operator can be evaluated between the states in (8.32) and those of the ground-state band. In general, one has

$$
\begin{aligned}
&\langle n_\mathrm{d}+1, v, \tilde{n}_\Delta, L_\mathrm{d}; n_\mathrm{f}=0; L=L_\mathrm{d} \| [d^\dagger \times \tilde{f}]^{(1)} \| n_\mathrm{d}, v', \tilde{n}'_\Delta, L'_\mathrm{d}; n'_\mathrm{f}=1, L'_\mathrm{f}=3; L'\rangle \\
&\quad = \sqrt{3}\sqrt{(2L'+1)}\,(-)^{L_\mathrm{d}+L'_\mathrm{d}} \begin{Bmatrix} L' & L'_\mathrm{d} & 3 \\ 2 & 1 & L_\mathrm{d} \end{Bmatrix} \\
&\qquad \times \langle n_\mathrm{d}+1, v, \tilde{n}_\Delta, L_\mathrm{d} \| d^\dagger \| n_\mathrm{d}, v', \tilde{n}'_\Delta, L'_\mathrm{d}\rangle.
\end{aligned}
\tag{8.46}
$$

Using the matrix elements of the $d^\dagger$ operator in Sect. 2.6.1, one obtains, denoting by F the E1 matrix elements,

$$
\begin{aligned}
L' &= 2n_\mathrm{d}+3 \to L=2n_\mathrm{d}+2; \quad F=\sqrt{[\tfrac{3}{7}(n_\mathrm{d}+1)(4n_\mathrm{d}+7)]}, \\
L' &= 2n_\mathrm{d}+2 \to L=2n_\mathrm{d}+2; \quad F=-\sqrt{[\tfrac{1}{7}n_\mathrm{d}(4n_\mathrm{d}+5)]}.
\end{aligned}
\tag{8.47}
$$

Thus

$$
\begin{aligned}
B(\mathrm{E1}; L' &= 2n_\mathrm{d}+3 \to L=2n_\mathrm{d}+2) = \gamma_1^2(\tfrac{3}{7})(n_\mathrm{d}+1), \\
B(\mathrm{E1}; L' &= 2n_\mathrm{d}+2 \to L=2n_\mathrm{d}+2) = \gamma_1^2(\tfrac{4}{7})(n_\mathrm{d}).
\end{aligned}
\tag{8.48}
$$

Similarly, one can evaluate the matrix elements of the E3 operator. Here, the important quantity to evaluate is the matrix element from the ground state to the 3^--states. There is only one 3^--state excited. The matrix element for this transition is (Arima and Iachello, 1978)

$$
\langle n_\mathrm{s}=N, n_\mathrm{d}=0, L=0 \| [s^\dagger \times \tilde{f}]^{(3)} \| n_\mathrm{s}=N-1, n_\mathrm{d}=0; n_\mathrm{f}=1; L'=3\rangle = \sqrt{N}\sqrt{7}.
\tag{8.49}
$$

Thus

$$B(\mathrm{E3};0_1^+\to 3_1^-)=7N\delta_3^2. \quad (8.50)$$

We next evaluate E2 transition matrix elements. If we neglect the contribution of the last term $[f^\dagger\times\tilde{f}]^{(2)}$ in (8.44), which is expected to be small, we find that only the first term contributes. Denoting by F' the reduced matrix elements of the operator $[s^\dagger\times\tilde{d}]^{(2)}$,

$$F'=\langle n_\mathrm{d},v,\tilde{n}_\Delta,L_\mathrm{d};n_\mathrm{f}=1,L_\mathrm{f}=3;L\|[s^\dagger\times\tilde{d}]^{(2)} \times\|n_\mathrm{d}+1,v',\tilde{n}'_\Delta,L'_\mathrm{d};n_\mathrm{f}=1,L_\mathrm{f}=3;L'\rangle, \quad (8.51)$$

we find

$$L'=2n_\mathrm{d}+5\to L=2n_\mathrm{d}+3,\quad F=\sqrt{[(N-n_\mathrm{d})(n_\mathrm{d}+1)(4n_\mathrm{d}+11)]},$$

$$L'=2n_\mathrm{d}+4\to L=2n_\mathrm{d}+2,\quad F=\sqrt{\left[(N-n_\mathrm{d})n_\mathrm{d}\frac{(2n_\mathrm{d}+5)(4n_\mathrm{d}+9)}{(2n_\mathrm{d}+3)}\right]},$$

$$L'=2n_\mathrm{d}+4\to L=2n_\mathrm{d}+3,\quad F=-\sqrt{\left[(N-n_\mathrm{d})\frac{3(4n_\mathrm{d}+9)}{(2n_\mathrm{d}+3)}\right]}, \quad (8.52)$$

from which one can obtain the corresponding $B(\mathrm{E2})$ values. For example,

$$B(\mathrm{E2};2n_\mathrm{d}+5\to 2n_\mathrm{d}+3)=\alpha_2^2(N-n_\mathrm{d})(n_\mathrm{d}+1). \quad (8.53)$$

It is particularly interesting to consider the $B(\mathrm{E1})/B(\mathrm{E2})$ branching ratios, since these are experimentally observable. They are given by

$$\frac{X\to Y}{X\to X}\quad\frac{B(\mathrm{E1};L'=2n_\mathrm{d}+5\to L=2n_\mathrm{d}+4)}{B(\mathrm{E2};L'=2n_\mathrm{d}+5\to L=2n_\mathrm{d}+3)}=\frac{3}{7}\frac{(n_\mathrm{d}+2)}{(N-n_\mathrm{d})(n_\mathrm{d}+1)}C,$$

$$n_\mathrm{d}=0,1,\ldots,N, \quad (8.54)$$

where $C=\gamma_1^2/\alpha_2^2$. Similarly, one obtains

$$\frac{X'\to Y}{X'\to X'}\quad\frac{B(\mathrm{E1};L'=2n_\mathrm{d}+4\to L=2n_\mathrm{d}+4)}{B(\mathrm{E2};L'=2n_\mathrm{d}+4\to L=2n_\mathrm{d}+2)}=\frac{1}{7}\frac{(n_\mathrm{d}+1)(2n_\mathrm{d}+3)}{(N-n_\mathrm{d})n_\mathrm{d}(2n_\mathrm{d}+5)}C,$$

$$n_\mathrm{d}=1,2,\ldots,N, \quad (8.55)$$

$$\frac{X'\to Y}{X'\to X}\quad\frac{B(\mathrm{E1};L'=2n_\mathrm{d}+4\to L=2n_\mathrm{d}+4)}{B(\mathrm{E2};L'=2n_\mathrm{d}+4\to L=2n_\mathrm{d}+3)}=\frac{1}{21}\frac{(n_\mathrm{d}+1)(2n_\mathrm{d}+3)}{(N-n_\mathrm{d})}C,$$

$$n_\mathrm{d}=0,1,\ldots,N, \quad (8.56)$$

where the Y, X, X' label the bands shown in Fig. 8.4.

Finally, one can also evaluate $B(\mathrm{M1})$ values. Using the M1 operator of

(8.44), with $\delta_1 = 0$, one obtains

$$X \to X' \quad B(\mathrm{M}1; L' = 2n_\mathrm{d} + 3 \to L = 2n_\mathrm{d} + 3) = \beta_1^2 \frac{3n_\mathrm{d}}{5(2n_\mathrm{d} + 3)}. \tag{8.57}$$

(*ii*) *Chain II*

The matrix elements of the E1 operator (8.43) can be evaluated between the states (8.36) and states of the ground-state band. One obtains

$$\begin{aligned} &\langle(\lambda = 2N, \mu = 0), K = 0, L = 0 \| [d^\dagger \times \tilde{f}]^{(1)} \| K_\mathrm{f}, L'' = 1\rangle \\ &= \sqrt{3} \begin{Bmatrix} 2 & 3 & 1 \\ 3 & 2 & 0 \end{Bmatrix} \langle(\lambda = 2N, \mu = 0), K = 0, L = 0 \| d^\dagger \| (\lambda' = 2N - 2, \mu' = 0), \\ &\qquad\qquad K' = 0, L' = 2\rangle \\ &\times \langle 0 \| \tilde{f} \| f\rangle \times \sqrt{\frac{5}{3}} \sqrt{\left(\frac{2}{1 + \delta_{K_\mathrm{f}0}}\right)} \langle 3\ K_\mathrm{f} 2\ 0 | 1\ K_\mathrm{f}\rangle, \end{aligned} \tag{8.58}$$

from which, inserting the appropriate values, one obtains,

$$\begin{aligned} B(\mathrm{E}1; K = 0, 0^+ \to K_\mathrm{f} = 0, 1^-) &= \frac{6}{35}\left(\frac{2N-2}{2N-1}\right) N\gamma_1^2, \\ B(\mathrm{E}1; K = 0, 0^+ \to K_\mathrm{f} = 1, 1^-) &= \frac{8}{35}\left(\frac{2N-2}{2N-1}\right) N\gamma_1^2. \end{aligned} \tag{8.59}$$

The matrix elements of the E3 operator can be obtained in a similar way. This operator has two terms. The important term is that with $[s^\dagger \times \tilde{f} + f^\dagger \times \tilde{s}]^{(3)}$. For this one obtains

$$\begin{aligned} &\langle(\lambda = 2N, \mu = 0), K = 0, L = 0 \| [s^\dagger \times \tilde{f}]^{(3)} \| K_\mathrm{f}, L'' = 3\rangle \\ &= \langle(\lambda = 2N, \mu = 0), K = 0, L = 0 \| s^\dagger \| (\lambda' = 2N - 2, \mu' = 0), K' = 0, L' = 0\rangle \\ &\times \langle 0 \| \tilde{f} \| f\rangle \sqrt{\frac{1}{7}} \sqrt{\left(\frac{2}{1 + \delta_{K_\mathrm{f}0}}\right)} \langle 3\ K_\mathrm{f}\ 0\ 0 | 3\ K_\mathrm{f}\rangle. \end{aligned} \tag{8.60}$$

Inserting the appropriate values, one finds

$$\begin{aligned} B(\mathrm{E}3; K = 0, 0^+ \to K = 0, 3^-) &= \frac{N}{3}\left(\frac{2N+1}{2N-1}\right)\delta_3^2, \\ B(\mathrm{E}3; K = 0, 0^+ \to K = 1, 3^-) &= 2\frac{N}{3}\left(\frac{2N+1}{2N-1}\right)\delta_3^2, \\ B(\mathrm{E}3; K = 0, 0^+ \to K = 2, 3^-) &= 2\frac{N}{3}\left(\frac{2N+1}{2N-1}\right)\delta_3^2, \\ B(\mathrm{E}3; K = 0, 0^+ \to K = 3, 3^-) &= 2\frac{N}{3}\left(\frac{2N+1}{2N-1}\right)\delta_3^2. \end{aligned} \tag{8.61}$$

The B(E3) strength is thus in this case divided in the ratios 1:2:2:2, with sum rule

$$\sum_{K_f} B(\mathrm{E3}; K=0, 0^+ \rightarrow K_f, 3^-) = 7N\frac{1}{3}\left(\frac{2N+1}{2N-1}\right)\delta_3^2, \tag{8.62}$$

to be compared with (8.50).

We consider next E2 transitions. It is convenient here to rewrite (8.44) in the form

$$T^{(\mathrm{E2})} = \alpha_2\hat{Q} + \alpha_2'\hat{Q}' + \delta_2\hat{Q}_f, \tag{8.63}$$

where the operators $\hat{Q}, \hat{Q}'$ and $\hat{Q}_f$ have been defined in Part I and (8.35). The reduced matrix elements of $\hat{Q}$ are given by

$$\langle K_f, L' \| \hat{Q} \| K_f, L\rangle = \sqrt{\tfrac{1}{8}}\,(-)^{L+L'}\sqrt{(2L+1)}\,\langle LK_f 2\ 0 | L'K_f\rangle(4N-1). \tag{8.64}$$

Those of $\hat{Q}_f$ by

$$\langle K_f, L' \| \hat{Q}_f \| K_f, L\rangle = \sqrt{\tfrac{1}{7}}\,(-)^{1+L}\sqrt{(2L+1)}\,\langle LK_f 2\ 0 | L'K_f\rangle \times \langle 3K_f 2\ 0 | 3K_f\rangle\langle f \| \hat{Q}_f \| f\rangle. \tag{8.65}$$

From these, one can obtain B(E2) values, if needed. For example, assuming that only the $\hat{Q}$ term is important, one obtains

$$B(\mathrm{E2}; K_f, L \rightarrow K_f, L') = \tfrac{1}{8}\langle LK_f 2\ 0 | L'K_f\rangle^2 (4N-1)^2\alpha_2^2. \tag{8.66}$$

Finally, one can calculate M1 transitions. For these it is convenient to rewrite the operator in (8.44) as

$$T^{(\mathrm{M1})} = \sqrt{\left(\frac{3}{4\pi}\right)}\,(g_B\hat{L} + g_F\hat{L}_f), \tag{8.67}$$

where g_B and g_F are the effective boson g-factors. Using similar techniques as those discussed above, one finds

$$\langle K_f, L \| T^{(\mathrm{M1})} \| K_f, L\rangle = \sqrt{\left(\frac{3}{4\pi}\right)}\sqrt{(2L+1)}\,[(g_F - g_B)K_f\langle LK_f 1\ 0 | LK_f\rangle + g_B\sqrt{[L(L+1)]}], \tag{8.68}$$

from which one can calculate B(M1) values, if needed. One can also calculate the magnetic moments and g-factors of the octupole states,

$$g(K_f, L) = \frac{\mu(K_f, L)}{L} = g_B + (g_F - g_B)\frac{K_f^2}{L(L+1)}. \tag{8.69}$$

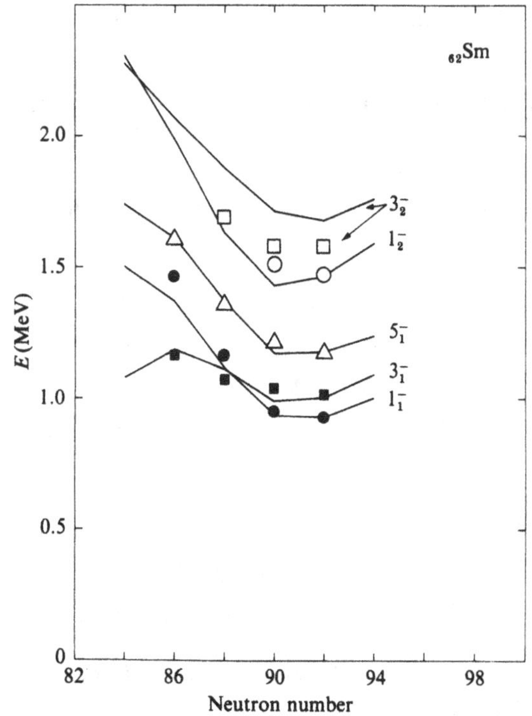

Fig. 8.6. Structure of the octupole states in transitional class A.

(iii) Chain III

Detailed calculations of electromagnetic matrix elements for octupole states in this chain have not been performed up to now.

8.5.3 Higher-order terms in the transition operators

The operators (8.43) and (8.44) describe the situation well in actual cases, with the exception of the E1 operator for which higher-order terms appear to be necessary. These higher-order terms have been written in a variety of ways. A form often used in calculations is (Scholten *et al.*, 1978)

$$T_\mu^{(\mathrm{E1})} = \gamma_1\{\hat{D}_\mu + x_1[\hat{Q}\times\hat{D}]_\mu^{(1)}\} + \gamma_1'\hat{n}_s\{\hat{D}_\mu + x_1[\hat{Q}\times\hat{D}]_\mu^{(1)}\}, \qquad (8.70)$$

where the dipole operator $\hat{D}$ is defined as

$$\hat{D} = [d^\dagger \times \tilde{f} + f^\dagger \times \tilde{d}]^{(1)}. \qquad (8.71)$$

The higher-order terms appear to arise from neglecting elementary bosons with $J^P = 1^-$ lying at higher excitation energies.

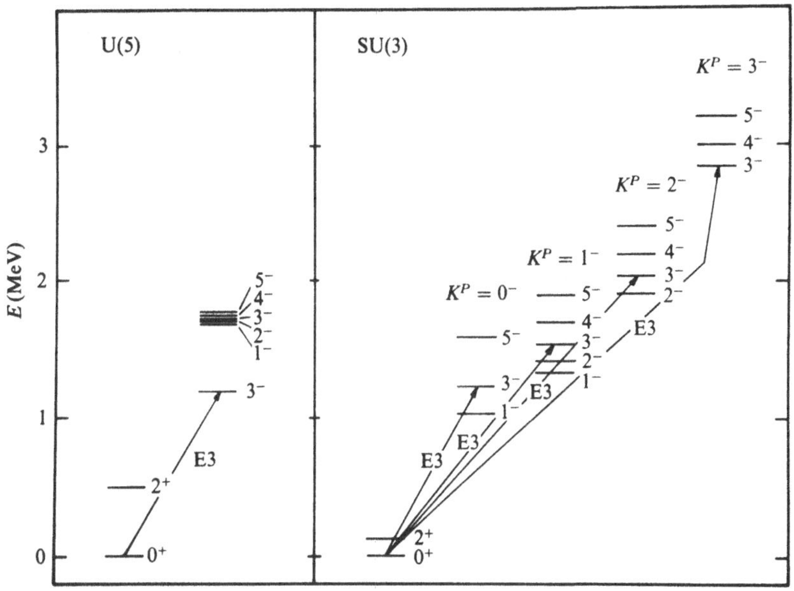

Fig. 8.7. Schematic representation of octupole excitations in chains I and II.

8.5.4 *Transitional classes*

An interesting application of the interacting boson model-1F has been made to the transitional class A (Scholten *et al.*, 1978). The Hamiltonian used is

$$H = H_{\rm sd}^{\rm (I)+(II)} + \varepsilon_{\rm f} + \theta_{\rm f} N + z(\hat{Q}_{\rm f} \cdot \hat{Q}). \tag{8.72}$$

The resulting spectra are compared with experiment in Fig. 8.6. In addition, electromagnetic properties of the states have been investigated. Particularly interesting are E3 and E1 properties. The expected behavior of E3 transitions in the two chains I and II is shown schematically in Fig. 8.7. As discussed in Sect. 8.5.2, only one state is excited in chain I while four states are expected to be excited in chain II. The experimental results are compared with the results of the calculations in Fig. 8.8. Finally, we show in Fig. 8.9 the expected pattern of E1 transitions from the ground state in chains I and II. In chain I the E1 matrix element leading from the ground state to the first excited 1^--state vanishes. The experimental and calculated values are compared in Fig. 8.10.

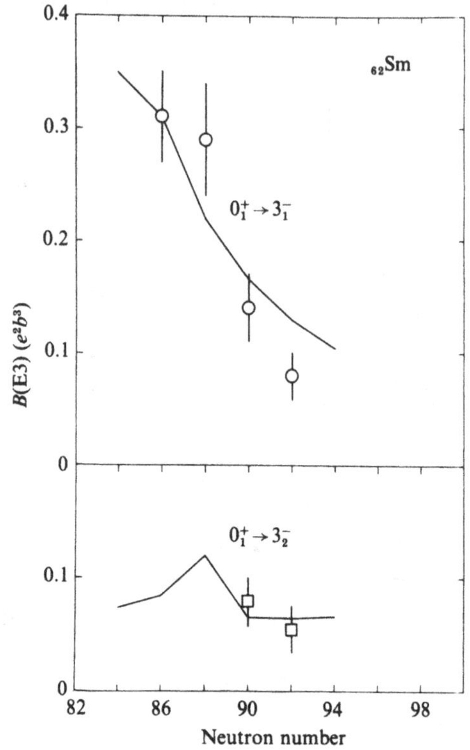

Fig. 8.8. Structure of E3 transitions in the transitional class A.

8.6 The interacting boson model-2CM

In the interacting boson model-2, bosons represent proton and neutron pairs in the valence shells. In some nuclei, it has been found important to include excitations of nucleons from the closed shells to the valence shells and/or excitations from the valence shells to the empty shells. Because of the neutron excess, this effect is mostly dramatically seen for protons and it is particularly pronounced where the energy gap between shells is not very large. Thus, if one considers closed shells at proton numbers 40 and 64, one must certainly include core excitations if one wants to describe the corresponding spectra accurately. However, even for proton numbers 50 and 82 the effects of excitations across the energy gaps have been observed experimentally.

The way in which core excitations are included within the framework of the interacting boson model-2, is a straightforward generalization of Part II and a consequence of the interpretation of bosons as nucleon pairs. In

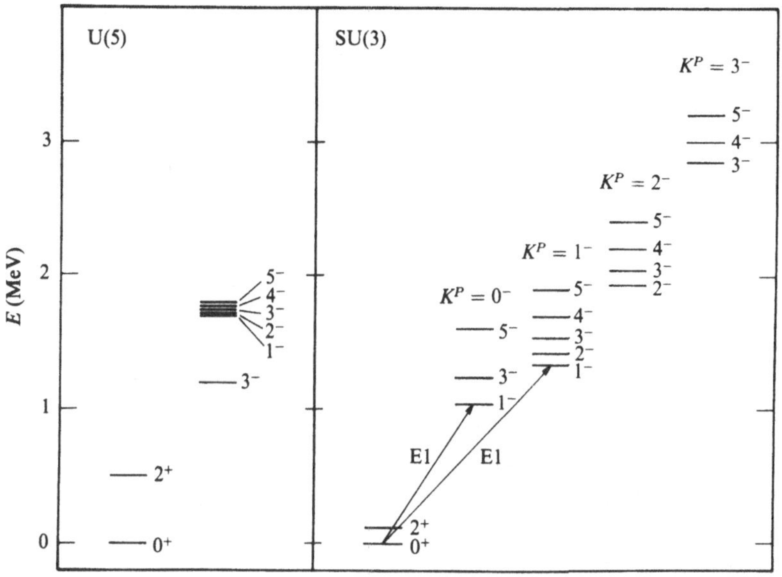

Fig. 8.9. Schematic representation of dipole excitations in chains I and II. For chain I, dipole excitations from the ground state are forbidden.

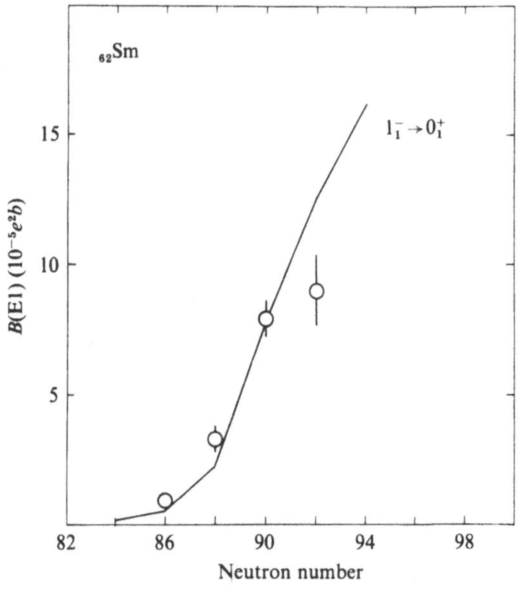

Fig. 8.10. Structure of E1 transitions in transitional class A.

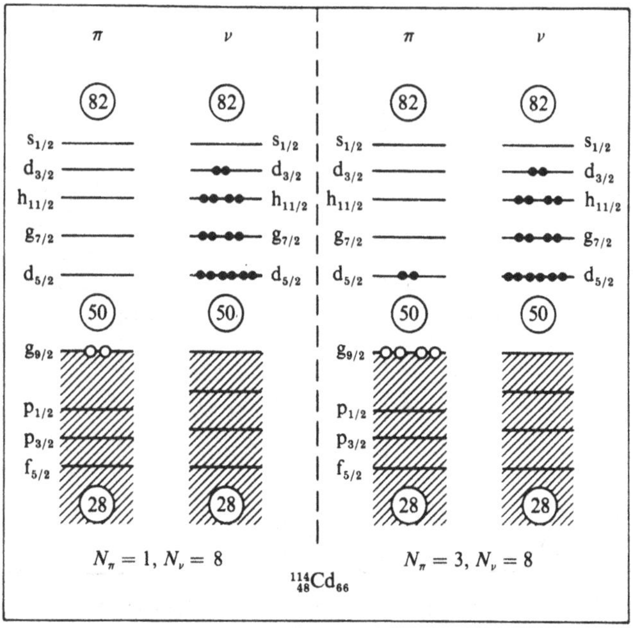

Fig. 8.11. Schematic illustration of the two shell-model configurations of importance in the description of $^{114}_{48}Cd_{66}$.

the spherical shell model, the low-lying excitations of interest here correspond to raising two particles from the shells below the gap to the shells above it, as shown schematically in Fig. 8.11 (two-particle, two-hole configurations). Thus, in treating properties of nuclei for which these excitations are important, one must consider two types of configurations: a normal configuration, built as in Part II (left part of Fig. 8.11), and an excited configuration in which two protons (or two neutrons) have been raised across the shell gap (right part of Fig. 8.11). For example, if two proton excitations across the shell gap at 50 are important, as it appears to be the case for $^{114}_{48}Cd_{66}$, one considers, in addition to the normal configuration, $N_\pi = \bar{1}$ and $N_\nu = 8$, the configuration $N_\pi = \bar{2}$, $N'_\pi = 1$, and $N_\nu = 8$ in which one proton pair has been moved from the shells below $Z = 50$ to those above it. The two configurations differ by an energy gap Δ_π and are coupled by the interaction

$$H_{\text{mix}} = \bar{\alpha}_\pi[s_\pi^\dagger \times s_\pi'^\dagger + \tilde{s}'_\pi \times \tilde{s}_\pi]_0^{(0)} + \bar{\beta}_\pi[d_\pi^\dagger \times d_\pi'^\dagger + \tilde{d}'_\pi \times \tilde{d}_\pi]_0^{(0)}, \quad (8.73)$$

for proton excitations (and a similar expression with π interchanged with ν for neutron excitations). H_{mix} creates a boson above $Z = 50$ ($s_\pi'^\dagger$, $d_\pi'^\dagger$) and a

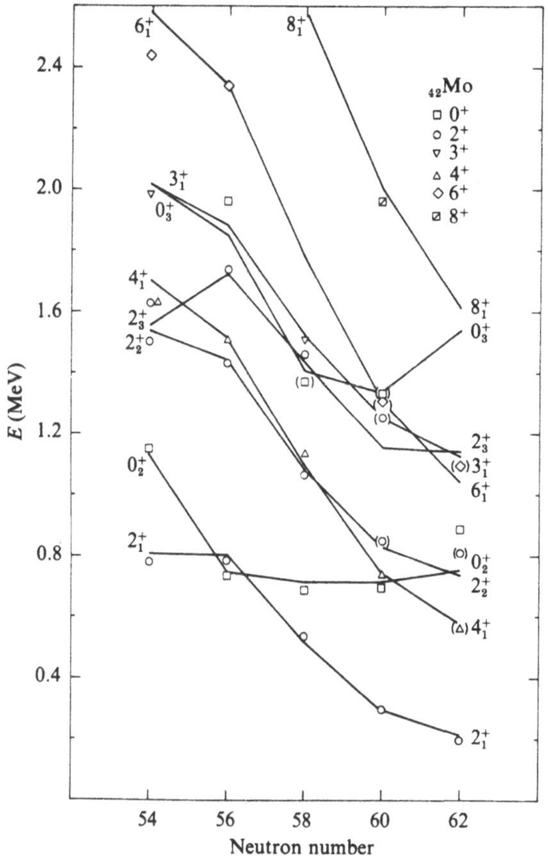

Fig. 8.12. Experimental (points) and calculated (lines) spectra of the $_{42}$Mo isotopes with $54 \leqslant$ neutron number $\leqslant 62$ (Sambataro and Molnar, 1982).

boson *hole* below $Z=50$ ($s_\pi^\dagger, d_\pi^\dagger$). In most calculations performed so far, the difference between primed and unprimed bosons has been neglected. Duval and Barrett (1982) have developed the appropriate formalism and discussed its main properties. Examples of configuration-mixing calculations within the framework of the interacting boson model-2 are found in the work of Sambataro and Molnar (1982) for $Z=42$, Duval and Barrett (1982) for $Z=80$, and Sambataro (1982) for $Z=48$. The results of the calculation of Sambataro and Molnar are shown in Fig. 8.12.

9
Group elements

9.1 Introduction

In the previous chapters, we have made use of two properties of interacting boson models: (i) their algebraic structure; and (ii) their geometric structure. There is a third property, (iii) their group structure, which is only now beginning to be exploited. This property is very important if one wants to study the response of a nucleus under multiple excitations. There are two processes here of considerable practical interest: (a) multiple Coulomb excitation of nuclei by means of heavy ions (Wenes, Yoshinaga and Dieperink, 1985); (b) multiple excitation of nuclei by energetic (~ 1 GeV) proton beams (Ginocchio *et al.*, 1986). In this chapter, we briefly discuss the basic problem that one has to solve and provide its solution in a selected number of cases.

9.2 Multiple excitations

The effects of strong external fields may, in many cases, be represented by an operator $\hat{\mathscr{U}}$ acting on the wave function of the target nucleus. Schematically this operator can be written as

$$\hat{\mathscr{U}}(\theta) = \mathrm{e}^{\mathrm{i}\,\theta T}, \tag{9.1}$$

where θ are some parameters, which may depend on time, and the Ts are transition operators. This form arises, for example, in the process shown in Fig. 9.1, in which a nucleus is Coulomb excited by a heavy ion passing nearby. For weak external fields, θ small, one can expand (9.1), obtaining

$$\hat{\mathscr{U}}(\theta) \approx 1 + \mathrm{i}\,\theta T, \tag{9.2}$$

thus recovering the transition operators discussed in Parts I and II. In general, however, one is faced with the evaluation of the matrix elements of $\hat{\mathscr{U}}$ between initial and final state

$$\mathscr{U}_{\mathrm{fi}}(\theta) = \langle f | \hat{\mathscr{U}}(\theta) | \mathrm{i} \rangle. \tag{9.3}$$

In models of the interacting boson type, the transition operators T are

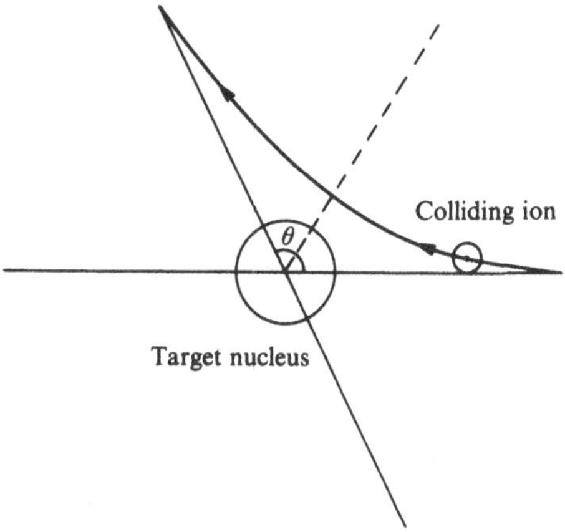

Fig. 9.1. Multiple Coulomb excitation of a nucleus by an impinging heavy ion.

assumed to be linear in the generators of a group G, Eqs. (1.24) and (4.24), and the states $|\mathrm{i}\rangle$ are assumed to be representations of the same group. Thus, the matrix elements (9.3) are nothing but the group elements and the problem of evaluating multiple excitation becomes that of finding the elements of the group G. Thus, in these applications, one makes use of the group itself rather than of its algebra.

9.3 Group elements

We begin by briefly reviewing the concept of group elements. Consider the rotation group O(3), with generators L_x, L_y, L_z. The group elements are the matrix elements

$$\langle L, M'|\mathrm{e}^{\mathrm{i}\sum_k \theta_k L_k}|L, M\rangle = \mathscr{D}^{(L)}_{M,M'}(\theta_k), \tag{9.4}$$

where $k=1,2,3$. These are usually called Wigner $\mathscr{D}$-functions (Wigner, 1959). In the case of the interacting boson model-1, the group G is U(6) and one needs the elements of

$$\hat{\mathscr{U}}(\theta_k) = \mathrm{e}^{\mathrm{i}\sum_{k=1}^{36} \theta_k G_k}, \tag{9.5}$$

where the G_k are the 36 generators of U(6) given in (2.1), and the θ_k are the corresponding 36 group-parameters. Since the group U(6) has three possible group chains (2.25), the matrix elements of $\hat{\mathscr{U}}$ can be evaluated in any of the three bases discussed in Sect. 2.4. For example, in basis I one has

$$\langle [N], n'_\mathrm{d}, v', \tilde{n}'_\Delta, L', M'_L|\hat{\mathscr{U}}(\theta_k)|[N], n_\mathrm{d}, v, \tilde{n}_\Delta, L, M_L\rangle$$
$$= \mathscr{D}^{[N]}_{n'_\mathrm{d}, v', \tilde{n}'_\Delta, L', M'_L; n_\mathrm{d}, v, \tilde{n}_\Delta, L, M_L}(\theta_k). \tag{9.6}$$

In general, for a unitary group U(n), one can write

$$\langle[\lambda_1,\lambda_2,\ldots,\lambda_n];\mu'_1,\ldots,\mu'_m|\hat{\mathscr{U}}(\theta_k)|[\lambda_1,\lambda_2,\ldots,\lambda_n];\mu_1,\ldots,\mu_m\rangle = \mathscr{D}^{[\lambda_1,\lambda_2,\ldots,\lambda_n]}_{\mu'_1,\ldots,\mu'_m;\mu_1,\ldots,\mu_m}(\theta_k), \quad (9.7)$$

where $k=1,\ldots,n$, and the μs are a complete set of labels. A general technique for finding the group elements of U(n) groups has been recently discussed and applied to U(4) by Balster, van Roosmalen and Dieperink (1983) and to U(6) by Wenes, Dieperink and van Roosmalen (1984). Here we shall confine ourselves to simple cases. We consider the matrix elements of the operator (Ginocchio *et al.*, 1986)

$$\hat{\mathscr{U}}(\alpha,\phi_m)=e^{i[\alpha(d_0^\dagger s+s^\dagger d_0)+\sum_{m=-2}^{+2}\phi_m d_m^\dagger d_m]}, \quad (9.8)$$

and begin by constructing the group elements (or representation matrices) for single boson states (Ginocchio *et al.*, 1986). One can see that, under the transformation (9.8), the boson operators $d_m^\dagger$ with $m\neq 0$ are diagonal, while the $s^\dagger$ and $d_0^\dagger$ operators transform into each other. It is convenient then to introduce linear combinations of the $s^\dagger$ and $d_0^\dagger$ operators,

$$\begin{aligned} B_+^\dagger &= (\cos u)s^\dagger+(\sin u)d_0^\dagger \\ B_-^\dagger &= -(\sin u)s^\dagger+(\cos u)d_0^\dagger \end{aligned} \quad (9.9)$$

with

$$\begin{aligned} \cos u &= \left(\frac{\phi-\phi_0}{2\phi}\right)^{\frac{1}{2}}, \\ \sin u &= \left(\frac{\phi+\phi_0}{2\phi}\right)^{\frac{1}{2}}, \\ \phi &= (\alpha^2+\phi_0^2)^{\frac{1}{2}}. \end{aligned} \quad (9.10)$$

One then obtains

$$\begin{aligned} \hat{\mathscr{U}}s^\dagger\hat{\mathscr{U}}^{-1} &= e^{i\phi_0}[Ws^\dagger+Xd_0^\dagger], \\ \hat{\mathscr{U}}d_0^\dagger\hat{\mathscr{U}}^{-1} &= e^{i\phi_0}[Xs^\dagger+W'd_0^\dagger], \\ \hat{\mathscr{U}}d_\mu^\dagger\hat{\mathscr{U}}^{-1} &= e^{i\phi_\mu}d_\mu^\dagger, \quad \mu\neq 0, \end{aligned} \quad (9.11)$$

where

$$\begin{aligned} W &= \cos\phi - i(\phi_0/\phi)\sin\phi, \\ W' &= \cos\phi + i(\phi_0/\phi)\sin\phi, \\ X &= i\,\alpha(\sin\phi)/\phi. \end{aligned} \quad (9.12)$$

Table 9.1. *Coefficients $A_{v\tilde{n}_\Delta L}$ for lowest values of v*

v	$\tilde{n}_\Delta$	L	A
0	0	0	1
1	0	2	1
2	0	2	$-(1/7)^{\frac{1}{2}}$
2	0	4	$(9/35)^{\frac{1}{2}}$
3	0	6	$(3/77)^{\frac{1}{2}}$
3	0	4	$-3(2/385)^{\frac{1}{2}}$
3	0	3	0
3	1	0	$-(1/105)^{\frac{1}{2}}$

Equations (9.11) and (9.12) allow one to evaluate the group elements for the three chains discussed in Part I.

9.3.1 Chain I

States here can be written as

$$|[N], n_d, v, \tilde{n}_\Delta, L, M_L\rangle = \eta_{Nn_dv}(s^\dagger)^{N-n_d}(d^\dagger \cdot d^\dagger)^{(n_d-v)/2}|[v], v, v, \tilde{n}_\Delta, L, M_L\rangle,$$

$$\eta_{Nn_dv} = \sqrt{\left[\frac{(2v+3)!!}{(N-n_d)!\,(n_d+v+3)!!\,(n_d-v)!!}\right]}.$$

The ground state is

$$|\text{g.s.}\rangle = |[N], n_d=0, v=0, \tilde{n}_\Delta=0, L=0, M_L=0\rangle = \frac{(s^\dagger)^N}{\sqrt{N!}}|0\rangle. \quad (9.13)$$

The group elements between the ground state and any excited state are given by

$$\langle [N], n_d, v, \tilde{n}_\Delta, L, M_L=0|\hat{\mathscr{U}}|[N], n_d=0, v=0, \tilde{n}_\Delta=0, L=0, M_L=0\rangle$$
$$= \sqrt{\left[\frac{N!\,(2v+3)!!}{(N-n_d)!\,(n_d+v+3)!!\,(n_d-v)!!}\right]} e^{iN\phi_0} W^{N-n_d} X^{n_d} A_{v\tilde{n}_\Delta L}, \quad (9.14)$$

where $A_{v\tilde{n}_\Delta L}$ is a constant that depends only on v, $\tilde{n}_\Delta$ and L. This constant is given in Table 9.1 for the lowest values of v.

9.3.2 Chain II

In order to obtain the group elements here, it is convenient to introduce the intrinsic state (3.17). For prolate deformations ($\gamma=0°$) and SU(3) symmetry, the intrinsic ground state is

$$|[N], \beta=\sqrt{2}, \gamma=0°\rangle=(s^{\dagger}+\sqrt{2}\, d_0^{\dagger})^N|0\rangle. \tag{9.15}$$

States of the ground band ($\lambda=2N$, $\mu=0$) can be obtained from (9.15) by angular momentum projection,

$$|[N], (2N,0), K=0, L, M_L\rangle$$
$$=\frac{(2L+1)}{8\pi^2\Lambda_L}\int \mathrm{d}\Omega \mathscr{R}(\Omega)\mathscr{D}^{(L)}_{M_L,0}(s^{\dagger}+\sqrt{2}\, d_0^{\dagger})^N|0\rangle, \tag{9.16}$$

where $\mathscr{R}(\Omega)$ is a general rotation parametrized by the Euler angles Ω, and $\mathscr{D}^{(L)}_{M_L,0}(\Omega)$ the corresponding Wigner $\mathscr{D}$-function. The normalization Λ_L is given by

$$\Lambda_L=\sqrt{\left[\frac{3^N(2L+1)N!\,(2N)!}{(2N-L)!!\,(2N+L+1)!!}\right]}. \tag{9.17}$$

If one assumes, as in Part I, that the transition operator is a generator of SU(3), $\hat{Q}$, then

$$\phi=\frac{3}{2\sqrt{2}}\alpha;\quad \phi_0=\tfrac{1}{3}\phi;\quad \phi_1=\phi_{-1}=\phi_0;\quad \phi_2=\phi_{-2}=-2\phi_0, \tag{9.18}$$

and

$$W=\cos\phi-\frac{\mathrm{i}}{3}\sin\phi,$$

$$W'=\cos\phi+\frac{\mathrm{i}}{3}\sin\phi, \tag{9.19}$$

$$X=\frac{2\sqrt{2}}{3}\mathrm{i}\sin\phi.$$

Use of the previous formulas and some algebra produces the group elements between the ground state and excited states in terms of hypergeometric function

$$\langle[N], (2N,0), K=0, L, M_L=0|\hat{\mathscr{U}}|[N], (2N,0), K=0, L=0, M_L=0\rangle$$
$$=\sqrt{\left[\frac{2^{L/2}(2L+1)N!\,(2N+L+1)!!}{(2N+1)!!\,(N-L/2)!}\right]}\frac{(L-1)!!}{(2L+1)!!}$$
$$\times \mathrm{e}^{-2\mathrm{i}N\phi_0}(\mathrm{e}^{6\mathrm{i}\phi_0}-1)^{(L/2)}F\left[-\left(N-\frac{L}{2}\right), \frac{L+1}{2}; L+\tfrac{3}{2}; 1-\mathrm{e}^{6\mathrm{i}\phi_0}\right]. \tag{9.20}$$

This expression can also be rewritten in terms of Jacobi polynomials. Ginocchio *et al.* (1986) have also evaluated the corresponding formulas for oblate deformations ($\gamma=60°$).

9.3.3 Chain III

States in this chain can be written as

$$|[N],\sigma,\tau,\tilde{v}_\Delta,L,M_L\rangle=\bar{\eta}_{N\sigma}(s^\dagger\cdot s^\dagger-d^\dagger\cdot d^\dagger)^{(N-\sigma)/2}|[\sigma],\sigma,\tau,\tilde{v}_\Delta,L,M_L\rangle,$$

$$\bar{\eta}_{N\sigma}=\sqrt{\left[\frac{(2\sigma+4)!!}{(N+\sigma+4)!!\,(N-\sigma)!!}\right]}. \quad (9.21)$$

Those with $N=\sigma$ can in turn be written as

$$|[\sigma],\sigma,\tau,\tilde{v}_\Delta,L,M_L\rangle=\sum_k D_k(\sigma,\tau)(s^\dagger)^{\sigma-\tau-2k}(s^\dagger\cdot s^\dagger-d^\dagger\cdot d^\dagger)^k$$
$$\times|[\tau],\tau,\tau,\tilde{v}_\Delta,L,M_L\rangle, \quad (9.22)$$

$$D_k(\sigma,\tau)=\sqrt{\left[\frac{2^{\sigma+1}(\sigma-\tau)!\,(2\tau+3)!!}{(\sigma+1)!\,(\sigma+\tau+3)!}\right]}\left(-\frac{1}{4}\right)^k\frac{(\sigma+1-k)!}{(\sigma-\tau-2k)!\,k!}.$$

If we assume, as discussed in Part I, that the transition operator is a generator of O(6), then $\phi_m=0$ in (9.8). Hence the angle ϕ becomes equal to α and

$$W=W'=\cos\alpha,$$
$$X=\mathrm{i}\sin\alpha. \quad (9.23)$$

Since $(d_0^\dagger s+s^\dagger d_0)$ is a generator of O(6), the operator $\hat{\mathscr{U}}$ cannot connect different O(6) representations. Thus

$$\langle[N],\sigma',\tau',\tilde{v}'_\Delta,L',M_L|\hat{\mathscr{U}}|[N],\sigma,\tau,\tilde{v}_\Delta,L,M_L\rangle$$
$$=\delta_{\sigma',\sigma}\langle[N],\sigma,\tau',v'_\Delta,L',M_L|\hat{\mathscr{U}}|[N],\sigma,\tau,\tilde{v}_\Delta,L,M_L\rangle. \quad (9.24)$$

The ground state is, in this case,

$$|\text{g.s.}\rangle=|[N],N,\tau=0,\tilde{v}_\Delta=0,L=0,M_L=0\rangle, \quad (9.25)$$

and the group elements between the ground state and the excited states are

$$\langle[N],N,\tau,\tilde{v}_\Delta,L,M_L|\hat{\mathscr{U}}|[N],N,\tau=0,\tilde{v}_\Delta=0,L,M_L\rangle$$
$$=\sqrt{\left[\frac{3\cdot 2^{N+1}N!}{(N+3)(N+2)}\right]}A_{\tau\tilde{v}_\Delta L}X^\tau\sum_k D_k(N,\tau)(\cos\alpha)^{N-\tau-2k}. \quad (9.26)$$

These matrix elements can also be rewritten in a variety of ways.

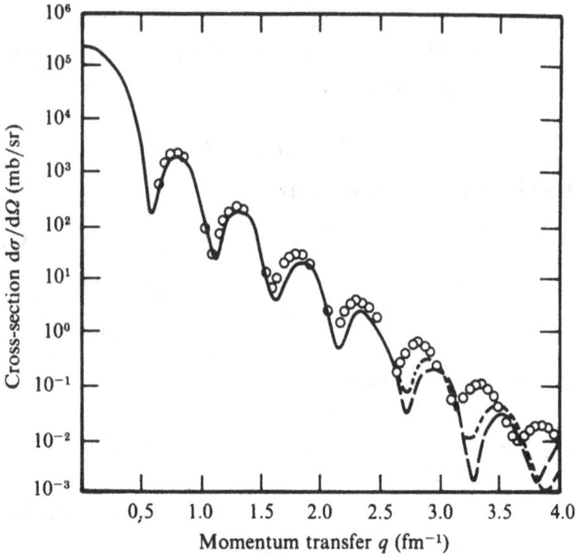

Fig. 9.2. Calculated elastic cross-section of an 800 MeV proton from a nucleus with SU(3) symmetry. The dashed line is the cross-section without and the solid line with channel coupling. Beyond a momentum transfer of 2.5 fm^{-1} the coupled-channel calculation is sensitive to the oblate or prolate shape of the nucleus and hence the solid line there shifts into a short-dashed and a long-dashed line, respectively (Ginocchio *et al.*, 1986).

9.4 Applications

Ginocchio *et al.* (1986) have applied the previous formulas to the study of elastic and inelastic scattering of energetic protons (800 MeV) off nuclei. Their aim is to study the contribution of multiple scattering to the elastic and inelastic cross-sections. An example of their calculation is shown in Fig. 9.2.

Wenes, Yoshinaga and Dieperink (1985) have extended calculations of group elements to cases without dynamic symmetry by writing a computer program for their numerical evaluation. This allows one to perform calculations in transitional nuclei and thus study the effects of multiple excitations in a large number of nuclei.

References

Akiyama, A. (1985), *Nucl. Phys.* **A433**, 369.
Arima, A. (1984), *Nucl. Phys.* **A421**, 63.
Arima, A. and Iachello, F. (1975), *Phys. Rev. Letters* **35**, 1069.
Arima, A. and Iachello, F. (1976), *Ann. Phys.* (NY) **99**, 253.
Arima, A. and Iachello, F. (1978), *Ann. Phys.* (NY) **111**, 201.
Arima, A. and Iachello, F. (1979), *Ann. Phys.* (NY) **123**, 468.
Arima, A. and Iachello, F. (1984), *in* J. W. Negele and E. Vogt (eds.), *Advances in Nuclear Physics*, Vol. 13, Plenum.
Arima, A., Otsuka, T., Iachello, F. and Talmi, I. (1977), *Phys. Letters* **66B**, 205.
Balster, G. J., van Roosmalen, O. S. and Dieperink, A. E. L. (1983), *J. Math. Phys.* **24**, 1392.
Barfield, A. F., Barrett, B. R., Sage, K. A. and Duval, P. D. (1983), *Z. Physik* **A311**, 205.
Bazantay, J. P., Cavedon, J. M., Clemens, J. C., Frois, B., Goutte, D., Huet, M., Leconte, P., Mizuno, Y., Phan, X. H., Platchkov, S. K., Boeglin, W. and Sick, I. (1985), *Phys. Rev. Letters* **54**, 643.
Bijker, R. (1985), *Phys. Rev.* **C32**, 1442.
Bijker, R., Dieperink, A. E. L., Scholten, O. and Spanhoff, R. (1980), *Nucl. Phys.* **A344**, 207.
Bohle, D., Küchler, G., Richter, A. and Steffen, W. (1984), *Phys. Letters* **148B**, 260.
Bohle, D., Richter, A., Steffen, W., Dieperink, A. E. L., LoIudice, N., Palumbo, F. and Scholten, O. (1984), *Phys. Letters* **137B**, 27.
Bohr, A. (1951), *Phys. Rev.* **81**, 134.
Bohr, A. (1952), *Mat. Fys. Medd. Dan. Vid. Selsk.* **26**, No. 14.
Bohr, A. and Mottelson, B. R. (1953), *Mat. Fys. Dan. Vid. Selsk.* **27**, No. 16.
Bohr, A. and Mottelson, B. R. (1975), *Nuclear Structure*, Vol. 2, Benjamin.
Bohr, A. and Mottelson, B. R. (1980), *Physics Scripta* **22**, 468.
Borghols, W. T. A., Blasi, N., Bijker, R., Harakeh, M. N., de Jager, C. W., van der Laan, J. B., de Vries, H. and van der Werf, S. Y. (1985), *Phys. Letters* **152B**, 330.
Brink, D. M. and Satchler, G. R. (1968), *Angular Momentum*, Oxford University Press.
Castaños, O., Chacon, E., Frank, A. and Moshinsky, M. (1979), *Journ. Math. Physics* **20**, 35.
Casten, R. F. (1981), *in* F. Iachello (ed.), *Interacting Bose–Fermi Systems in Nuclei*, Plenum.
Casten, R. F. (1985a), *Phys. Letters* **152B**, 145.
Casten, R. F. (1985b), *Phys. Rev. Letters* **54**, 1991.
Casten, R. F. and Cizewski, J. A. (1978), *Nucl. Phys.* **A309**, 477.
Casten, R. F. and von Brentano, P. (1985), *Phys. Letters* **152B**, 22.
Casten, R. F., von Brentano, P. and Haque, A. M. I. (1985), *Phys. Rev.* **C31**, 1991.
Casten, R. F., von Brentano, P., Heyde, K., van Isacker, P. and Jolie, J. (1985), *Nucl. Phys.* **A439**, 289.
Chen, H. T. and Arima, A. (1983), *Phys. Rev. Letters* **51**, 447.
Chen, X., Zhang, M., Sun, H. and Han, Q. (1982), *Scientia Sinica* **A25**, 834.

Cizewski, J. A. (1981), *in* F. Iachello (ed.), *Interacting Bose–Fermi Systems in Nuclei*, Plenum.

Cizewski, J. A., Casten, R. F., Smith, G. J., Stelts, M. L., Kane, W. R., Börner, H. G. and Davidson, W. F. (1978), *Phys. Rev. Letters* **40**, 167.

Cooper, L. N. (1956), *Phys. Rev.* **104**, 1189.

de Jager, C. W. (1984), *in* O. Scholten (ed.), *Interacting Boson–Boson and Boson–Fermion Systems*, World Scientific.

De Shalit, A. and Talmi, I. (1963), *Nuclear Shell Theory*, Academic Press.

Dewald, A., Kaup, U., Gast, W., Gelberg, A., Schuh, H. W., Zell, K. O. and von Brentano, P. (1982), *Phys. Rev.* **C25**, 226.

Dieperink, A. E. L. (1979), *in* F. Iachello (ed.), *Interacting Bosons in Nuclei*, Plenum.

Dieperink, A. E. L. (1983), *in* D. Wilkinson (ed.), *Collective Bands in Nuclei*, *Progress in Particle and Nuclear Physics*, Vol. 9, Pergamon.

Dieperink, A. E. L. and Bijker, R. (1982), *Phys. Letters* **116B**, 77.

Dieperink, A. E. L., Iachello, F., Rinat, A. and Creswell, C. (1978), *Phys. Letters* **76B**, 135.

Dieperink, A. E. L., Scholten, O. and Iachello, F. (1980), *Phys. Rev. Letters* **44**, 1747.

Duval, P. D. and Barrett, B. R. (1981a), *Phys. Rev.* **C23**, 492.

Duval, P. D. and Barrett, B. R. (1981b), *Phys. Rev.* **C24**, 1272.

Duval, P. D. and Barrett, B. R. (1981c), *Phys. Letters* **100B**, 223.

Duval, P. D. and Barrett, B. R. (1982), *Nucl. Phys.* **A376**, 213.

Duval, P. D., Goutte, D. and Vergnes, M. (1983), *Phys. Letters* **124B**, 297.

Dyson, J. F. (1956), *Phys. Rev.* **102**, 1217, 1230.

Elliott, J. P. (1958), *Proc. Roy. Soc.* **A245**, 128, 562.

Elliott, J. P. and Evans, J. A. (1981), *Phys. Letters* **101B**, 216.

Elliott, J. P. and White, A. P. (1980), *Phys. Letters* **97B**, 169.

Feng, D. H., Gilmore, R. and Deans, S. R. (1981), *Phys. Rev.* **C23**, 1254.

Feshbach, H. and Iachello, F. (1973), *Phys. Letters* **45B**, 7.

Feshbach, H. and Iachello, F. (1974), *Ann. Phys.* (NY) **84**, 211.

Gilmore, R. (1974), *Lie Groups, Lie Algebras and Some of Their Applications*, Wiley.

Gilmore, R. (1979), *J. Math. Phys.* **20**, 891.

Gilmore, R. and Feng, D. H. (1978), *Nucl. Phys.* **A301**, 189.

Ginocchio, J. and Kirson, M. W. (1980a), *Phys. Rev. Letters* **44**, 1744.

Ginocchio, J. and Kirson, M. W. (1980b), *Nucl. Phys.* **A350**, 31.

Ginocchio, J. N., Otsuka, T., Amado, R. D. and Sparrow, D. A. (1986), *Phys. Rev.* **C33**, 247.

Goutte, D. (1984), *in* O. Scholten (ed.), *Interacting Boson–Boson and Boson–Fermion Systems*, World Scientific.

Góźdź, A. and Szpikowski, S. (1978), *Annales Universitatis Lublin* **33**, 1.

Góźdź, A. and Szpikowski, S. (1980), *Nukleonica* **25**, 1055.

Greiner, W. (1965), *Phys. Rev. Letters* **14**, 599.

Greiner, W. (1966), *Nucl. Phys.* **80**, 417.

Hamermesh, M. (1962), *Group Theory and Its Application to Physical Problems*, Addison-Wesley.

Hamilton, W. D., Irbäck, A. and Elliott, J. P. (1984), *Phys. Rev. Letters* **53**, 2469.

Harter, H., Gelberg, A. and von Brentano, P. (1985), *Phys. Letters* **157B**, 1.

Haxel, O., Jensen, J. H. and Suess, H. E. (1949), *Phys. Rev.* **75**, 1766.

Heyde, K., van Isacker, P., Waroquier, M., Wenes, G., Gigase, Y. and Stachel, J. (1983), *Nucl. Phys.* **A398**, 235.

Holstein, T. and Primakoff, H. (1940), *Phys. Rev.* **58**, 1098.

Iachello, F. (1969), *The Interacting Boson Model*, PhD. Thesis, Massachusetts Institute of Technology, Cambridge, Mass.

Iachello, F. (1980a), *in* G. Bertsch and D. Kurath (eds.), *Nuclear Spectroscopy, Lecture Notes in Physics*, Vol. 119, Springer Verlag.
Iachello, F. (1980b), *Phys. Rev. Letters* **44**, 772.
Iachello, F. (1981), *Nucl. Phys.* **A358**, 89c.
Iachello, F. (1984), *Phys. Rev. Letters* **53**, 1427.
Iachello, F. (1985), *in* M. Vallieres and B. H. Wildenthal (eds.), *Nuclear Shell Models*, World Scientific.
Iachello, F. and Scholten, O. (1979), *Phys. Rev. Letters* **43**, 679.
Janssen, D., Jolos, R. V. and Dönau, F. (1974), *Nucl. Phys.* **A224**, 93.
Kaup, U. and Gelberg, A. (1979), *Z. Physik* **A293**, 311.
Kaup, U., Mönkemeyer, C. and von Brentano, P. (1983), *Z. Physik* **A310**, 129.
Klein, A. and Vallieres, M. (1981), *Phys. Rev. Letters* **46**, 586.
Kota, V. K. B. (1984), *in* O. Scholten (ed.), *Interacting Boson–Boson and Boson–Fermion Systems*, World Scientific.
Kyrchev, G. and Paar, V. (1986), *Ann. Phys.* (NY) **170**, 257.
Kyrchev, G. and Paar, V. (1983), *Nucl. Phys.* **A395**, 61.
Leviatan, A. (1984), *Phys. Letters* **143B**, 25.
Leviatan, A. (1985), *Z. Physik* **A321**, 467.
Lin, Y. S. (1982), *Phys. Energiae Fortis et Phys. Nucl.* **6**, 77.
Lin, Y. S. (1983), *Chinese Phys.* **3**, 86.
Lipas, P. (1962), *Nucl. Phys.* **39**, 468.
Lipas, P. (1984), *in* O. Scholten (ed.), *Interacting Boson-Boson and Boson-Fermion Systems*, World Scientific.
LoIudice, N. and Palumbo, F. (1978), *Phys. Rev. Letters* **41**, 1532.
Mayer, M. G. (1949), *Phys. Rev.* **75**, 1969.
Moinester, M. A., Alster, J., Azuelos, G. and Dieperink, A. E. L. (1982), *Nucl. Phys.* **A383**, 264.
Novoselski, A. (1984), *Computer Program BOSON*, Weizmann Institute of Science, Rehovot, Israel.
Novoselski, A. (1985), *Phys. Letters* **155B**, 299.
Novoselski, A. and Talmi, I. (1985), *Phys. Letters* **160B**, 13.
Nwachuku, C. O. and Rashid, M. A. (1977), *J. Math. Phys.* **18**, 1387.
Otsuka, T. (1977), *Computer Program NPBOS*, University of Tokyo, Japan.
Otsuka, T., Arima, A., Iachello, F. and Talmi, I. (1978), *Phys. Letters* **76B**, 139.
Otsuka, T. and Ginocchio, J. N. (1985), *Phys. Rev. Letters* **55**, 276.
Pittel, S., Dukelsky, J., Perazzo, R. P. J. and Sofia, H. M. (1984), *Phys. Letters* **144B**,
Popov, V. S. and Perelomov, A. M. (1967), *Sov. J. Nucl. Phys.* **5**, 489.
Puddu, G., Scholten, O. and Otsuka, T. (1980), *Nucl. Phys.* **A348**, 109.
Rainwater, J. (1950), *Phys. Rev.* **79**, 432.
Ratna Raju, R. D. (1981), *Phys. Rev.* **C23**, 518.
Ratna Raju, R. D. (1982), *J. Phys.* **G8**, 1663.
Riedinger, L. L., Johnson, N. R. and Hamilton, J. M. (1969), *Phys. Rev.* **179**, 1214.
Rohozinski, S. G. (1978), *J. Phys.* **G4**, 1075.
Sambataro, M. (1982), *Nucl. Phys.* **A380**, 365.
Sambataro, M. and Dieperink, A. E. L. (1981), *Phys. Letters* **107B**, 249.
Sambataro, M. and Molnar, G. (1982), *Nucl. Phys.* **A376**, 201.
Schaaser, H. and Brink, D. (1984), *Phys. Letters* **143B**, 269.
Scholten, O. (1976), *Computer Program PHINT*, University of Groningen, The Netherlands.
Scholten, O. (1980), *The Interacting Boson Model and Some Applications*, PhD. Thesis, University of Groningen, The Netherlands.
Scholten, O., Heyde, K., van Isacker, P., Jolie, J., Moreau, J., Waroquier, M. and Sau, J. (1985a), *Nucl. Phys.* **A438**, 41.

Scholten, O., Heyde, K., van Isacker, P. and Otsuka, T. (1985b), *Phys. Rev.* **C32**, 1729.
Scholten, O., Iachello, F. and Arima, A. (1978), *Ann. Phys.* (NY) **115**, 325.
Schwinger, J. (1965), *in* L. C. Biedenharn and H. van Dam (eds.), *Quantum Theory of Angular Momentum*, Academic Press.
Shlomo, S. and Talmi, I. (1972), *Nucl. Phys.* **A198**, 81.
Stachel, J., van Isacker, P. and Heyde, K. (1982), *Phys. Rev.* **C25**, 650.
Sun, H., Han, Q., Chen, X. and Zhang, M. (1982), *Scientia Sinica* **A25**, 1165.
Szpikowski, S. and Góźdź, A. (1980), *Nucl. Phys.* **A340**, 76.
Talmi, I. (1971), *Nucl. Phys.* **A172**, 1.
Talmi, I. (1973), *Nuovo Cimento* **3**, 85.
van der Laan, J. B., Burghardt, A. J. C., de Jager, C. W. and de Vries, H. (1985), *Phys. Letters* **153B**, 130.
van Isacker, P. (1983), *Phys. Rev.* **C27**, 2447.
van Isacker, P. (1984), *Het Interagerend Boson Model*, PhD. Thesis, University of Ghent, Belgium.
van Isacker, P. and Chen, J. Q. (1981), *Phys. Rev.* **C24**, 684.
van Isacker, P., Frank, A. and Dukelsky, J. (1985), *Phys. Rev.* **C31**, 671.
van Isacker, P., Frank, A. and Sun, H. Z. (1984), *Ann. Phys.* (NY) **157**, 183.
van Isacker, P., Heyde, K., Jolie, J. and Sevin, A. (1986), *Ann. Phys.* (NY) **171**, 253
van Isacker, P., Heyde, K., Waroquier, M. and Wenes, G. (1981), *Phys. Letters* **104B**, 5.
van Isacker, P., Heyde, K., Waroquier, M. and Wenes, G. (1982), *Nucl. Phys.* **A380**, 383.
van Isacker, P. and Lipas, P. D. (1985), *Phys. Rev.* **C31**, 1546.
van Isacker, P. and Puddu, G. (1980), *Nucl. Phys.* **A348**, 125.
van Roosmalen, O. (1982), *Algebraic Description of Nuclear and Molecular Rotation–Vibration Spectra*, PhD. Thesis, University of Groningen, The Netherlands.
van Roosmalen, O., Iachello, F., Levine, R. D. and Dieperink, A. E. L. (1983), *J. Chem. Phys.* **79**, 2515.
Vergados, J. D. (1968), *Nucl. Phys.* **A111**, 681.
Vergnes, M. (1981), *in* F. Iachello (ed.), *Interacting Bose–Fermi Systems in Nuclei*, Plenum.
Warner, D. D. (1981), *Phys. Rev. Letters* **47**, 1819.
Warner, D. D. and Casten, R. F. (1983), *Phys. Rev.* **C28**, 1798.
Warner, D. D., Casten, R. F. and Davidson, W. F. (1981), *Phys. Rev. Lett.* **45**, 1761.
Weiguny, A. (1981), *Z. Physik* **A301**, 335.
Wenes, G., Dieperink, A. E. L. and van Roosmalen, O. S. (1984), *Nucl. Phys.* **A424**, 81.
Wenes, G., Yoshinaga, N. and Dieperink, A. E. L. (1985), *Nucl. Phys.* **A443**, 472.
Wigner, E. (1937), *Phys. Rev.* **51**, 106.
Wigner, E. (1959), *Group Theory and Its Application to the Quantum Mechanics of Atomic Spectra*, Academic Press.
Wilets, J. and Jean, M. (1956), *Phys. Rev.* **102**, 788.
Wolf, A., Berant, Z., Warner, D. D., Gill, R. L., Schmid, M., Chrien, R. E., Peaslee, G., Yamamoto, H., Hill, J. C., Wohn, F. K., Chung, C. and Walters, W. B. (1983), *Phys. Letters* **123B**, 165.
Wu, H. C. (1982), *Phys. Letters* **110B**, 1.
Wu, H. C. and Zhou, X. Q. (1984), *Nucl. Phys.* **A417**, 67.
Wybourne, B. G. (1974), *Classical Groups for Physicists*, Wiley.
Yoshida, N., Arima, A. and Otsuka, T. (1982), *Phys. Letters* **114B**, 86.
Yoshinaga, N., Akiyama, Y., and Arima, A. (1986), *Phys. Rev. Lett.* **56**, 1116.
Zhang, M., Chen, X., Sun, H. and Han, Q. (1982), *Scientia Sinica* **A25**, 952.

Index